NEW TOPOLOGICAL INVARIANTS FOR REAL- AND ANGLE-VALUED MAPS

An Alternative to Morse–Novikov Theory

NEW TOPOLOGICAL INVARIANTS FOR REAL- AND ANGLE-VALUED MAPS

An Alternative to Morse–Novikov Theory

Dan Burghelea

Ohio State University, USA

World Scientific

NEW JERSEY · LONDON · SINGAPORE · BEIJING · SHANGHAI · HONG KONG · TAIPEI · CHENNAI · TOKYO

Published by

World Scientific Publishing Co. Pte. Ltd.

5 Toh Tuck Link, Singapore 596224

USA office: 27 Warren Street, Suite 401-402, Hackensack, NJ 07601

UK office: 57 Shelton Street, Covent Garden, London WC2H 9HE

Library of Congress Cataloging-in-Publication Data

Names: Burghelea, Dan, author.

Title: New topological invariants for real- and angle-valued maps :
  an alternative to Morse-Novikov theory / by Dan Burghelea (Ohio State University, USA).

Description: New Jersey : World Scientific, 2017. |
  Includes bibliographical references and index.

Identifiers: LCCN 2017030717 | ISBN 9789814618243 (hardcover : alk. paper)

Subjects: LCSH: Manifolds (Mathematics) | Mappings (Mathematics) | Topological spaces.

Classification: LCC QA613.2 .B87 2017 | DDC 514/.34--dc23

LC record available at https://lccn.loc.gov/2017030717

British Library Cataloguing-in-Publication Data

A catalogue record for this book is available from the British Library.

To Ani and Gabriela

Acknowledgments

This book was finalized during a visit at Max Planck Institute for Mathematics (Bonn, Germany) and at Mathematical Institute of the Romanian Academy (Bucharest, Romania) (2016). I thank both institutions for their financial support. I am thankful to colleagues who were attending lectures on various parts of the material helping to bring improvements to this text and in particular to Stefan Haller, coauthor of many of the results presented in this work, for some critical observations. I also thank Ana Burghelea for technical support and Ms. Kwong Lai Fun, editor for World Scientific, for her patience in working with a disorganized author and help in editing the manuscript in the present book format.

My warmest thanks go to the copy editor (who does not want his name mentioned) for helping to convert this work into a hopefully readable text; without his help the final version of this book would have been considerably delayed.

Preface

This book proposes an alternative to Morse-Novikov theory referred below as AMN-theory and is primarily about the new topological invariants for real- and angle-valued maps which appear in the context of this theory.

The invariants discussed are on one side analogues of rest points, instantons between rest points and closed trajectories of vector fields and on the other side refine basic topological invariants like Betti numbers, homology and monodromy.

They are associated to tame maps defined on compact locally contractible metric spaces, considerably more general than Morse maps defined on compact smooth manifolds, are computable by computer implementable algorithms when the map is simplicial and defined on a finite simplicial complex and have remarkable robustness and topological properties.

They carry information about the rest points and trajectories of some classes of flows that admit Lyapunov maps relating dynamics to topology in the same spirit Morse–Novikov theory does.

Contents

List of Figures

Chapter 1

Preview

1.1 Introduction

This book proposes an alternative to Morse-Novikov theory, referred to below as the AMN theory, and is primarily about the new topological invariants for real- and angle-valued maps that appear in the context of this theory.

The invariants discussed are analogues of the rest points, instantons[1], and closed trajectories of *locally conservative vector fields*[2] on the one hand, and refined basic topological invariants like Betti numbers, homology, and monodromy on the other hand[3].

These invariants are associated to tame maps defined on compact locally contractible metric spaces, considerably more general than the Morse maps defined on compact smooth manifolds, are computable by computer-implementable algorithms, when the map is simplicial and defined on a finite simplicial complex, and enjoy remarkable robustness and topological properties.

Moreover, these invariants carry information about the rest points and trajectories of flows that admit Lyapunov maps relating dynamics to topology in the same spirit as the Morse-Novikov theory does.

To better understand the motivation for and what AMN theory is, we begin with a few observations about what classical Morse-Novikov theory, a topic in Differential and Algebraic Topology, is.

---

[1]Isolated trajectories between rest points.

[2]Locally the gradient of a smooth function with respect to some metric g on the manifold.

[3]When f is angle-valued, homology means Novikov homology and Betti numbers mean Novikov-Betti numbers, cf. definitions in Section 2.2.

1.2   Morse-Novikov theory

Morse theory considers smooth real-valued maps f defined on smooth manifolds M and relates the set of critical points of f (when f is a Morse real-valued map) and the isolated trajectories between rest points (instantons) of a vector field X which has f as a Lyapunov function[4] to the topology of the manifold M. This is done for a large class of smooth vector fields with Lyapunov function which is generic with respect to the C^r-topology, $r \geq 1$.

Figure 1.1 depicts a surface and a Morse map, the projection on the line below the surface. The red curves on the surface represent isolated trajectories of the gradient with respect to the metric induced by regarding the surface as embedded in the three-dimensional space, as suggested by the figure.

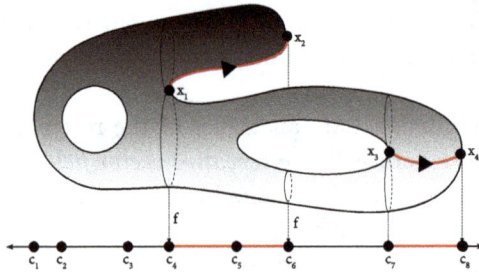

Fig. 1.1   A surface with a Morse function

The Novikov theory considers angle-valued maps f, or more generally degree-one closed differential forms ω, instead of real-valued maps. It relates the critical points of the map f (or the zeros of the 1-form ω), the isolated trajectories between rest points, called in this book *instantons*, and the isolated closed trajectories of a vector field for which f or ω is Lyapunov function or 1-form [5], to the topology of the pair consisting of the manifold and the homotopy class of f, or the cohomology class of ω.

One should note that the class of vector fields which admit a Lyapunov closed 1-form (i.e., locally conservative vector fields) is considerably larger than the class of vector fields which admit a Lyapunov real-valued function (i.e., conservative vector fields). A vector field from the former class may

---

[4]A smooth function $f : M \to \mathbb{R}$ such that $X(f)(x) < 0$ iff $df(x) \neq 0$.
[5]i.e., $\omega(X)(x) < 0$ iff $X(x) \neq 0$.

have closed trajectories, while a vector field from the latter class can not. Note also that a smooth vector field which admits a Lyapunov closed 1-form, also admits a Lyapunov angle-valued map [Burghelea, D., Haller, S. (2008a)].

Figure 1.2 below depicts a surface and a Morse angle-valued map, the central projection on the dotted circle. The red curves on the surface indicate instantons and the green curve a closed trajectory for the gradient of the angle-valued map with respect to the metric induced by regarding the surface as embedded in the three-dimensional space.

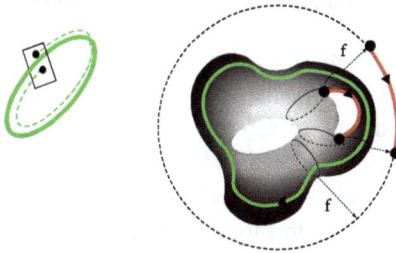

Fig. 1.2 A surface with a Morse angle-valued map

The relationship between the dynamical elements of the vector field and the topology of the manifold with an angle-valued map provided by Morse-Novikov theory, can be used in both directions. In one, the critical points of the Morse maps or the rest points, instantons, and closed trajectories of vector fields provide a lot of information about the topology of the underlying space (in the present case a manifold). In the opposite direction, knowledge of the topology of the underlying space provides substantial constraints on the number of these dynamical elements, and it leads to results on existence and information about their quantity and qualitative properties. The last feature is very appealing outside mathematics. Morse theory came recently to the attention of fields of science outside traditional mathematics, such as Data Analysis, Shape Recognition, Computer Science, and Physics, cf. [Carlsson, G. (2009)], [Allili, M., Corriveau, D. (2007)].

To motivate our proposal for an Alternative to Morse–Novikov theory (AMN-theory) note that:

• The dynamical elements in Morse-Novikov theory, namely, the rest points, instantons, and closed trajectories, are robust to C^r-perturbations,

$r \geq 1$, of the dynamics (vector field) and of the Lyapunov map, but not to C^0-perturbations, the only ones that can be tested experimentally.

• The possible infiniteness of the number of instantons and of closed trajectories makes the relation with topology considerably more subtle and practically much harder to conclude existence and evaluate the size of the sets of such elements (cf [Burghelea, D., Haller, S. (2008a)] [Hutchings, M. (2002)], [Hutchings, M., Lee, Y.J. (1999)], [Pajitnov, A.V. (2006)] for work about these aspects).

• Many compact spaces of interest are not manifolds, and even when they are, many of the real- or angle-valued maps of interest do not satisfy the hypotheses under which the Morse-Novikov theory is applicable.

The proposed *Alternative to Morse-Novikov theory* addresses at least in part some of these drawbacks (reduced generality of spaces and maps, lack of finiteness, and lack of C^0-robustness), providing at the same time some new perspective on and refinements of classical topological invariants related to dynamics. Some of these refinements are of mathematical interest and have implications outside Morse-Novikov theory.

There are other mathematical developments which address these drawbacks, like the Morse theory on stratified spaces developed by Goretski and MacPherson [Goresky, M., MacPherson, R. (1988)], or the combinatorial Morse-Novikov theory developed by Robin Forman [Forman, R. (1998)], [Forman, R. (2002)], but these developments go in different directions.

1.3   The AMN theory

The AMN theory considers tame real- or angle-valued continuos maps f defined on a compact ANR X in the presence of a field κ, the field of coefficients for homology. The reader unfamiliar with the notions of ANR and tame map should think of spaces homeomorphic to finite simplicial complexes or to smooth manifolds as examples of ANR's, and simplicial maps or smooth maps which in some local coordinates are polynomials[6], as examples of tame maps. A definition is provided in Section 2.2.

Instead of critical points, the AMN theory considers *critical values* for f real-valued, or *critical angles* for f angle-valued. They are real numbers or angles [7] for which the topology of the level changes in any neighborhood of the critical value or of the critical angle, respectively.

---

[6]In particular, maps with all critical points nondegenerate.
[7]Real numbers mod 2π.

Instead of critical points and instantons between critical points, the AMN theory defines and calculates *barcodes*. These are intervals of four types: closed $[a, b], a \leq b$, open (a, b), closed-open $[a, b)$, and open-closed $(a, b]$, $a < b$, with both a and b critical values for f real-valued map and lifts in $\mathbb{R}$ of critical angles θ_1 and θ_2 for angle-valued maps. In the second case the barcodes are actually equivalence classes of such intervals, with two intervals equivalent if they can be identified by a translation of $2\pi k$, equivalently, intervals whose left endpoint belongs to $[0, 2\pi)$. As it will be shown in the book, only the mixed barcodes are related to the instantons; the closed and open barcodes are related to the homology of the underlying space.

Instead of closed trajectories the AMN theory defines and calculates the Jordan cells or the Jordan blocks whose direct sums calculate the *homological monodromies*. They can be regarded as analogues of Poincaré return maps for closed trajectories. However, the relation between these two is considerably subtler and involves the concept of *torsion*, cf. [Hutchings, M. (2002)], [Burghelea, D., Haller, S. (2008a)], [Burghelea, D., Haller, S. (2008b)], [Burghelea, D. (2011)].

The left side of Fig. 1.2 indicates the Poincaré return map for a closed trajectory; it is the linear isomorphism of the normal space to a trajectory at a point on the trajectory, in our figure suggested by the dotted curve around of the closed trajectory.

The barcodes are finite collections of intervals with multiplicities, $\mathcal{B}_r^c(f)$, $\mathcal{B}_r^o(f)$, $\mathcal{B}_r^{c,o}(f)$, and $\mathcal{B}_r^{o,c}(f)$ for any $r = 0, 1, 2, \ldots, \dim X$. As indicated in Chapter 8, for f a Morse real- or angle-valued map the barcodes recover the number of critical points and the rank of the *boundary map* in the Morse complex or the Novikov complex of f,[8] rank determined by the algebraic counting of instantons.

In view of the properties of closed and open barcodes, it is convenient to put together the sets $B_r^c(f)$ and $B_{r-1}^o(f)$ as a collection of points in the complex plane defining the *configuration* δ_r^f of points with multiplicity in $\mathbb{C}$ for f real-valued, and of points with multiplicity in $\mathbb{C} \setminus 0$ for f angle-valued. For real-valued maps, the closed r-barcode $[a, b]$ will correspond to the complex number $a + ib$, and the open $(r - 1)$-barcode (a, b) to the complex number $b + ia$, cf. Fig. 1.3a. For angle-valued maps the closed r-barcode $[\theta', \theta'']$ will correspond to the complex number $e^{(\theta'' - \theta') + i\theta'}$, and the open $(r - 1)$-barcode (θ', θ'') to the complex number $e^{(\theta' - \theta'') + i\theta''}$, cf. Fig.

---

[8]Cf. Section 8.1 for the description of the Morse complex.

1.3b. The black circles indicate closed r-barcodes, the empty circles open $(r-1)$-barcodes. All these points have multiplicity ≥ 1, the multiplicity of the corresponding barcode.

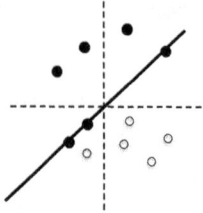

Fig. 1.3a: Configuration for real-value map   Fig. 1.3b: Configuration for angle-value map

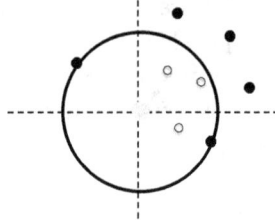

Fig. 1.3   Configurations δ_r^f

The *convenience* comes from the following facts.

• The cardinality of the support of δ_r^f is constant in f, and is equal to the r-th Betti number cf. Theorems 5.1, and 5.4.

• The configurations δ_r^f can be refined to configurations $\widehat{\delta}_r^f$ of κ-vector spaces when f real-valued, and of free $\kappa[t^{-1}, t]$-modules when f is angle-valued, with the property that $\dim \widehat{\delta}_r^f = \delta_r^f$, respectively rank $\widehat{\delta}_r^f = \delta_r^f$; both assignments $f \rightsquigarrow \delta_r^f$ and $f \rightsquigarrow \widehat{\delta}_r^f$ are continuous, cf. Theorem 5.2.

• Continuous perturbation of the map f results in continuous variation of the points in the support of δ_r^f; points in the support close to the diagonal (in the case when f is real-valued), or to the unit circle (in the case when f is angle-valued) can move from above and on the diagonal, respectively outside and on the unit circle, when they represent closed r-barcodes, to below the diagonal, respectively inside the unit circle, when they represent open $(r-1)$-barcodes and viceversa.

• When X is a closed topological manifold, the Poincaré Duality between Betti numbers and between homology vector spaces is refined to a Poincaré Duality between configurations $\delta_r(f)$ and between configurations $\widehat{\delta}_r(f)$, cf. Theorems 5.3, 5.6.

Note when f is real-valued the homology involved is the standard homology with coefficients in the field κ, which consists of κ-vector spaces, and the Betti numbers are the dimensions of these vector spaces. When f

is angle-valued the homology involved is the Novikov homology associated with X and the homotopy class of f, which in this book consists of free $\kappa[t^{-1}, t]$-modules, cf. Section 2.2 for definitions. The Novikov-Betti numbers are the ranks of these free modules. If instead of the ring $\kappa[t^{-1}, t]$ of Laurent polynomials with coefficients in κ one considers the field $\kappa[t^{-1}, t]]$ of Laurent power series with coefficients in κ, then the Novikov homology consists of vector spaces over $\kappa[t^{-1}, t]]$ and the Novikov-Betti numbers are the dimensions of these vector spaces.

In the same manner, but so far only for aesthetic reasons, one can put together the sets $B_r^{c,o}(f)$ and $B_r^{o,c}(f)$ as a configuration $\gamma_r(f)$ of points with multiplicity located in $\mathbb{C} \setminus \Delta$, with Δ the diagonal $\{z \in \mathbb{C} | z = i\bar{z}\}$, for f real-valued, and in $\mathbb{C} \setminus (0 \sqcup S^1)$ for f angle-valued. The cardinality of the support of $\gamma_r(f)$ is not constant, but a weaker robustness discovered by the authors of [Cohen-Steiner, D., Edelsbrunner, H., Harer, J. (2007)] still holds, and so does a Poincaré Duality. This Poincaré Duality is however quite different from the Poincaré Duality between the configurations $\delta_r(f)$, cf. Theorems 6.2, 6.4 versus Theorems 5.3, 5.6.

The r-monodromy is a *similarity class* of invertible matrices with coefficients in κ, equivalently, of pairs (V_r, T_r) with V_r a finite-dimensional κ-vector space and $T_r : V_r \to V_r$ a linear isomorphism. This similarity class is completely determined by the collection of *Jordan cells* $\mathcal{J}_r(f)$ or, equivalently, by a characteristic polynomial (with coefficients in κ) and its canonical divisors

$$D_{n_r}^{T_r}(z) | D_{n_r - 1}^{T_r}(z) | \cdots | D_1^{T_r}(z)$$

cf. Subsection 2.1.1 for definition. Recall that a Jordan cell is a $k \times k$ matrix $T(\lambda; k)$ with $\lambda \in \bar{\kappa} \setminus 0$ on the diagonal, 1 above the diagonal, and all the other entries equal to zero. Recall also that $\bar{\kappa}$ denotes the algebraic closure of κ.

The barcodes and Jordan cells

(1) are *computer friendly*, i.e., computable by computer-implementable algorithms, cf. [Burghelea, D., Dey, T (2013)] and [Burghelea, D. (2015b)],

(2) each of the sets $\mathcal{B}_r^c(f)$, $\mathcal{B}_r^o(f)$, $\mathcal{B}_r^{c,o}(f)$, $\mathcal{B}_r^{o,c}(f)$, $\mathcal{J}_r(f)$ are finite when X is compact,

(3) are defined for a considerably larger class of spaces and maps than manifolds and Morse maps,

(4) refine familiar topological invariants like the Betti numbers and the Novikov-Betti numbers, and calculate the monodromy.

In Chapter 4, Subsection 4.2.2, a chain complex of κ-vector spaces for a f tame real-valued map, and of $\kappa[t^{-1}, t]]$-vector spaces[9] for f a tame angle-valued map is derived from the collections $\mathcal{B}_r^c(f)$, $\mathcal{B}_r^o(f)$, and $\mathcal{B}_r^{c,o}(f)$. In case X is a closed smooth manifold and f is a Morse map, this complex is isomorphic to the Morse complex or to the Novikov complex associated with f and a vector field which has f as a Lyapunov map[10]. For a Morse map f the Morse complex, respectively the Novikov complex, carries significant information about the dynamical elements of the vector field, such as rest points and instantons. One expects that the chain complexes proposed in Chapter 4 for any tame map and discussed in more details in Chapter 8 retain these features for flows on a compact ANR which have f as a Lyapunov map.

It is also remarkable that most of the basic algebraic topology invariants[11] which can be recovered from critical points, instantons, and closed trajectories, such as the Betti numbers, the Novikov-Betti numbers with respect to any field, the Alexander function, can be equally well recovered from barcodes and Jordan blocks. Theorems 4.1 and 5.1 provide such results.

In this book the barcodes and Jordan blocks are defined in two different ways. In Chapter 4 the configurations $\delta_r(f)$ and $\gamma_r(f)$ as well as the Jordan cells are derived via graph representations. In Chapters 5 and 6 the configurations δ_r^f, γ_r^f and their refinements $\widehat{\delta}_r^f$ are defined in the spirit of measure theory as densities of measures, and in Chapter 7 the Jordan cells $\mathcal{J}_r(f)$ are derived via the linear algebra of linear relations.

Note that in Chapters 5 and 7 the configurations $\delta_r(f)$ and the Jordan cells $\mathcal{J}_r(f)$ are ultimately associated to any continuous map f. This, however, is not the case for the configurations γ_r^f.

1.4   Contents of the book

The book has nine chapters including Chapter 1, this Preview.

Chapter 2, Preliminary, reviews for the reader's convenience the linear

---

[9]The field of Laurent power series

[10]This shows that the Morse or the Novikov complex is independent on the vector field, a result proved in [Cornea, O., Ranicki, A. (2003)] for the integral Morse complex.

[11]The torsion of the integral homology as a finite abelian group cannot, but its cardinality can.

algebra of matrices and of linear relations, as well as a few concepts and results in topology needed in this book. We suggest the reader take first a superficial look at this material and return to the appropriate subsection of this chapter whenever necessary. For example, the algorithm for the calculation of the regular part of a linear relation[12] might be better to consider after reading Chapter 7.

The chapter contains a section on Linear Relations which follows closely [Burghelea, D., Haller, S. (2015)] and [Burghelea, D. (2015a)], and in particular an algorithm for the calculation of the regular part of a linear relation between two equal finite-dimensional vector spaces, and a subsection on Fredholm maps used in Chapter 6.

Chapter 3 contains the needed material about the representations of the two graphs used in this paper, $\mathcal{Z}$ and G_{2m}, as well as an algorithm for the decomposition of such representations into indecomposable components, proposed in [Burghelea, D., Dey, T (2013)]. The presentation follows closely [Burghelea, D., Haller, S. (2015)].

Chapter 4 provides the definitions of barcodes and Jordan blocks for a tame map via graph representations and the computation of various types of homologies of the source space of the tame map in terms of barcodes and Jordan blocks. This part follows largely [Burghelea, D., Haller, S. (2015)]. We sketch an algorithm proposed in [Burghelea, D., Dey, T (2013)] to derive the representations $\rho_r(f)$ for an angle-valued map, then implicitly for a real-valued map from an input "simplicial complex and a simplicial map". In the case of real-valued maps the practitioners can also use the algorithms proposed in [Cohen-Steiner, D., Edelsbrunner, H., Morozov, D. (2006)] for the calculation of the barcodes of Zigzag Persistence.

In additions, this chapter introduces the AM chain complex (Alternative Morse complex) and the AN chain complex (Alternative Novikov complex) for any tame map which, when applied to a Morse map is isomorphic to the Morse complex, respectively the Novikov complex, tensored by the field of coefficients.

Chapter 5 follows [Burghelea, D., Haller, S. (2015)], [Burghelea, D. (2015b)], and [Burghelea, D. (2016a)]. It provides an alternative treatment of closed and open barcodes; one defines the configuration $\delta_r(f)$ as well as the refinement configuration $\widehat{\delta}_r(f)$, and one establishes two of the important results, the *stability property* and the Poincaré Duality for the configurations $\delta_r(f)$ and $\widehat{\delta}_r(f)$.

---

[12]Concept introduced by S. Haller

Chapter 6 provides the definition of the configuration $\gamma_r(f)$ with the help of the Fredholm cross-ratio, a concept introduced in Subsection 2.1.2. Here we also prove the Cohen-Steiner-Edelsbrunner-Harer stability theorem and establish a Poincaré Duality theorem for the configurations $\gamma_r(f)$. Note that in the case of real-valued maps the configuration $\gamma_r(f)$ is simply a juxtaposition of the persistence diagrams (in the sense of [Edelsbrunner, H., Letscher, D., Zomorodian, A. (2002)]) without the points at infinity of the maps f and $-f$.

Chapter 7 introduces the topological r-monodromies and the sets of Jordan cells $\mathcal{J}_r(f)$ with the help of linear relations and proves their homotopy invariance based on properties of linear relations. The homotopy invariance of $\mathcal{J}_r(f)$ follows also from the homological definition of monodromy and of Theorem 4.1. The present definition as well as the definition of the Jordan cells based on graph representations make the monodromies, and in view of Theorem 4.1, also the Novikov-Betti numbers, computer friendly. The treatment based on linear relations suggests new methods and therefore new algorithms for the calculation of the Alexander polynomial of a knot and of the Alexander rational function for a pair (X, ξ), $\xi \in H^1(X; \mathbb{Z})$. The definition of monodromy we propose under the name of the geometric monodromy is given first for ξ representable by a continuous map which has at least one weakly regular value, and subsequently extended to arbitrary continuous maps using a few results in infinite-dimensional topology of compact Hilbert cube manifolds.

Chapter 8 reviews the definition of the classical Morse and Novikov complexes associated with a Morse-Smale vector field which has a real- and respectively angle-valued map f as Lyapunov map. The key result is that when tensored by a field κ, this complex is actually isomorphic to the AM or AN complex of f defined in terms of barcodes in Chapter 4. The AM and AN complexes depend only on the Morse map, since they are defined in terms of barcodes. This result justifies in part the name "Alternative to Morse Novikov theory" in the title of this book; it derives one of the key objects of classical Morse-Novikov theory in terms of some of the new invariants discussed in this book, the barcodes. The chapter also contains some computations of relevance for the algebraic topology of complements of complex algebraic hypersurfaces, most of them known but only for the field $\kappa = \mathbb{C}$ and derived by different methods.

Finally, Chapter 9 discusses the relation with *Persistence Theory* at various stages of development and makes a number of observations and analogies, including a measure-theoretic perspective on the configurations

δ_r^f and γ_r^f which might provide a new direction of research.

The material presented here is essentially my work partly in collaboration with Tamal Dey and Stefan Haller, as specified in each chapter whenever the case. A large part of this book follows parts of this work published or unpublished, but largely available in preprints.

The AMN theory is a new territory expecting explorations which I hope will reward the researcher's efforts with new mathematics and possible applications outside mathematics.

Concerning the book.

- I tried to make the reading of each chapter possible without detailed knowledge of the previous material and with minimal references to the previous notations and definitions. To achieve this each chapter begins with a brief summary of the relevant concepts and notations. Avoidable repetitions do occur unpleasantly often but, I hope, the reading is nevertheless easier.

- In order to lighten heavy notations, various specifications such as indices, exponents, or others notational additions are sometimes discarded when there is no danger of ambiguity or confusion.

- In general I use "$=$" for equality or for canonical isomorphism and "$\simeq$" for existing but unspecified isomorphism. We also use "$\sqcup$" for disjoint union and "$\sharp$" to denote cardinality.

- There is an unfortunate inconsistency of notations. I use $\kappa[t^{-1}, t]$ for the Laurent polynomials with coefficients in κ and $\kappa[S]$ for the κ-vector space generated by a set S. Both are standard mathematical notations for these concepts. Fortunately, they appear in situations where there is no risk of confusion.

- We will use the standard abbreviations:
  - **f.g** for finitely generated
  - **p.l** for *piecewise linear*

Chapter 2

Preparatory Material

2.1 Linear algebra

2.1.1 *Matrices*

The material presented in this subsection follows very closely [Gelfand, I. M. (1961)], Chapter 3, Section 21.

An $n \times m$ matrix is a table

$$A = \begin{pmatrix} a_{1,1} & a_{1,2} & a_{1,3} & \cdots & a_{1,m} \\ a_{2,1} & a_{2,2} & a_{2,3} & \cdots & a_{2,m} \\ a_{3,1} & a_{3,2} & a_{3,3} & \cdots & a_{3,m} \\ \vdots & \ddots & \ddots & \ddots & \vdots \\ a_{n,1} & a_{n1,2} & a_{n,3} & \cdots & a_{n,m} \end{pmatrix}$$

whose entries $a_{i,j}$ are elements in a given commutative ring with unit R.

In this section R will be either a field κ or $\kappa[t^{-1}, t]]$ (Laurent power series), or the ring $\kappa[t]$ (polynomials), or the ring $\kappa[t^{-1}, t]$ (Laurent polynomials) with coefficients in κ. [1]

Two $n \times m$ matrices, A with entries $a_{i,j}$ and B with entries $b_{i,j}$ can be added, the result being the matrix $C = A + B$ with entries $c_{i,j} = a_{i,j} + b_{i,j}$.

An $n \times m$ matrix A and an $m \times p$ matrix B can be composed (multiplied), yielding the $n \times p$ matrix $C = A \cdot B$ with entries $c_{i,j} = \sum_{r=1,\cdots m} a_{i,r} b_{r,j}$.

A matrix A can be regarded as an R-linear map $T : R^n \to R^m$, and any linear map $T : V \to W$, with V and W free R-modules of rank n and m, and with bases $e_1, e_2, \ldots, e_n$ for V and $f_1, f_2, \ldots, f_m$ for W, can be written

---

[1] An expression $\sum_{i \in \mathbb{Z}} a_i t^i$ with all but finitely many (all but finitely many negative) coefficients equal to zero is called a Laurent polynomial (Laurent power series). If all but finitely many coefficients and all negative coefficients are zero, then the expression is a polynomial.

as a matrix A with entries $a_{i,j}$ given by

$$T(e_i) = \sum_{j=1,\dots,m} a_{i,j} f_j.$$

The reader is assumed to be familiar with the standard concepts and notations: $\ker T$, $\mathrm{coker}\, T$, $\mathrm{img}\, T$, direct sum ($\oplus$).

If $n = m$, then the matrix A is referred below as an n-square matrix or an $n \times n$ matrix, and for a square matrix A, its *determinant* $\det(A) \in R$ is defined by

$$\det A = \sum_{\sigma} \mathrm{sign}(\sigma) a_{1,\sigma(1)} a_{2,\sigma(2)} \cdots a_{n,\sigma(n)},$$

where the sum is over all permutations of the set $\{1, 2, \dots, n\}$ and $\mathrm{sign}(\sigma)$ denotes the sign of the permutation σ.

One denotes by $E \equiv Id_n$ the matrix with entries $e_{i,j} = \delta_{i,j}{}^2$. Clearly, $\det(E) = 1$. Endowed with the operations of addition and multiplication, the set of all $n \times n$ matrices with coefficients in R is a ring with unit, the matrix E (often written simply Id). A matrix A is invertible iff $\det(A)$ is an invertible element in R.

Given an $n \times n$ matrix, one considers its $(k \times k)$ submatrices, i.e., arrays with entries the elements $a_{i,j}$ located at the intersection of k different rows $1 \leq i_1 < i_2 \cdots i_k \leq n$ and k different columns $1 \leq j_1 < j_2 < \cdots < j_k \leq n$, and one refers to their determinants as k-minors.

Definition 2.1. Two $n \times n$ matrices A and B are said to be:

equivalent, if there exists invertible matrices C and D such that $A = C \cdot B \cdot D$, and

similar, if there exists the invertible matrix C such that $A = C \cdot B \cdot C^{-1}$ (equivalently, $A \cdot C = C \cdot B$).

For matrices with coefficients in a field we will use in this book two equivalent collections of invariants to characterize similar square matrices; the first is based on a collection of polynomials associated to the matrix, called *canonical divisors,* while the second is based on Jordan cells. To describe them some definitions and observations are needed.

Recall that the expression $p(x) = a_0 x^k + a_1 x^{k-1} + \cdots + a_k$, $a_i \in \kappa$, $i = 0, 1, \dots k$, $a_0 \neq 0$, is called a *polynomial of degree k* with coefficients in the field κ and when, in addition $a_0 = 1$, a *monic polynomial* of degree k.

${}^2 \delta_{i,j} = 0$ if $i \neq j$, and $\delta_{\mathrm{i,i}} = 1$

Given a finite collection of polynomials $p_1(x), p_2(x), \ldots, p_r(x)$, their greatest common divisor, denoted by

$$\text{g.c.d}(p_1(x), p_2(x), \ldots, p_r(x)),$$

is the monic polynomial of highest degree which divides all polynomials $p_i(x)$, $i = 1, 2, \ldots, r$. The g.c.d always exists, cf. [Lang, S. (2002)]. Note that given two polynomials $p(x)$ and $q(x)$, their g.c.d$(p(x), q(x)) = 1$ iff there exists polynomials $\alpha(x)$ and $\beta(x)$ such that $1 = \alpha(x)p(x) + \beta(x)q(x)$.

One writes $A(x)$ for an $n \times n$ matrix with coefficients in $\kappa[x]$, the ring of polynomials in one variable x with coefficients in κ, when at least one of the entries has degree larger than 0. When all entries are polynomials of degree zero, hence elements in κ, the matrix is called *constant* and one writes simply A, B, etc. The matrix $A(x)$ is called a *polynomial matrix*.

Note that any $A(x)$ can be written $A(x) = A_0 x^n + A_1 x^{n-1} + \cdots + A_{n-1}x + A_n$ with A_i constant matrices and $A_0 \neq 0$, in which case $A(x)$ is referred to as a polynomial matrix of *degree n*. The simplest example of matrix with polynomial entries is $A(x) = xE - A$.

Let $A(x)$ be an $n \times n$ polynomial matrix.

Denote by

- $D_1(A(x)))$ the g.c.d. of the 1×1 minors of $A(x)$,
- $D_2(A(x)))$ the g.c.d. of the 2×2 minors of $A(x)$,
- ⋯⋯⋯⋯
- $D_{n-1}(A(x)))$ the g.c.d. of the $(n-1) \times (n-1)$ minors of $A(x)$,
- $D_n(A(x) = \det A(x)$. In view of the definition of the determinant, $D_1(A(x))$ divides $D_2(A(x))$, which in turn divides $D_3(A(x))$, and continuing each $D_{i-1}(A(x)$ divides $D_i(A(x))$, $i \leq n$; we set
- $E_k(A(x)) := D_k(A(x))/D_{k-1}(A(x))$.

The polynomial $D_n(A(x))$ is called the **characteristic polynomial**, the collection $D_n(A(x)), D_{n-1}(A(x)), \cdots, D_1(A(x))$ the **canonical divisors** of the characteristic polynomial and the polynomials $E_k(A(x))$ the **elementary divisors** of the characteristic polynomial.

Definition 2.2. Two polynomial matrices $A(x)$ and $B(x)$ are said to be *equivalent* if they are equivalent as matrices with coefficients in $\kappa[x]$ (i.e., there exists polynomial matrices $C(x)$ and $D(x)$, invertible as matrices with entries in $\kappa[x]$, such that $A(x) = C(x) \cdot B(x) \cdot D(x)$.

Theorem 2.1. (Canonical Smith form)

1.   *If $A(x)$ and $B(x)$ are equivalent polynomial matrices, then $D_k(A(x)) = D_k(B(x))$.*

2.   *Any $n \times n$ matrix $A(x)$ with entries in $\kappa[x]$ is equivalent to the polynomial matrix*

$$A = \begin{pmatrix} E_1(A(x)) & 0 & 0 & \cdots & 0 \\ 0 & E_2(A(x)) & 0 & \cdots & 0 \\ 0 & 0 & E_3(A(x)) & \cdots & 0 \\ \vdots & \vdots & \vdots & \ddots & \vdots \\ 0 & 0 & 0 & \cdots & E_n(A(x)) \end{pmatrix}.$$

3.   *If A and B are two constant matrices (with entries in κ), then the matrices $xE - A$ and $xE - B$ are equivalent iff A and B are similar.*

Item 2. follows from Item 1. For a proof of Item 1. we refer to [Gelfand, I. M. (1961)]. To check Item 3., suppose that $xE - A = P(x)(xE - B)Q(x)$, with $P(x) = \sum_{0 \le i \le p} P_i x^i$ and $Q(x) = \sum_{0 \le j \le q} Q_j x^j$ invertible matrices with entries in $\kappa[x]$, P_i and Q_j constant matrices. Then necessarily $xE - A = P_0(xE - B)Q_0$ with P_0 and Q_0 invertible, whence $P_0 \cdot Q_0 = E$ and $A = P_0 B Q_0$. Therefore, A is similar to B. The converse is obvious.

When $A(x) = xE - A$ with A an $n \times n$ matrix, we refer to the polynomials $D_n(A(x)), D_{n-1}(A(x)), \cdots, D_1(A(x))$ as the **characteristic polynomial** and the **canonical divisors** of the characteristic polynomial of the matrix A, and denote them by $D_k^A(x)$, $k = n, n - 1, \ldots, 1$. We also refer to the polynomials $E_k(A(x))$ as the elementary divisors of the characteristic polynomial $D_n^A(x)$ and denote them by $E_k^A(x)$. As an immediate corollary of Theorem 2.1 one has:

Theorem 2.2. *Two square $n \times n$ matrices are similar iff their elementary divisors and then their canonical divisors, are the same.*

Since the elementary divisors of the characteristic polynomials of two similar matrices coincide, the characteristic polynomial with its canonical divisors, or equivalently the elementary divisors, can be associated to any linear map $T : V \to V$, with V a finite-dimensional vector space, with no reference to matrix representations of T.

Moreover, since the polynomials $D_k^A(x)$ remain the same when a matrix A with entries in κ is regarded as a matrix with entries in the algebraic closure $\bar{\kappa}$ of κ, or any other field extending κ, one obtains the following useful corollary.

Corollary 2.1. *Let $\kappa \subset \bar{\kappa}$. Two matrices A and B with entries in the field κ are similar iff they are similar as matrices with entries in $\bar{\kappa}$.*

Definition 2.3. An $n \times n$ matrix A is *decomposable* if it is similar to a block matrix $\begin{pmatrix} A_1 & 0 \\ 0 & A_2 \end{pmatrix}$ with A_1 a $k \times k$ matrix with $1 \le k \le n-1$; equivalently, if A is similar to the direct sum of two nontrivial square matrices, A_1 and A_2.

A square matrix is *indecomposable* if it is not decomposable. An indecomposable matrix is called a *Jordan block*.

Any matrix A is similar to a block diagonal matrix

$$\begin{pmatrix} A_1 & 0 & \cdots & 0 & 0 \\ 0 & A_2 & \cdots & 0 & 0 \\ \vdots & \vdots & \ddots & \vdots & \vdots \\ 0 & 0 & \cdots & A_{p-1} & 0 \\ 0 & 0 & \cdots & 0 & A_p \end{pmatrix}$$

with A_i, $i = 1, \ldots, p$ indecomposable matrices.

For $\lambda \in \kappa$ and $k \in \mathbb{N}_{>1}$ one denotes

$$T(\lambda, k) = \begin{pmatrix} \lambda & 1 & 0 & \cdots & 0 & 0 \\ 0 & \lambda & 1 & \cdots & 0 & 0 \\ 0 & 0 & \lambda & \ldots & 0 & 0 \\ \vdots & \vdots & \vdots & \ddots & \vdots & \vdots \\ 0 & 0 & 0 & \cdots & \lambda & 1 \\ 0 & 0 & 0 & \cdots & 0 & \lambda \end{pmatrix}, \tag{2.1}$$

and one refers to the matrix $T(\lambda, k)$ and $T(\lambda, 1) = (\lambda)$ as a *Jordan cell*. It is not hard to see that any Jordan cell is indecomposable, cf. [Gelfand, I. M. (1961)] Chapter 3, Section 18.

Proposition 2.1. *If $\lambda \in \kappa$ is an eigenvalue[3] of an $n \times n$ indecomposable matrix A (Jordan block), then A is similar to the Jordan cell $T(\lambda, k)$.*

Proof. Let $T : \kappa^n \to \kappa^n$ be the linear map represented by A. If λ has multiplicity strictly less than n, then A is similar to the matrix $\begin{pmatrix} A_1 & 0 \\ 0 & A_2 \end{pmatrix}$, with A_1 a matrix representing T restricted to

$$\bigcup_{r \ge 1} \ker(T - \lambda E)^r,$$

---

[3]This means that $\dim \ker(A - \lambda E) \ne 0$.

the so-called generalized eigenspace of λ, and A_2 the linear map induced by T on

$$\kappa^n / \bigcup_{r \geq 1} \ker(T - \lambda E)^r,$$

hence A is not indecomposable.

Since under the hypothesis $\bigcup_{r \geq 1} \ker(T - \lambda E)^r = \kappa^n$, one can choose a basid $v_1, v_2, \ldots, v_n$, such that $(T - \lambda E)(v_1) = 0$ and $(T - \lambda E)(v_{i+1}) = v_i$, $i > 1$. We leave the verification of this last statement to the reader. Clearly, with respect to this basis T is given by the matrix $T(\lambda, n)$. The vector v_1 is called the *leading vector*. $\qquad \square$

Combining the *similarity of any matrix with a block diagonal matrix with indecomposable matrices on the diagonal* stated above with Proposition 2.1 and Theorems 2.2 one derives the following result.

Theorem 2.3. (Jordan canonical form theorem)
Any $n \times n$ matrix with n eigenvalues, in particular any $n \times n$ matrix with entries in an algebraically closed field is similar to a direct sum of Jordan cells

$$\begin{pmatrix} J_1 & 0 & \cdots & 0 & 0 \\ 0 & J_2 & \cdots & 0 & 0 \\ \vdots & \vdots & \ddots & \vdots & \vdots \\ 0 & 0 & \cdots & J_{r-1} & 0 \\ 0 & 0 & \cdots & 0 & J_r \end{pmatrix}. \tag{2.2}$$

Moreover, the collection of these Jordan cells is unique up to a permutation, providing a complete set of invariants for the similarity class of A.

A direct proof of the Jordan canonical form (or decomposition) theorem which can be converted into an algorithm for the calculation of the Jordan cells is given in [Gelfand, I. M. (1961)], Chapter 3, Section 18. For the benefit of potential practitioners we include this proof in Appendix, Subsection 2.1.3.

It is straightforward from the definitions to derive from the collection $\mathcal{J}(A)$ of Jordan cells of A the characteristic polynomial, the canonical divisors, and the elementary divisors, cf. [Gelfand, I. M. (1961)], Section 20.

If $\mathcal{J}(A)$ consists of:
p pairs (λ_1, n_i), $(n_1 \geq n_2 \geq \cdots \geq n_p)$,

q pairs (λ_2, m_i), $(m_1 \geq m_2 \geq \cdots \geq m_q)$,
etc, then

$$D_n^A = (x - \lambda_1)^{n_1 + n_2 + n_3 + \cdots + n_p}(x - \lambda_2)^{m_1 + m_2 + m_3 + \cdots + m_q} \cdots ,$$
$$D_{n-1}^A = (x - \lambda_1)^{n_2 + n_3 + \cdots + n_p}(x - \lambda_2)^{m_2 + m_3 + \cdots + m_q} \cdots ,$$
$$D_{n-2}^A = (x - \lambda_1)^{n_3 + \cdots + n_p}(x - \lambda_2)^{m_3 + \cdots + m_q} \cdots ,$$
$$\cdots\cdots$$

and

$$E_n^A = (x - \lambda_1)^{n_1}(x - \lambda_2)^{m_1} \cdots ,$$
$$E_{n-1}^A = (x - \lambda_1)^{n_2}(x - \lambda_2)^{m_2} \cdots ,$$
$$E_{n-2}^A = (x - \lambda_1)^{n_3}(x - \lambda_2)^{m_3} \cdots ,$$
$$\cdots\cdots$$

Echelon form

Let κ be a field and let

$$M = \begin{pmatrix} a_{1,1} & a_{1,2} & a_{1,3} & \cdots & a_{1,m} \\ a_{2,1} & a_{2,2} & a_{2,3} & \cdots & a_{2,m} \\ a_{3,1} & a_{3,2} & a_{3,3} & \cdots & a_{3,m} \\ \vdots & \ddots & \ddots & \ddots & \vdots \\ a_{n,1} & a_{n1,2} & a_{n,3} & \cdots & a_{n,m} \end{pmatrix} .$$

be an $n \times m$ matrix with entries in κ.

It is useful to think of M as defining a linear map $M : V = \kappa^m \to W = \kappa^n$, and think of any linear map $M : V \to W$ in the presence of a basis $\{e_1, e_2, \ldots, e_m\}$ of V and a basis $\{f_1, f_2, \ldots, f_n\}$ of W as a matrix with $M(e_i) = \sum_j a_{ji} f_j$.

The transpose M^T of the matrix M is the $m \times n$ matrix whose rows are the columns of M and columns are the rows of M; M^T can be viewed as the linear map $T^* : W^* \to V^*$ induced by $T : V \to W$ by *passing to the dual spaces*, and described in terms of the dual bases $\{f_1^*, f_2^*, \ldots, f_n^*\}$ and $\{e_1^*, e_2^*, \ldots, e_j^*\}$.

A row of the matrix M is a *zero-row* if all its entries are zero. The *leading entry* and the *ending entry* in a row are the first and respectively the last nonzero entries.

A column of the matrix M is a *zero-column* if all its entries are zero. The *leading entry* and the *ending entry* in a column are the first and respectively the last nonzero entries.

Definition 2.4.

- The matrix M is in *row echelon form*, written REF, if (1) and (2) below hold, and in *reduced row echelon form*, written RREF, if (1), (2), (3) and (4) hold:

  (1) all zero rows are below nonzero ones,
  (2) for each row the leading entry lies to the right of the leading entry of the previous row,
  (3) for each nonzero row the leading term is 1,
  (4) if a column has an entry 1, then all other entries are zero.

  The following matrix is in reduced row echelon form:

$$M = \begin{pmatrix} 0\,0\,1\,0\,*\,*\,0\,* \\ 0\,0\,0\,1\,*\,*\,0\,* \\ 0\,0\,0\,0\,0\,0\,1\,* \\ 0\,0\,0\,0\,0\,0\,0\,0 \\ 0\,0\,0\,0\,0\,0\,0\,0 \end{pmatrix}.$$

- The matrix M is in *column echelon form*, written CEF, if (1) and (2) below hold, and in *reduced column echelon form*, written RCEF, if (1), (2), (3) and (4) hold:

  (1) all zero columns lie to the right of nonzero ones,
  (2) for each column the leading entry lies below the leading entry of the previous column,
  (3) for each nonzero column the leading term is 1,
  (4) if a row has an entry 1, then all other entries are zero.

  A matrix M is in CEF (RCEF) form iff its transpose M^T is in REF (respectively, RREF) form.

The following matrix is in reduced row echelon form:

$$
M = \begin{pmatrix}
0 & 0 & 0 & 0 & 0 \\
0 & 0 & 0 & 0 & 0 \\
1 & 0 & 0 & 0 & 0 \\
0 & 1 & 0 & 0 & 0 \\
* & * & 0 & 0 & 0 \\
* & * & 0 & 0 & 0 \\
0 & 0 & 1 & 0 & 0 \\
* & * & * & 0 & 0
\end{pmatrix} .
$$

Proposition 2.2.

(i) *For any $n \times m$ matrix M one can produce an invertible $n \times n$ matrix C such that the composition $C \cdot M$ is in* RREF. *Such a matrix will be denoted by $L(M)$.*

(ii) *For any $n \times m$ matrix matrix M one can produce an invertible $m \times m$ matrix C' such that the composition $M \cdot C'$ is in* RCEF. *Such matrix will be denoted by $R(M)$.*

For the proof it suffices to check item (i). Item (ii) follows by applying item (i) to M^T.

The construction of $L(M)$ is based on the *Gauss elimination* procedure (cf. [Gelfand, I. M. (1961)]) consisting in operations of permuting rows, multiplying rows by a nonzero element of κ, and replacing a row by itself plus a multiple of another row. Each such operation is realizable by left multiplication by a matrix of type $T(i, j)$, $T(i; \lambda)$, $T(i, j; \lambda)$, $1 \le i < j \le n$, with

$$
T(i; \lambda)_{r,s} = \begin{cases}
1, & \text{if } r = s \ne i, \\
\lambda, & \text{if } r = s = i, \\
0, & \text{otherwise},
\end{cases}
$$

$$
T(i, j)_{r,s} = \begin{cases}
1, & \text{if } r = s \ne i, j, \\
1, & \text{if } r = i,\ s = j \text{ or } r = j,\ s = i, \\
0, & \text{otherwise},
\end{cases}
$$

$$
T(i, j; \lambda)_{r,s} = \begin{cases}
1, & \text{if } r = s, \\
\lambda, & \text{if } r = i, s = j, \\
0, & \text{otherwise}.
\end{cases}
$$

All basic software which carries linear algebra packages contain sub-packages which input a matrix and output its reduced row/column echelon form, as well as the matrix $L(M)$ or $R(M)$.

2.1.2   *Fredholm maps and Fredholm cross-ratio*

Let κ be a fixed field. A linear map $f : A \to B$ between two vector spaces A and B is called *Fredholm* if the dimensions of its kernel and cockernel are finite. Recall that

$$\ker f := \{x \in A \mid f(x) = 0\},$$
$$\operatorname{img} f := \{y \in B \mid y = f(x), x \in A\},$$
$$\operatorname{coker} f := B/\operatorname{img} f.$$

For A a κ-vector space one denotes by $A^* := \hom(A, \kappa)$ the dual space of A (and similarly for B^*), and for a linear map $f : A \to B$ one denotes by $f^* : B^* \to A^*$ the induced linear map, the dual of f. Note that $\ker(f^*)$ is canonically isomorphic to $(\operatorname{coker} f)^*$.

For $f : A \to B$ a Fredholm map one defines the *index* of f, denoted by $\operatorname{ind} f$, as the number

$$\operatorname{ind} f := \dim \ker f - \dim \operatorname{coker} f.$$

The following properties are straightforward consequences of the definition.

(1) If the spaces A and B are finite-dimensional, then

$$\operatorname{ind} f = \dim A - \dim B.$$

(2) If $f : A \to B$ and $g : B \to C$ are two Fredholm linear maps, then $g \cdot f$ is Fredholm and

$$\operatorname{ind}(g \cdot f) = \operatorname{ind} f + \operatorname{ind} g.$$

Let $\alpha : A \to B$, $\beta : B \to C$, $\gamma : C \to D$ be Fredholm maps. To these three Fredholm maps we associate the following two diagrams, $\widehat{\mathcal{D}}$ and $\underline{\mathcal{D}}$:

$$\widehat{\mathcal{D}} \equiv \begin{cases} \ker(\gamma\beta\alpha) \xrightarrow{\;j_2\;} \ker(\gamma\beta) \\ \quad\uparrow{\scriptstyle i_1} \qquad\qquad \uparrow{\scriptstyle i_2} \\ \ker(\beta\alpha) \xrightarrow{\;j_1\;} \ker(\beta) \end{cases} \qquad \underline{\mathcal{D}} \equiv \begin{cases} \operatorname{coker}(\gamma\beta\alpha) \xrightarrow{\;j'_2\;} \operatorname{coker}(\gamma\beta) \\ \quad\uparrow{\scriptstyle i'_1} \qquad\qquad \uparrow{\scriptstyle i'_2} \\ \operatorname{coker}(\beta\alpha) \xrightarrow{\;j'_1\;} \operatorname{coker}(\beta) \end{cases}$$

with i_1 and i_2 injective and j_1' and j_2' surjective, as well as the following two finite-dimensional vector spaces[4]:

$$\widehat{\omega}(\alpha, \beta, \gamma) = \operatorname{coker} \widehat{\mathcal{D}} := \operatorname{coker}(j_1 \cup_{\ker(\beta\alpha)} i_1 \to \ker(\gamma\beta)),$$

a quotient space of $\ker(\gamma\beta)$, and

$$\underline{\omega}(\alpha, \beta, \gamma) = \ker \underline{\mathcal{D}} := \ker(\operatorname{coker}(\beta\alpha) \to i_2' \times_{\operatorname{coker}(\gamma\beta)} j_2'),$$

a subspace of $\operatorname{coker}(\beta\alpha)$. In view of Theorem 2.4 below the dimensions of these two spaces are equal and called the *Fredholm cross-ratio*[5].

Note that for $\alpha : A \to B$ Fredholm map the diagram

$$
\begin{array}{ccccccc}
(\ker \alpha)^* & \longleftarrow & A^* & \longleftarrow & B^* & \longleftarrow & (\operatorname{coker}(\alpha))^* \\
\uparrow & & \| & & \| & & \uparrow \\
\operatorname{coker}(\alpha^*) & \longleftarrow & A^* & \longleftarrow & B^* & \longleftarrow & \ker(\alpha^*)
\end{array}
$$

provides the canonical isomorphisms

$$\ker(\alpha^*) \to (\operatorname{coker}\alpha)^* \quad \text{and} \quad \operatorname{coker}(\alpha^*) \to (\ker\alpha)^*,$$

which in turn extend to the canonical isomorphisms

$$\underline{\omega}(\gamma^*, \beta^*, \alpha^*) \to (\widehat{\omega}(\alpha, \beta, \gamma)) \quad \text{and} \quad \widehat{\omega}(\gamma^*, \beta^*, \alpha^*) \to (\underline{\omega}(\alpha, \beta, \gamma)).$$

Note also that the assignments $(\alpha, \beta, \gamma) \rightsquigarrow \widehat{\omega}(\alpha, \beta, \gamma)$ and $(\alpha, \beta, \gamma) \rightsquigarrow \underline{\omega}(\alpha, \beta, \gamma)$ are functorial and enjoy the following properties:

Theorem 2.4.

1. *For α, β, γ Fredholm maps, it holds that*

$$\dim \widehat{\omega}(\alpha, \beta, \gamma) = \dim \underline{\omega}(\gamma^*, \beta^*, \alpha^*) =$$

$$\dim \ker(\gamma \cdot \beta) + \dim \ker(\beta \cdot \alpha) - \dim \ker(\gamma \cdot \beta \cdot \alpha) - \dim \ker(\beta) =$$

$$\dim \operatorname{coker}(\beta \cdot \alpha) + \dim \operatorname{coker}(\gamma \cdot \beta) - \dim \operatorname{coker}(\gamma \cdot \beta \cdot \alpha) - \dim \operatorname{coker}(\beta).$$

2. *Consider the diagram*

$$
\begin{array}{ccccccc}
M & \xrightarrow{=} & M & \xrightarrow{=} & M & \xrightarrow{=} & M \\
\uparrow\scriptstyle a & & \uparrow\scriptstyle b & & \uparrow\scriptstyle c & & \uparrow\scriptstyle d \\
A & \xrightarrow{\alpha} & B & \xrightarrow{\beta} & C & \xrightarrow{\gamma} & D \\
\uparrow & & \uparrow & & \uparrow & & \uparrow \\
A' & \xrightarrow{\alpha'} & B' & \xrightarrow{\beta'} & C' & \xrightarrow{\gamma'} & D' \\
\uparrow\scriptstyle a'' & & \uparrow\scriptstyle b'' & & \uparrow\scriptstyle c'' & & \uparrow\scriptstyle d'' \\
N & \xrightarrow{=} & N & \xrightarrow{=} & N & \xrightarrow{=} & N
\end{array}
\qquad (2.3)
$$

---

[4]Recall that for $\alpha_1 : A \to B_1$ and $\alpha_2 \to B_2$, the push-out $\alpha_1 \cup_A \alpha_2$ is defined as $B_1 \oplus B_2 / \{(\alpha_1(a), -\alpha_2(a)) \mid a \in A\}$, and for $\beta_1 : B_1 \to C$ and $\beta_2 : B_2 \to C$ the pull-back $\beta_1 \times_C \beta_2$ is defined as $\{(b_1, b_2) \in B_1 \times B_2 \mid \beta_1(b_1) = \beta_2(b_2)\}$.

[5]By analogy with the familiar *cross-ratio* of four points on a line

in which the columns are exact sequences and α, β, γ and α', β', γ' are Fredholm maps. Then the linear maps

$$\widehat{\omega}(\alpha', \beta', \gamma') \to \widehat{\omega}(\alpha, \beta, \gamma)$$

and

$$\underline{\omega}(\alpha, \beta, \gamma) \to \underline{\omega}(\alpha, \beta, \gamma)$$

are isomorphisms.

    3. *For α, β, γ Fredholm maps one has the following:*

      (i) *A factorization $\alpha = \alpha_2 \cdot \alpha_1$, with $\alpha_1 : A \to A'$ and $\alpha_2 : A' \to B$ Fredholm maps, induces the short exact sequences*

$$0 \to \widehat{\omega}(\alpha_1, \beta\alpha_2, \gamma) \to \widehat{\omega}(\alpha, \beta, \gamma) \to \widehat{\omega}(\alpha_2, \beta, \gamma) \to 0,$$
$$0 \to \underline{\omega}(\alpha_1, \beta\alpha_2, \gamma) \to \underline{\omega}(\alpha, \beta, \gamma) \to \underline{\omega}(\alpha_2, \beta, \gamma) \to 0. \tag{2.4}$$

      (ii) *A factorization $\beta = \beta_2 \cdot \beta_1$, with $\beta_1 : B \to B'$ and $\beta_2 : B' \to B$ Fredholm maps, induces the short exact sequences*

$$0 \to \widehat{\omega}(\alpha, \beta_1, \beta_2) \to \widehat{\omega}(\alpha, \beta_1, \gamma\beta_2) \to \widehat{\omega}(\alpha, \beta, \gamma) \to 0,$$
$$0 \to \underline{\omega}(\alpha, \beta_1, \beta_2) \to \underline{\omega}(\alpha, \beta_1, \gamma\beta_2) \to \underline{\omega}(\alpha, \beta, \gamma) \to 0. \tag{2.5}$$

      (iii) *A factorization $\gamma = \gamma_2 \cdot \gamma_1$, with $\gamma_1 : C \to C'$ and $\gamma_2 : C' \to D$ Fredholm maps, induces the short exact sequences*

$$0 \to \widehat{\omega}(\alpha, \beta, \gamma_1) \to \widehat{\omega}(\alpha, \beta, \gamma) \to \widehat{\omega}(\alpha, \gamma_1\beta, \gamma_2) \to 0,$$
$$0 \to \underline{\omega}(\alpha, \beta, \gamma_1) \to \underline{\omega}(\alpha, \beta, \gamma) \to \underline{\omega}(\alpha, \gamma_1\beta, \gamma_2) \to 0. \tag{2.6}$$

Proof. Item 1. It suffices to check the statements for $\widehat{\omega}(\alpha, \beta, \gamma)$; the companion statements for $\underline{\omega}(\alpha, \beta, \gamma)$ follow from the natural isomorphism

$$\underline{\omega}(\alpha, \beta, \gamma)^* = (\widehat{\omega}(\gamma^*, \beta^*, \alpha^*)).$$

The verification of this assertion will be better conceptualized when one relies on the following observations.

Consider the commutative diagram $\mathbb{D}$ of finite-dimensional vector spaces

$$\begin{array}{ccc} A_2 & \xrightarrow{\; i_2^{A,B} \;} & B_2 \\ {\scriptstyle j_A} \Big\uparrow & {\scriptstyle k} \nearrow & \Big\uparrow {\scriptstyle j_B} \\ A_1 & \xrightarrow[\; i_1^{A,B} \;]{} & B_1 \end{array} \tag{2.7}$$

which satisfies:

      (1) j_A and j_B are injective,

(2) $j_A : \ker(i_1^{A,B}) \to \ker(i_2^{A,B})$ is an isomorphism (the diagram is cartesian),

(3) $i_2^{A,B}(A_2) \cap j_B(B_1) = k(A_1)$.

Note that the diagram $\widehat{\mathcal{D}}$,

$$\begin{array}{ccc} \ker(\gamma\beta\alpha) & \xrightarrow{i_2} & \ker(\gamma\beta)] \ , \\ \Big\uparrow{\scriptstyle j_1} & \nearrow{\scriptstyle \kappa} & \Big\uparrow{\scriptstyle j_2} \\ \ker(\beta\alpha) & \xrightarrow{i_1} & \ker(\beta) \end{array}$$

satisfies (1),(2),(3), and $\widehat{\omega}(\alpha,\beta,\gamma) = \ker(\gamma\beta)/(\mathrm{img}\, i_2 + \mathrm{img}\, j_2)$.

In view of the injectivity of j_A, $\dim\left(j_A \cup_{A_1} i_1^{A,B}\right) = \dim A_2 + \dim B_1 - \dim A_1$, and in view of properties (2) and (3) the linear map $A_2 \cup_{i_1^{A,B}} B_1 \to B_2$ induced by this diagram is injective[6]. These facts imply that

$$\dim(\widehat{\omega}(\alpha,\beta,\gamma)) = \dim A_1 + \dim B_2 - \dim A_2 - \dim B_1 \qquad (2.8)$$

and then Item 1. in Theorem 2.4 follows.

Item 2. To prove it, notice that:

(i) If $N = 0$, then the statement follows from the isomorphisms

$$\begin{aligned} \ker(\beta) &\simeq \ker(\beta'), \\ \ker(\beta\gamma) &\simeq \ker(\beta'\gamma') \\ \ker(\alpha\beta) &\simeq \ker(\alpha'\beta'), \\ \ker(\alpha\beta\gamma) &\simeq \ker(\alpha'\beta'\gamma'). \end{aligned} \qquad (2.9)$$

(ii) If $M = 0$, then the statement follows from the isomorphisms

$$\begin{aligned} \mathrm{coker}(\beta) &\simeq \mathrm{coker}(\beta'), \\ \mathrm{coker}(\beta\gamma) &\simeq \mathrm{coker}(\beta'\gamma'), \\ \mathrm{coker}(\alpha\beta) &\simeq \mathrm{coker}(\alpha'\beta'), \\ \mathrm{coker}(\alpha\beta\gamma) &\simeq \mathrm{coker}(\alpha'\beta'\gamma'). \end{aligned} \qquad (2.10)$$

---

[6]Indeed, $i_2(a_2) + j_B(b_1) = 0$ implies by (3) above that $i_2(a_2) = -j_B(b_1) = \kappa(\alpha)$, $\alpha \in A_1$. Since $i_2(a_2) - j_A(\alpha)) = 0$, by (2) above there exists $\beta \in A_1$ such that $i_2(j_A(\alpha+\beta)) = a_2$, therefore $a_2 = (a_2 + j_A(\beta), b_1)$ represents the same element as (a_2, b_1) in $j_A \cup_{A_1} i_2$, the difference between these two being $j_A(\beta), -i_1(\beta)$, hence the desired injectivity is established.

Consider the diagram

$$A \xrightarrow{\alpha} B \xrightarrow{\beta} C \xrightarrow{\gamma} D$$

$$\ker a \xrightarrow{\alpha''} \ker b \xrightarrow{\beta''} \ker c \xrightarrow{\gamma''} \ker d$$

$$A' \xrightarrow{\alpha'} B' \xrightarrow{\beta'} C' \xrightarrow{\gamma'} D'$$

and the linear maps

$$\widehat{\omega}(\alpha', \beta', \gamma') \to \widehat{\omega}(\alpha'', \beta'', \gamma'') \to \widehat{\omega}(\alpha, \beta, \gamma).$$

The first arrow is an isomorphism by (ii) above the second by (i) above. This establishes Item 2.

To prove Item 3 consider the diagram

$$(2.11)$$

and make the following

Observation 2.1. Suppose that each of the three diagrams $\mathbb{B}_1, \mathbb{B}_2, \mathbb{B}$, associated with (2.11),

$\mathbb{B}_1$ with vertices A_1, A_2, B_1, B_2,
$\mathbb{B}_2$ with vertices B_1, B_2, C_1, C_2, and
$\mathbb{B}$ with vertices A_1, A_2, C_1, C_2
satisfy the properties (1) (2) (3) of the diagram (2.7). Then (2.11) induces the exact sequence

with i induced by i_2^B, well defined because $\mathrm{img}(i_2^B \cdot j_B) \subseteq \mathrm{img}\, j_C$, and p the projection induced by the inclusion $(i_2(A_2) + j_C(C_1)) \subseteq (i_2^B(B_2) + j_C(C_1))$.

Clearly p is surjective and $p \cdot i = 0$. Property (3) in diagram (2.7) implies i injective. Calculation of dimensions based on (2.8) implies that the sequence is exact.

A similar observation holds for the diagram

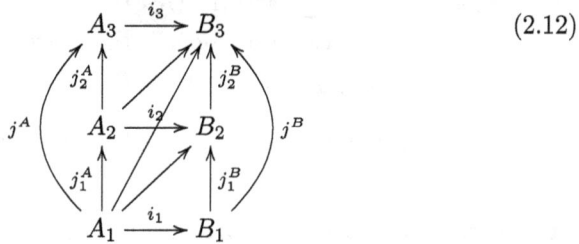

$$\tag{2.12}$$

Observation 2.2. Suppose that each of the three diagrams $\mathbb{B}_1, \mathbb{B}_2, \mathbb{B}$, associated with (2.12),

$\mathbb{B}_1$ with vertices A_2, A_3, B_2, B_3,

$\mathbb{B}_2$ with vertices A_1, A_2, B_1, B_2, and

$\mathbb{B}$ with vertices A_1, A_3, B_1, B_2

satisfies the properties (1) (2) (3) of the diagram (2.7). Then (2.12) induces the exact sequence

Observation 2.1 applied to diagram (2.11) with

$$A_1 = \ker(\beta\alpha), \quad A_2 = \ker(\gamma\beta\alpha),$$
$$B_1 = \ker(\beta\alpha_2), \quad B_2 = \ker(\gamma\beta\alpha_2),$$
$$C_1 = \ker(\beta), \quad C_2 = \ker(\gamma\beta),$$

verifies Item 3 (i) in Theorem 2.4.

Observation 2.2 applied to diagram (2.12) with

$$A_1 = \ker(\beta_1\alpha), \quad B_1 = \ker(\beta_1),$$
$$A_2 = \ker \beta\alpha), \quad B_2 = \ker(\beta),$$
$$A_3 = \ker(\gamma\beta\alpha), \quad B_3 = \ker(\gamma\beta)$$

verifies item 3 (ii) in Theorem 2.4 and applied to diagram (2.12) with

$$A_1 = \ker(\beta\alpha), \quad B_1 = \ker(\beta),$$
$$A_2 = \ker(\gamma_1\beta\alpha), \quad B_2 = \ker(\gamma_1\beta),$$
$$A_3 = \ker(\gamma\beta\alpha), \quad B_3 = \ker(\gamma\beta),$$

verifies Item 3 (iii) in Theorem 2.4.   □

The following factorization statement is quite technical, but will be used as such in Chapter 6.

Proposition 2.3. *Let $\alpha : A \to B$, $\beta : B \to C$, $\gamma : C \to D$, and $\delta : D \to F$ be Fredholm maps. Suppose that one has the factorization of $\alpha, \beta, \gamma, \delta$ as the composition of Fredholm maps*

(1) $\alpha = \alpha_3 \cdot \alpha_2 \cdot \alpha_1,$
(2) $\beta = \beta_3 \cdot \beta_2 \cdot \beta_1,$
(3) $\gamma = \gamma_3 \cdot \gamma_2 \cdot \gamma_1,$

and suppose that the compositions $\alpha_2 \cdot \alpha_1$, $\beta_2 \cdot \beta_1$, $\gamma_2 \cdot \gamma_1$ are isomorphisms and δ is injective.
    Define

$$\alpha' := \beta_1 \cdot \alpha_3 \cdot \alpha_2,$$
$$\beta' := \gamma_1 \cdot \beta_3 \cdot \beta_2, \tag{2.13}$$
$$\gamma' := \delta \cdot \gamma_3 \cdot \gamma_2.$$

Then the induced map $\widehat{\omega}(\alpha, \beta, \gamma) \to \widehat{\omega}(\alpha', \beta', \gamma')$ is an isomorphism.

Proof. First note that the linear map $\widehat{\omega}(\alpha, \beta, \gamma) \to \widehat{\omega}(\alpha', \beta'\gamma')$ is injective. Indeed, consider an inverse $\delta_2 : F \to D$ of δ and denote

$$\alpha'' := \beta_2 \cdot \beta_1 \cdot \alpha_3,$$
$$\beta'' := \gamma_2 \cdot \gamma_1 \cdot \beta_3, \tag{2.14}$$
$$\gamma'' := \delta_2 \cdot \delta \cdot \gamma_3.$$

Clearly, by the naturality of the construction $\widehat{\omega}$ one has the linear maps

$$\widehat{\omega}(\alpha, \beta, \gamma) \to \widehat{\omega}(\alpha', \beta', \gamma') \to \widehat{\omega}(\alpha'', \beta'', \gamma'')$$

whose composition, in view of the hypothesis, is an isomorphism. Therefore, it suffices to check the equality $\dim \widehat{\omega}(\alpha, \beta, \gamma) = \dim \widehat{\omega}(\alpha', \beta', \gamma')$.

By the hypothesis, $\alpha_1, \beta_1, \gamma_1, \delta_1$ are injective and $\alpha_2, \beta_2, \gamma_2$, are surjective. Then we have

$$\dim \ker(\gamma \cdot \beta \cdot \alpha) = \begin{cases} \dim \ker(\delta_1 \cdot \gamma \cdot \beta \cdot \alpha) = \dim \ker(\delta_1 \cdot \gamma \cdot \beta \cdot \alpha_3) = \\ \dim \ker(\delta_1 \cdot \gamma \cdot \beta \cdot \alpha_3 \cdot \alpha_2) - \dim \ker(\alpha_2) = \\ \dim \ker(\gamma' \cdot \beta' \cdot \alpha') - \dim \ker \alpha_2, \end{cases}$$

$$(2.15)$$

$$\dim \ker(\beta) = \begin{cases} \dim \ker(\gamma_1 \cdot \beta) = \dim \ker(\gamma_1 \cdot \beta_3 = \\ \dim \ker(\gamma_1 \cdot \beta_3 \cdot \beta_2) - \dim \ker(\beta_2) = \\ \dim \ker(\beta') - \dim \ker \beta_2, \end{cases} \qquad (2.16)$$

$$\dim \ker(\gamma \cdot \beta) = \begin{cases} \dim \ker(\delta_1 \gamma \cdot \beta) = \dim \ker(\delta_1 \gamma \cdot \beta_3 = \\ \dim \ker(\delta_1 \gamma \cdot \beta_3 \beta_2) - \dim \ker(\beta_2) = \\ \dim \ker(\gamma' \cdot \beta') - \dim \ker \beta_2, \end{cases} \qquad (2.17)$$

$$\dim \ker(\beta \cdot \alpha) = \begin{cases} \dim \ker(\gamma_1 \beta \cdot \alpha) = \dim \ker(\gamma_1 \beta \cdot \alpha_3) = \\ \dim \ker(\gamma_1 \cdot \beta \cdot \alpha_3 \cdot \alpha_2) - \dim \ker(\alpha_2) = \\ \dim \ker(\beta' \cdot \alpha') - \dim \ker \alpha_2. \end{cases} \qquad (2.18)$$

Adding the final calculations in (2.15), (2.16), (2.17), and (2.18) the assertion follows. $\square$

2.1.3 *Appendix*

Proof of the Jordan canonical form theorem, the existence part (cf. [Gelfand, I. M. (1961)])

We start with a linear transformation $A : V \to V$ with $\dim V < \infty$ and want to find a base in V s.t A has a matrix representation of the form (2.2) with J_i Jordan cells. We prove the statement by induction on the dimension of V.

For this purpose we start with an eigenvalue $\lambda \in \kappa$ and consider $V_1 := \ker(A - \lambda E)$ and $V' = (A - \lambda E)(V)$. Let $r = \dim V_1$. Both subspaces V_1 and V' are clearly invariant under A and the restrictions of A to these spaces have $\dim V_1$ and $\dim V'$ eigenvalues in κ. If $V_1 \cap V' = 0$ then the result follows from the induction hypothesis on V' and the observation that

$A : V_1 \to V_1$ is a direct sum of Jordan cells of size 1 and eigenvalue λ. If $V_1 = V$, then A is a direct sum of Jordan cells of size 1.

Suppose that $W = V_1 \cap V'$ has positive dimension p, $p < \dim V$, hence $\dim V' = p + r$ and $\dim V_1 = p + q$, $q \geq 0$. Note that the restriction of A to $V_1 \cap V'$ has $\dim(V_1 \cap V')$ eigenvalues in κ. Choose a basis $e_1, e_2, \ldots, e_{p+r}$ of V' such that $A : V' \to V'$ has Jordan form. Then p elements of this basis, say $e_1, \ldots, e_p$, are the eigenvectors of Jordan cells of eigenvalue λ and then elements in V_1. If $q > 0$, then complete this set of vectors to a basis of V_1 by choosing additional vectors $f_1, f_2, \ldots, f_q$. As noticed, the vectors $e_1, \ldots, e_p$ are the eigenvectors of different Jordan cells (of size $n_1, n_2, \ldots, n_p$), hence the basis $e_1, \cdots, e_{p+r}$ contains p different elements $e_{j_1}, \ldots, e_{j_p}$ which are lead vectors for these Jordan cells.

Choose $h_1, \ldots, h_p \in V$ such that $(A - \lambda E)h_k = e_{j_k}$. Note that the vectors $f_1, \ldots, f_q, h_1, \ldots, h_p, e_1, \ldots, e_{p+r}$ constitute a basis for V, and with respect to this basis A has Jordan form. The Jordan cells corresponding to eigenvalues different from λ for $A : V' \to V'$ are the same as for $A : V \to V$, and as long as the eigenvalue λ is concerned, there are exactly q Jordan cells of size one with eigenvectors $f_1, \ldots, f_q$ and exactly p Jordan cells of sizes $n_1 + 1, n_2 + 1, \ldots, n_p + 1$ with eigenvectors $e_1, \ldots, e_p$ and leading vectors $h_1, \ldots h_p$. This completes the proof of the existence part.

2.2   Linear relations

2.2.1   *General considerations*

This section follows closely [Burghelea, D., Haller, S. (2015)] and [Burghelea, D. (2015a)]. Fix a field κ.

A *linear relation* from V to W, two finite-dimensional vector spaces over κ, can be considered as a linear subspace $R \subseteq V \times W$. In symbols, we write $R \colon V \rightsquigarrow W$. For $v \in V$ and $w \in W$ we write vRw if $(v, w) \in R$.

Every linear map $f : V \to W$ can be regarded as a linear relation $R(f) \colon V \rightsquigarrow W$, setting $vR(f)w$ if $f(v) = w$. As a subspace of $V \times W$, $R(f)$ is the graph of the linear map f. To simplify notation, we will often write $f \colon V \rightsquigarrow W$ instead of $R(f) \colon V \rightsquigarrow W$.

If U is another vector space and $S \colon W \rightsquigarrow U$ is a linear relation, then the composition $SR \colon V \rightsquigarrow U$ is the linear relation defined by $v(SR)u$ iff there exists $w \in W$ such that vRw and wSu. Clearly, this is an associative composition generalizing the ordinary composition of linear maps. For the identical relation from V to V we take $\Delta_V \subset V \times V := R_{E_V}$. The familiar category with the finite-dimensional vector spaces as objects and the linear maps as morphisms extends to the category of finite-dimensional vector spaces and linear relations. In this last category the direct sum of two objects $R_i \colon V_i \rightsquigarrow W_i$, $i = 1, 2$ is the linear relation $R_1 \oplus R_2 \colon V_1 \times V_2 \rightsquigarrow W_1 \times W_2$ defined by $(v_1, v_2)(R_1 \oplus R_2)(w_1, w_2)$ iff $v_1 R_1 w_1$ and $v_2 R_2 w_2$.

If $R \colon V \rightsquigarrow W$ is a linear relation then one defines a linear relation $R^\dagger \colon W \rightsquigarrow V$ by $w R^\dagger v$ iff $v R w$. Clearly, $R^{\dagger\dagger} = R$, $(SR)^\dagger = R^\dagger S^\dagger$ and $(R_1 \oplus R_2)^\dagger = R_1^\dagger \oplus R_2^\dagger$.

Two linear maps $a_1 : U \to V_1$ and $a_2 : U \to V_2$ define a linear relation $R^{(a_1, a_2)} \colon V_1 \rightsquigarrow V_2$ by the rule

$$R^{(a_1, a_2)} := \{(v_1, v_2) \mid \exists u \in U, a_1(u) = v_1, a_2(u) = (v_2)\}.$$

For simplicity we will often write $a_2 a_1^\dagger$ instead of $R^{(a_1, a_2)}$. Clearly, $(a_2 a_1^\dagger)^\dagger = a_1 a_2^\dagger \colon V_2 \rightsquigarrow V_1$.

Further, two linear maps $\alpha_1 : V_1 \to W$ and $\alpha_2 : V_2 \to W$ define a linear relation $R(\alpha_1, \alpha_2) \colon V_1 \rightsquigarrow V_2$ by the rule

$$R(\alpha_1, \alpha_2) := \{(v_1, v_2) \mid \alpha_1(v_1) = \alpha_2(v_2)\}.$$

For simplicity we will often write $\alpha_2^\dagger \alpha_1$ instead of $R(\alpha_1, \alpha_2)$. Clearly $(\alpha_2^\dagger \alpha_1)^\dagger = \alpha_1^\dagger \alpha_2 \colon V_2 \rightsquigarrow V_1$.

It is not hard to see that any linear relation $R \subset V_1 \times V_2$ can be realized as $R(\alpha_1, \alpha_2)$ for some linear maps $\alpha_1 : V_1 \to W, \alpha_2 : V_2 \to W$, or as $R^{(a_1, a_2)}$ for some linear maps $a_1 : U \to V_1, a_2 : U \to V_2$ [7].

As mentioned above, the category with the finite-dimensional vector spaces as objects and the linear relations as morphisms is abelian and extends the category of finite-dimensional vector spaces and linear maps. This category is equipped with the involution $\dagger$ on the set of morphisms which extends the *inverse of a linear isomorphism*. In particular, the Krull-Remack-Schmidt theorem holds, and therefore any linear relation decomposes as (is isomorphic to) a direct sum of indecomposable relations. The collection of these indecomposable relations is unique, but not the decomposition.

---

[7]For $R \subseteq V_1 \times V_2$, $i = 1, 2$ let

(i) $\mathrm{pr}_i : V_1 \times V_2 \to V_i$ be the projections $\mathrm{pr}_i : (v_1, v_2) = v_i$, and $\mathrm{in}_i : V_i \to V_1 \times V_2$ be the injections $\mathrm{in}_1(v_1) = (v_1, 0)$, $\mathrm{in}_2(v_2) = (0, v_2)$,

(ii) $\alpha_i : R \to V_i$ be the restrictions of pr_i to R and $\alpha : R \to V_1 \times V_2$ defined by $\alpha = (\alpha_1, -\alpha_2)$,

(iii) $\mathrm{pr} : V_1 \times V_2 \to W = V_1 \times V_2 / \mathrm{img}(\alpha)$ be the canonical projection,

(iv) $a_i : V_i \to W$ be the composition $a_i = \mathrm{pr} \cdot \mathrm{in}_i$.

Clearly $R = R(a_1, a_2) = R^{(a_1, a_2)}$.

A linear relation $R\colon V \rightsquigarrow W$ gives rise to the following subspaces:

$$\mathrm{dom}(R) := \{v \in V \mid \exists w \in W : vRw\} = \mathrm{pr}_V(R),$$
$$\mathrm{img}(R) := \{w \in W \mid \exists v \in V : vRw\} = \mathrm{pr}_W(R),$$
$$\ker(R) := \{v \in V \mid vR0\} \cong (V \times 0) \cap R,$$
$$\mathrm{mul}(R) := \{w \in W \mid 0Rw\} \cong (0 \times W) \cap R.$$

Here pr_V and pr_W denote the projections of $V \times W$ on V and W, respectively. One has

Observation 2.3.

(1) $\ker(R) \subseteq \mathrm{dom}(R) \subseteq V$ and $W \supseteq \mathrm{img}(R) \supseteq \mathrm{mul}(R)$.
(2) $\ker(R^\dagger) = \mathrm{img}(R)$ and $\mathrm{dom}(R^\dagger) = \mathrm{img}(R)$.
(3) $\dim \mathrm{dom}(R) + \dim \ker(R^\dagger) = \dim(R) = \dim(R^\dagger) = \dim \mathrm{dom}(R^\dagger) + \dim \ker(R)$.

Therefore one readily verifies:

Lemma 2.1. *Given a linear relation $R\colon V \rightsquigarrow W$, it holds that*

(1) $R = R(f)$ *for a linear map* $f : V \to W$ *iff* $\mathrm{dom}(R) = V$ *and* $\mathrm{mul}(R) = 0$;
(2) $R = R(\theta)$ *for a linear isomorphism* $\theta : V \to W$ *iff* $\mathrm{dom}(R) = V$, $\mathrm{img}(R) = W$, $\mathrm{mul}(R) = 0$, *and* $\ker(R) = 0$;
(3) $R = R(\theta)$ *for a linear isomorphism* $\theta : V \to V$ *iff* $\mathrm{dom}(R) = V$ *and* $\ker(R) = 0$.

Item (1) is straightforward, Item (2) follows from Item (1) and Observation 2.3 (2), and Item (3) follows from item (2) and Observation 2.3 (3).

For a linear relation $R\colon V \rightsquigarrow V$, we introduce the following subspaces:

$$K_+ := \{v \in V \mid \exists k\, \exists v_i \in V : vRv_1Rv_2R\cdots Rv_kR0\} = \bigcup_n \ker(R^n),$$

$$K_- := \{v \in V \mid \exists k\, \exists v_i \in V : 0Rv_{-k}\cdots Rv_{-2}Rv_{-1}Rv\} = \bigcup_n \ker((R^\dagger)^n),$$

$$D_+ := \{v \in V \mid \exists v_i \in V : vRv_1Rv_2Rv_3R\cdots\} = \bigcap_n \mathrm{dom}(R^n),$$

$$D_- := \{v \in V \mid \exists v_i \in V : \cdots Rv_{-3}Rv_{-2}Rv_{-1}Rv\} = \bigcap_n \ker((R^\dagger)^n),$$

$$D := D_- \cap D_+ = \{v \in V \mid \exists v_i \in V : \cdots Rv_{-2}Rv_{-1}RvRv_1Rv_2R\cdots\}.$$

Clearly, $K_- \subseteq D_- \subseteq V \supseteq D_+ \supseteq K_+$. Note also that when we pass from R to $R^\dagger$, the roles of $+$ and $-$ get interchanged.

Consider

$$V_{\text{reg}} := \frac{D}{(K_- + K_+) \cap D}$$

and introduce the relation $R_{\text{reg}} \colon V_{\text{reg}} \rightsquigarrow V_{\text{reg}}$ defined by the composition

$$V_{\text{reg}} = \frac{D}{(K_- + K_+) \cap D} \xrightarrow{\pi^\dagger} D \xrightarrow{\iota} V \xrightarrow{R} V \xrightarrow{\iota^\dagger} D \xrightarrow{\pi} \frac{D}{(K_- + K_+) \cap D} = V_{\text{reg}}$$

where ι and π denote the canonical inclusion and projection, respectively, viewed as relations.

In other words, two elements in V_{reg} are related by R_{reg} iff they admit representatives in D which are related by R. We refer to R_{reg} as the *regular part* of R.

It is convenient to denote by R_D the linear relation $R_D := \iota^\dagger R \iota \colon D \rightsquigarrow D$, in which case $R_{\text{reg}} = \pi R_D \pi^\dagger$.

Examples:

(1) If $f : V \to V$ is a linear map, then one has

$$V_{\text{reg}} = V \Big/ \Big(\bigcup_n \ker(f^n) \Big) \quad \text{and} \quad R(f)_{\text{reg}} = R(f_{\text{reg}}),$$

where f_{reg} is the linear map induced on the quotient V_{reg}.

(2) If $\alpha, \beta : V \to W$ are two linear isomorphisms, then $R(\alpha, \beta)_{\text{reg}} = R(\beta^{-1}\alpha)$.

(3) If $V = \kappa^{n+1}$, $W = \kappa^n$,

$$\alpha = \begin{pmatrix} 0 & 1 & 0 & \cdots & 0 \\ 0 & 0 & 1 & \cdots & 0 \\ & & \cdots & & \\ 0 & 0 & 0 & \cdots & 1 \end{pmatrix} \quad \text{and} \quad \beta = \begin{pmatrix} 1 & 0 & 0 & \cdots & 0 & 0 \\ 0 & 1 & 0 & \cdots & 0 & 0 \\ & & \cdots & & & \\ 0 & 0 & 0 & \cdots & 1 & 0 \end{pmatrix},$$

then the relations $R(\alpha, \beta) \colon V \rightsquigarrow V$ and $R(\alpha^T, \beta^T)$ have $V_{\text{reg}} = 0$.

Items (1) and (2) are straightforward. To check Item (3) note that for $R(\alpha, \beta)$ with α and β as above one has $D = K_+ = K_-$.

Proposition 2.4. *The relation* $R_{\text{reg}} \colon V_{\text{reg}} \rightsquigarrow V_{\text{reg}}$ *is an isomorphism of vector spaces.*

Proof. First note that:

(i) if xRy and $x, y \in D$, then there exists $x_i \in V$, $i \in \mathbb{Z}$, such that $x_i R x_{i+1}$, $x_0 = x$, and $x_1 = y$;

(ii) if $y \in K^+ + K^-$, then there exists a positive integer k and vectors $x_i^+, x_{-i}^- \in V$, $i = 0, 1, \ldots, k$, such that

(a) $y = x_0^+ + x_0^-$,

(b) $x_k^+ = x_{-k}^- = 0$,

(b) $x_i^+ R\, x_{i+1}^+$ and $x_{-i-1}^- R x_{-i}^-$ for all $i \leq k - 1$.

In view of Lemma 2.1, it suffices to check that $\mathrm{dom}(R_{\mathrm{reg}}) = V_{\mathrm{reg}}$ and $\ker(R_{\mathrm{reg}}) = 0$. The first is obvious since π is surjective and by (i) above $D = \mathrm{img}(R_D)$. To check the second equality observe that, by (ii) above, $x \in D$ and xRy with $y = x_1^+ + x_1^-$ and $x_1^\pm \in K_\pm$ imply the existence of additional elements

(1) $x_i \in V$, $i \in \mathbb{Z}$, with $x_i R x_{i+1}$, $x_0 = x$;

(2) $x_{-i}^- \in V$, $i = 0, 1, \ldots, k$, with $x_{-r-1}^- R x_{-r}^-$ for $-1 \leq r \leq k - 1$, and $x_{-k}^- = 0$;

(3) $x_i^+ \in V$, $i = 2, \ldots, k$, with $x_r^+ R x_{r+1}^+$, $1 \leq r \leq k - 1$ and $x_k^+ = 0$.

Then $(x - x_0^-)R(y - x_1^-) \in K_+ \subset D_+$ and $x - x_0^- \in D_-$, and so $x - x_0^- \in D_- \cap D_+ \cap K_+ = D \cap K_+$. Hence, $x_0^- = -x + (x - x_0^-) \in D$, and since $x_0^- \in K_-$ one has $x_0^- \in D \cap K_-$. Therefore

$$x = x_0^- + (x - x_0^-) \in D \cap K_- + D \cap K_+ \subseteq D \cap (K_+ + K_-).$$

This implies $\ker(R_{\mathrm{reg}}) = 0$. $\qquad\qquad\square$

Theorem 2.5.

(i) *If* $R: V \rightsquigarrow V$, *then* $(R^\dagger)_{\mathrm{reg}} = (R_{\mathrm{reg}})^\dagger = (R_{\mathrm{reg}})^{-1}$. [8]

(ii) *If* $R_i: V_i \rightsquigarrow V_i$, $i = 1, 2$, *then* $(R_1 \oplus R_2)_{\mathrm{reg}} = (R_1)_{\mathrm{reg}} \oplus (R_2)_{\mathrm{reg}}$.

(iii) *If* $R_1: V_1 \rightsquigarrow V_2$ *and* $R_2: V_2 \rightsquigarrow V_1$, *then* $(R_2 \cdot R_1)_{\mathrm{reg}}$ *is similar to* $(R_1 \cdot R_2)_{\mathrm{reg}}$,

Proof. Items (i) and (ii) follow from the definitions. To check item (iii) one needs a few additional considerations.

Definition 2.5. For a relation $R: V \rightsquigarrow V$, a collection $\{\underline{v}\} := \{v_i \in V, i \in \mathbb{Z} : v_i R v_{i+1}\}$ is called a R-*compatible sequence*. The compatible sequence $\{\underline{v}\}$ is said to be *zero at* $+\infty$ (respectively, *at* $-\infty$) if $v_i = 0$ for $i \geq N$ (respectively, $i \leq -N$), with N some positive integer.

[8] Here we regard R_{reg} as a linear isomorphism.

Clearly, all v_i's in a compatible sequence belong to D. One denotes by $[v_i]$ their class in V_{reg}. Based on this definition, we make the following

Observation 2.4.

(1) $v \in D$ iff there exists an R-compatible sequence $\{\underline{v}\}$ with $v_0 = v$;

(2) $v \in D \cap K_+$ iff there exists an R-compatible sequence $\{\underline{v}\}$ zero at $+\infty$ with $v_0 = v$;

(3) $v \in D \cap K_-$ iff there exists an R-compatible sequence $\{\underline{v}\}$ zero at $-\infty$ with $v_0 = v$. The above remains true for any specified i instead of $i = 0$;

(4) If $\{v_i\}$ is a compatible R-sequence, then $[v_i] = [v_{i+1}]$.

Definition 2.6. For two relations $R_1 : V \rightsquigarrow W$, $R_2 : W \rightsquigarrow V$ a pair of two sequences $\{\underline{v}, \underline{w}\} := \{v_i \in V, w_i \in W, i \in \mathbb{Z}\}$ is called a compatible (R_1, R_2)-pair if $v_i R_1 w_i$ and $w_i R_2 v_{i+1}$. One says that the pair is zero at ∞ or at $-\infty$ if for some N one has $v_i = 0$, and $w_i = 0$ for all $i \geq N$ or for $i \leq -N$.

Based on this definition and on Observation 2.4, we further make

Observation 2.5.

Suppose $S_1 = R_2 R_1$ and $S_2 = R_1 R_2$.

(1) If $\{\underline{v}, \underline{w}\}$ is a compatible (R_1, R_2) pair, then $\{v_i\}$ is an S_1-compatible sequence and $\{w_i\}$ is an S_2-compatible sequence.

(2) (i) If $v \in V_1$, then $v \in D(S_1)$ iff there exists a compatible pair $\{\underline{v}, \underline{w}\}$ with $v_0 = v$.

(ii) If $w \in V_2$ then $w \in D(S_2)$ iff there exists a compatible pair $\{\underline{v}, \underline{w}\}$ with $w_0 = w$.

(iii) If $v R_1 w, v \in D(S_1), w \in D(S_2)$, then there exists a compatible pair $\{\underline{v}, \underline{w}\}$ with $v_0 = v$ and $w_0 = w$. The same remains true for any i instead of $i = 0$.

(3) If $v \in V_1$, then $v \in D(S_1) \cap K(S_1)_\pm$ iff there exists a compatible pair $\{\underline{v}, \underline{w}\}$ with $v_0 = v$ which is zero at $\pm\infty$.

(4) If $w \in V_2$, then $v \in D(S_2) \cap K(S_2)_\pm$ iff there exists a compatible pair $\{\underline{v}, \underline{w}\}$ with $w_0 = w$ which is zero at $\pm\infty$ [9].

To prove item (iii) in Theorem 2.5 one produces two linear maps $\tilde{\omega}_1 : D(S_1) \to W_{\text{reg}}$ and $\tilde{\omega}_2 : D(S_2) \to V_{\text{reg}}$, such that:

---

[9] The notation $D(R)$, $K(R)$, etc, specifies that D and K refer to the relation in the parentheses.

(1) $\widetilde{\omega}_1$ and $\widetilde{\omega}_2$ factor through $\omega_1 : V_{\mathrm{reg}} \to W_{\mathrm{reg}}$ and $\omega_2 : W_{\mathrm{reg}} \to V_{\mathrm{reg}}$.

(2) $\omega_1 \cdot \omega_2 = id$, $\omega_2 \cdot \omega_1 = id$.

(3) $R(\omega_1) \cdot (S_1)_{\mathrm{reg}} = (S_2)_{\mathrm{reg}} \cdot \mathbb{R}(\omega_1)$.

To construct $\widetilde{\omega}_1$ choose a compatible pair $\{\underline{v}, \underline{w}\}$ with $v_0 = v$ and define $\omega(v)$ to be $[w_0]$. Observation 2.5 implies that this assignment is well defined and factors through a linear map $\omega_1 : V_{\mathrm{reg}} \to W_{\mathrm{reg}}$. Similarly, one defines $\widetilde{\omega}_2$ and verifies that if factors through a linear map $\omega_2 : W_{\mathrm{reg}} \to V_{\mathrm{reg}}$, i.e., (1) holds. Observation 2.5 implies the independence of the chosen compatible pairs, as well as the equalities $\omega_2 \cdot \omega_1 = id$ and $\omega_1 \cdot \omega_2 = id$, which establishes item (iii).

$\square$

The following proposition is a consequence of finite-dimensionality.

Proposition 2.5. *Suppose* $R \colon V \rightsquigarrow V$ *is a linear relation on a finite dimensional vector space. Then*

$$D_+ = D + K_+, \quad D_- = K_- + D \tag{2.19}$$

and

$$K_- \cap D_+ = K_- \cap K_+ = D_- \cap K_+. \tag{2.20}$$

For the proof we first establish two lemmas.

Lemma 2.2. *Suppose* $R \colon V \rightsquigarrow W$ *is a linear relation between vector spaces such that* $\dim V = \dim W < \infty$. *Then the following are equivalent:*

(i) R *is an isomorphism.*

(ii) $\mathrm{dom}(R) = V$ *and* $\ker(R) = 0$.

(iii) $\mathrm{dom}(R^\dagger) = W$ *and* $\ker(R^\dagger) = 0$.

Proof. The statement follows from the dimension formula

$$\dim \mathrm{dom}(R) + \dim \ker(R^\dagger) = \dim(R) = \dim \mathrm{dom}(R^\dagger) + \dim \ker(R)$$

and Lemma 2.1. $\square$

Lemma 2.3. *If* V *is finite-dimensional, then the composition of relations*

$$D_+/K_+ \xrightarrow{\pi^\dagger} D_+ \xrightarrow{\iota} V \xrightarrow{R^k} V \xrightarrow{\iota^\dagger} D_+ \xrightarrow{\pi} D_+/K_+,$$

where ι *and* π *denote the canonical inclusion and projection, respectively, is a linear isomorphism, for all* $k \geq 0$. *Analogously, the relation induced by* R^k *on* D_-/K_- *is an isomorphism, for all* $k \geq 0$. *Moreover, for sufficiently large* k,

$$D_- = \mathrm{dom}((R^\dagger)^k) \quad \text{and} \quad D_+ = \mathrm{dom}(R^k).$$

Proof. One verifies that $\mathrm{dom}(\pi\iota^\dagger R^k \iota\pi^\dagger) = D_+/K_+$ and $\ker(\pi\iota^\dagger R^k \iota\pi^\dagger) = 0$. The first assertion thus follows from Lemma 2.1 above. Considering $R^\dagger$, we obtain the second statement. Clearly, $\mathrm{dom}(R^k) \supseteq \mathrm{dom}(R^{k+1})$ for all $k \geq 0$. Since V is finite-dimensional, we must have $\mathrm{dom}(R^k) = \mathrm{dom}(R^{k+1})$ for sufficiently large k. Given $v \in \mathrm{dom}(R^k)$, we thus find $v_1 \in \mathrm{dom}(R^k)$ such that vRv_1. Proceeding inductively, we construct $v_i \in \mathrm{dom}((R^\dagger)^k)$ such that $vRv_1Rv_2R\cdots$, whence $v \in D_+$. This shows $\mathrm{dom}(R^k) \subseteq D_+$ for sufficiently large k. As the converse inclusion is obvious we get $D_+ = \mathrm{dom}(R^k)$. Considering $R^\dagger$, we obtain the last statement. $\square$

Proof of Proposition 2.5. From Lemma 2.3 we get $\mathrm{img}(\pi\iota^\dagger R^k) = D_+/K_+$, whence $D_+ \subseteq \mathrm{img}(R^k) + K_+$, for every $k \geq 0$, and thus $D_+ \subseteq D_- + K_+$. This implies $D_+ = D + K_+$. Considering $R^\dagger$, we obtain the other equality in (2.19). From Lemma 2.3 we also get $\ker((\pi\iota^\dagger R^k))^\dagger) = 0$, whence $\ker((R^\dagger)^k) \cap D_+ \subseteq K_+$, for every $k \geq 0$. This gives $K_- \cap D_+ = K_- \cap K_+$. Considering $R^\dagger$, we get the other equality in (2.20). $\square$

Proposition 2.6. *Let* $R\colon V \rightsquigarrow V$ *be a linear relation on a finite-dimensional vector space over an algebraically closed field and let* $R \cong R_1 \oplus \cdots \oplus R_N$ *denote a decomposition into indecomposable linear relations. Then* $R_{\mathrm{reg}} \cong (R_1)_{\mathrm{reg}} \oplus \cdots \oplus (R_N)_{\mathrm{reg}}$.

Proof. In view of Theorem 2.4, it suffices to show that:

(1) If $R\colon V \rightsquigarrow V$ is an isomorphism of vector spaces, then $V_{\mathrm{reg}} = V$ and $R_{\mathrm{reg}} = R$.

(2) If $R\colon V \rightsquigarrow V$ is an indecomposable linear relation on a finite-dimensional vector space which is not a linear isomorphism, then $V_{\mathrm{reg}} = 0$.

The first statement is obvious, since in this case we have $K_- = K_+ = 0$ and $D = D_- = D_+ = V$.

To check the second, one needs a description of indecomposable linear relations, which because any relation $R\colon V \rightsquigarrow V$ appears as $R(a,b)$ with $a, b : V \to W$ linear maps, is provided by the following theorem due to Kronecker. This theorem establishes item (2) above. $\square$

Theorem 2.6. (*Kronecker*) *Suppose* $a : V \to W$ *and* $b : V \to W$ *are two linear maps, where* V *and* W *are vector spaces over an algebraically closed field* κ, *with* $\dim V = n$ *and* $\dim W = m$, *such that the linear relation*

$R(a,b)$ *is indecomposable. Then there exists linear isomorphisms* $\alpha : \kappa^n \to V$ *and* $\beta : \kappa^m \to W$ *such that either one of the following possibilities holds:*

(1) a *is an isomorphism and* $\beta^{-1} \cdot a \cdot \alpha = Id$, $\beta^{-1} \cdot b \cdot \alpha = T(\lambda, n)$, $\lambda \in \kappa$.

(2) b *is an isomorphism and* $\beta^{-1} \cdot b \cdot \alpha = Id$, $\beta^{-1} \cdot a \cdot \alpha = T(\lambda, n)$, $\lambda \in \kappa$.

(3) $n = m + 1$ *and*

$$\beta^{-1} \cdot a \cdot \alpha = \begin{pmatrix} 0 & 1 & 0 \cdots 0 \\ 0 & 0 & 1 \cdots 0 \\ & & \cdots \\ 0 & 0 & 0 \cdots 1 \end{pmatrix},$$

$$\beta^{-1} \cdot b \cdot \alpha = \begin{pmatrix} 1 & 0 & 0 \cdots 0\, 0 \\ 0 & 1 & 0 \cdots 0\, 0 \\ & & \cdots \\ 0 & 0 & 0 \cdots 1\, 0 \end{pmatrix}.$$

(4) $n = m - 1$ *and*

$$\beta^{-1} \cdot a \cdot \alpha = \begin{pmatrix} 0 & 0 & 0 & \cdots & 0 \\ 1 & 0 & 0 & \cdots & 0 \\ 0 & 1 & 0 \cdots 0 & & 0 \\ & & \cdots \\ 0 & 0 & 0 & \cdots & 1, \end{pmatrix}$$

$$\beta^{-1} \cdot b \cdot \alpha = \begin{pmatrix} 1 & 0 & 0 & \cdots & 0 \\ 0 & 1 & 0 \cdots 0\, 0 \\ & & \cdots \\ 0 & 0 & 0 & \cdots & 1 \\ 0 & 0 & 0 & \cdots & 0 \end{pmatrix}.$$

In case κ is not algebraically closed the same conclusion holds with $T(\lambda, n)$ replaced by an indecomposable matrix.

The proof can be derived from Jordan canonical form theorem, Theorem 2.3.

Indeed, in the case (2) the relation is actually $R(b)$ and the statement follows from Theorem 2.3. Case (1) follows by passing from R to $R^\dagger$. Finally, cases (3) and (4) follows from the observation that $V_{reg} = 0$ since $V = D$ and either K_+ or K_- is actually V.

A direct proof of Kronecker Theorem can be also fond in [Benson, J.D. (1998)].

2.2.2   Calculation of $R(A, B)_{reg}$ (*an algorithm*)

For the algorithm described below we need the following technical Proposition, the proof of which is given at the end of this subsection.

Proposition 2.7.

(i) *Consider the diagram*

$$V \xrightarrow{\ \alpha\ } W \xleftarrow{\ \beta\ } V \qquad (2.21)$$

$$\subseteq \uparrow \qquad \subseteq \uparrow \qquad \subseteq \uparrow$$

$$V' \xrightarrow{\ \alpha'\ } W' \xleftarrow{\ \beta'\ } V'$$

and suppose that $W' \supseteq \text{img}\,\alpha \cap \text{img}\,\beta$, $V' = \alpha^{-1}(W') \cap \beta^{-1}(W')$, *and* α' *and* β' *are the restrictions of* α *and* β. *Then* $R(\alpha, \beta)_{reg} = R(\alpha', \beta')_{reg}$.

(ii) *Consider the diagram*

$$V \xrightarrow{\ \alpha\ } W \xleftarrow{\ \beta\ } V \qquad (2.22)$$

$$\downarrow p' \qquad \downarrow p \qquad \downarrow p'$$

$$V' \xrightarrow{\ \alpha'\ } W' \xleftarrow{\ \beta'\ } V'$$

with both α *and* β *surjective.*
Define $V' := V/\ker\alpha$, $W' := W/\beta(\ker\alpha)$, *and let* $p : W \to W'$, $p' : V \to V'$ *be the canonical quotient maps and* $\overline{\alpha} : V' \to W$ *be the linear maps induced by* α, $\alpha' = p \cdot \overline{\alpha}$, *and* β' *be the linear map induced by* β *by passing to the quotient. Then* $R(\alpha, \beta)_{reg} = R(\alpha', \beta')_{reg}$.

The algorithm presented below inputs two $m \times n$ matrices (A, B) defining a linear relation $R(A, B)$ and outputs two invertible $k \times k$, $k \leq \inf\{m, n\}$, matrices (A', B') such that $R(A, B)_{reg}$ and $R(A', B')_{reg}$ are similar. It is based on three modifications, T_1, T_2, T_3, described below. The simplest way to perform these modification is to use familiar procedures of bringing a matrix to row or column echelon form, (REF) or (CEF), but less is actually needed.

Modification $T_1(A, B) = (A', B')$:
Produces the invertible $m \times m$ matrix C and the invertible $n \times n$ matrix D such that

$$CAD = \begin{pmatrix} A_{11} & A_{12} \\ 0 & 0 \end{pmatrix} \quad \text{and} \quad CBD = \begin{pmatrix} B_{11} & B_{12} \\ B_{2.1} & 0 \end{pmatrix}.$$

Precisely, one constructs first C which puts A in REF such that

$$CA = \begin{pmatrix} A_1 \\ 0 \end{pmatrix} \quad \text{and makes} \quad CB = \begin{pmatrix} B_1 \\ B_2 \end{pmatrix}.$$

Next, one constructs D which puts B_2 in CEF. Precisely,
$B_2 D = \begin{pmatrix} B_{21} & 0 \end{pmatrix}$.

Clearly CAD and CBD are as desired above.

Take $A' = A_{12}$ and $B' = B_{12}$.

In view of Proposition 2.7 (i), one has $R(A, B)_{\text{reg}} = R(A', B')_{\text{reg}}$.

Modification $T_2(A, B) = (A', B')$:

Produces the invertible $m \times m$ matrix C and the invertible $n \times n$ matrix D such that

$$CAD = \begin{pmatrix} A_{11} & A_{12} \\ A_{21} & 0 \end{pmatrix} \quad \text{and} \quad CBD = \begin{pmatrix} B_{11} & B_{12} \\ 0 & 0 \end{pmatrix}.$$

Precisely, one constructs first C which puts B in REF such that

$$CB = \begin{pmatrix} B_1 \\ 0 \end{pmatrix} \quad \text{and makes} \quad CA = \begin{pmatrix} A_1 \\ A_2 \end{pmatrix}.$$

Next one constructs D which puts A_2 in CEF. Precisely $A_2 D = \begin{pmatrix} A_{21} & 0 \end{pmatrix}$.

Take $A' = A_{12}, B' = B_{12}$.

Clearly CAD and CBD are as wanted above.

In view of Proposition 2.7 (i) one has $R(A, B)_{\text{reg}} = R(A', B')_{\text{reg}}$.

Note that if A was surjective, then A' remains surjective.

Note also that if for the pair of matrices (A, B) one denotes by $t(A, B)$ the pair (B, A), then $T_2(A, B) = tT_1(t(A, B))$.

Modification $T_3(A, B) = (A', B')$:

Produces the invertible $n \times n$ matrix D and the $m \times m$ invertible matrix C such that

$$CAD = \begin{pmatrix} A_{11} & 0 \\ A_{21} & 0 \end{pmatrix} \quad \text{and} \quad CBD = \begin{pmatrix} B_{11} & B_{12} \\ B_{21} & 0 \end{pmatrix}.$$

Precisely, one constructs D first which puts A in CEF i.e.,
$AD = \begin{pmatrix} A_1 & 0 \end{pmatrix}$ and makes $BD = \begin{pmatrix} B_1 & B_2 \end{pmatrix}$.

Next one constructs C to put B_2 in REF. Precisely,

$$CB_2 = \begin{pmatrix} B_{21} \\ 0 \end{pmatrix}.$$

Take $A' = A_{21}$ and $B' = B_{21}$.

Clearly, CAD and CBD are as desired above.

In view of Proposition 2.7 (ii), one has $R(A, B)_{\text{reg}} = R(A', B')_{\text{reg}}$. Note that if both A and B were surjective, then A' and B' remain surjective.

Here is how the algorithm works.

- (I) Inspect A
  if surjective move to (II)
  else:
    - apply T_1 and obtain A' and B'
    - make $A = A'$ and $B = B'$ and
    - go to (I)
- (II) Inspect B
  if surjective move to (III)
  else :
    - apply T_2 and obtain A' and B'
    - make $A = A'$ and $B = B'$ and
    - go to (II)
  (Note that if A was surjective by applying T_2, A' remains surjective.)
- (III) Inspect A
  if injective go to (IV).
  else
    - apply T_3 and obtain A' and B'
    - make $A = A'$ and $B = B'$ and
    - go to (III)
- (IV) Calculate $B^{-1} \cdot A$.

The process stops when both A and B become invertible.

The existence and construction of the invertible matrices C and D is discussed in Proposition 2.2 in the previous section.

Note: All basic software which carries linear algebra packages contain subpackages which input a matrix and output its (reduced) row/column echelon form as well as the matrix C or D.

An example

We illustrate Step 2 of the algorithm with the matrices A and B provided by the example in Chapter 7 Section 7.3:

$$
A = \begin{pmatrix} 3 & 3 & 0 \\ 2 & 3 & -1 \\ 1 & 2 & 3 \\ 0 & 0 & 0 \end{pmatrix} \quad \text{and} \quad B = \begin{pmatrix} 0 & 0 & 0 \\ 0 & 1 & 0 \\ 0 & 0 & 1 \\ 0 & 0 & 0 \end{pmatrix}. \tag{2.23}
$$

Inspect A; since not surjective, apply T_1 and find $C = Id$ and $D = Id$.

Then

$$A' = \begin{pmatrix} 3 & 3 & 0 \\ 2 & 3 & -1 \\ 1 & 2 & 3 \end{pmatrix} \quad \text{and} \quad B' = \begin{pmatrix} 0 & 0 & 0 \\ 0 & 1 & 0 \\ 0 & 0 & 1 \end{pmatrix}. \tag{2.24}$$

Update

$$A = \begin{pmatrix} 3 & 3 & 0 \\ 2 & 3 & -1 \\ 1 & 2 & 3 \end{pmatrix} \quad \text{and} \quad B = \begin{pmatrix} 0 & 0 & 0 \\ 0 & 1 & 0 \\ 0 & 0 & 1 \end{pmatrix}. \tag{2.25}$$

Since A is surjective, inspect B. Since B is not surjective, apply T_2 and find

$$C = \begin{pmatrix} 0 & 1 & 0 \\ 0 & 0 & 1 \\ 1 & 0 & 0 \end{pmatrix}, \quad \text{and then} \quad D = \begin{pmatrix} 1 & -1 & 0 \\ 0 & 1 & 0 \\ 0 & 0 & 1 \end{pmatrix}.$$

Then

$$CAD = \begin{pmatrix} 2 & 1 & -1 \\ 1 & 1 & 3 \\ 3 & 0 & 0 \end{pmatrix} \quad \text{and} \quad CBD = \begin{pmatrix} 0 & 1 & 0 \\ 0 & 0 & 1 \\ 0 & 0 & 0 \end{pmatrix},$$

and

$$A' = \begin{pmatrix} 1 & -1 \\ 1 & 3 \end{pmatrix} \quad \text{and} \quad B' = \begin{pmatrix} 1 & 0 \\ 0 & 1 \end{pmatrix}. \tag{2.26}$$

Both A' and B' are invertible, so consider

$$B^{-1} \cdot A = \begin{pmatrix} 1 & -1 \\ 1 & 3 \end{pmatrix}.$$

Hence, $\mathcal{J}([R_{\text{reg}}(A, B)]) = \{(2, 2)\}$.

Proof of Proposition 2.7.

Item (i) follows by observing that D and $D \cap (K^+ + K^-)$ for both $R(\alpha, \beta)$ and $R(\alpha', \beta')$ are actually the same.

To check this, note that

- $v_0 \in D$ implies the existence of the sequence

$$\cdots \leftarrow v_{-1} \rightarrow w_0 \xleftarrow{\beta} v_0 \xrightarrow{\alpha} w_1 \xleftarrow{\beta} v_1 \rightarrow w_2 \leftarrow v_2 \rightarrow \cdots,$$

- $v_0^+ \in K_+$ implies the existence of the sequence

$$v_0^+ \xrightarrow{\alpha} w_1^+ \xleftarrow{\beta} v_1^+ \xrightarrow{\alpha} w_2^+ \leftarrow \cdots \rightarrow w_{k-1}^+ \xleftarrow{\beta} v_{k-1}^+ \xrightarrow{\alpha} w_k^+ \leftarrow 0$$

for some k;

- $v_0^- \in K_-$ implies the existence of the sequence

$$0 \rightarrow w_{-k}^- \xleftarrow{\beta} v_{-k}^- \xrightarrow{\alpha} w_{-(k-1)}^- \leftarrow \cdots \leftarrow v_{-1} \xrightarrow{\alpha} w_0^- \xleftarrow{\beta} v_0^-$$

for some k.

In view of the hypotheses of Proposition 2.7 (i),

(1) if $v_0 \in D$, then all $w_i \in W'$, and so $v_i \in V'$. Hence $D = D'$.
(2) if $v_0^+ \in K_+$, then all $w_i \in W'$, and so $v_i \in V'$ for $i \geq 1$.
(3) if $v_0^- \in K_-$, then all $w_i \in W'$, and so $v_i \in V'$ for $i \leq -1$.

Suppose $v_0 \in D \cap (K_+ + K_-)$. If $v_0 = v_0^- + v_0^+$, then $v_0^- = v_0 - v_0^+ \in D + K_+ = D_+$ by Proposition 2.5, hence $v_0^- \in K_- \cap D_+ = K_- \cap K_+ = K \subseteq D = D'$, again by Proposition 2.5. This combined with (3) implies $v_0^- \in K_-'$. Similarly, $v_0^+ \in K_+'$. Hence, $v_0 \in K_+' + K_-'$, and since $D = D'$, one has $v_0 \in D' \cap (K_+' + K_-')$, which establishes the statement.

To check Item (ii) observe that the diagram (2.22) induces the linear map $\pi : D/(D \cap (K_+ + K_-)) \rightarrow D'/(D' \cap (K_-' + K_+'))$. This map is obviously surjective since both pairs α, β and α', β' being surjective implies that $V = D$ and $V' = D'$.

To show that π is injective we will verify that $p'^{-1}(K_\pm') \subset K_\pm$. For this purpose we consider the diagram (2.22) with α' and β' as specified by the hypotheses of Proposition 2.7 (ii) and we use the following lemma.

Lemma 2.4. *If* $w \in W, w' \in W', v' \in V'$ *such that* $p(w) = w'$ *and* $\beta'(v') = w'$, *then there exists* $v \in V$ *such that* $\beta(v) = w$ *and* $p'(v) = v'$.

Proof. We first choose $\underline{v} \in V$ with the property $p'(\underline{v}) = v'$, and then observe that $p(w - \beta(\underline{v})) = 0$, hence in view of the definition of the diagram (2.22) $w - \beta(\underline{v}) = \beta(u), u \in \ker \alpha$ and correct $\bar{v}$ to v by taking $v = \underline{v} + u$. $\square$

Based on Lemma 2.4, observe that given a sequence $v_0', v_1', \ldots, v_k' \in V'$ and $v_0 \in V$ with the property that

$$\begin{aligned} \alpha'(v_{i-1}') &= \beta'(v_i'), \quad 1 \leq i \leq k \\ p(v_0) &= v_0' \end{aligned} \tag{2.27}$$

one can inductively produce $v_1, v_2, \ldots, v_k \in V$ such that

$$\alpha(v_{i-1}) = \beta(v_i),$$
$$p(v_i) = v_i'. \tag{2.28}$$

Indeed, suppose inductively that $v_1, v_2, \ldots, v_i$, $i \leq r$ satisfying properties (2.28) are already produced. Then apply Lemma 2.4 to $w = \alpha(v_i)$, $w' = \alpha'(v_i')$, and $v' = v_{r+1}'$, and obtain v_{r+1}.

To conclude that ${p'}^{-1}(K_+') \subset K_+$, one chooses the sequence $\{v_i'\}$ so as to have (for some k) $\alpha(v_k') = 0$, which means that $v_0' \in K_+'$. Then the vector v_k constructed as above lies in $\ker \alpha$, which means that $v_0 \in K_+$.

To conclude ${p'}^{-1}(K_-') \subset K_-$ one chooses a sequence $\{v_i'\}$ to have (for some k) $v' = v_k' \in K_-'$ and $v_0' = 0$. Now for $v_0 = 0$ one constructs the sequence $v_1, v_2, \ldots, v_k \in V$. Then $v_k \in K_-$, hence ${p'}^{-1}(K_-') \subset K_-$.

2.3   Topology

2.3.1   *ANRs, tameness, regular and critical values*

A space X is an ANR if any closed subset A of a metrizable space B homeomorphic to X has a neighborhood U which retracts to A, cf. [Hu, S.T. (1965)] Chapter 3. Any space homeomorphic to a locally-finite simplicial complex, or to a finite-dimensional topological manifold, or to an infinite-dimensional manifold, i.e., a paracompact separable Hausdorff space locally homeomorphic to the infinite-dimensional separable Hilbert space or to the Hilbert cube, the product of countable many copies of $[0, 1]$, is an ANR. Any ANR is locally contractible. It can be shown that a locally compact locally contractible separable metrizable space is an ANR, cf. [Hu, S.T. (1965)]. If A, B are two ANRs, with A a closed subset of B, then A is a neighborhood deformation retract of B, cf. [Hu, S.T. (1965)].

Let $f : X \to \mathbb{R}$ or $f : X \to \mathbb{S}^1$ be a proper continuous map with X an ANR. The properness of f implies that X is locally compact, and when the target is $\mathbb{S}^1$, X is compact.

Definition 2.7.

(1) A value $t \in \mathbb{R}$ or $\theta \in \mathbb{S}^1$ is called *weakly regular* if $f^{-1}(t)$, respectively $f^{-1}(\theta)$, is an ANR.

(2) The map $f : X \to \mathbb{R}$ or $f : X \to \mathbb{S}^1$ is *weakly tame* if all its values are weakly regular. Therefore, for any bounded or unbounded closed interval I, the space $f^{-1}(I)$ is an ANR.

(3) A value $s = t$ or θ is a *regular value*, if there exists an $\epsilon > 0$ such that for any $s' \in (s - \epsilon, s + \epsilon)$ the level $f^{-1}(s')$ is a deformation retract of $f^{-1}(\theta - \epsilon, \theta + \epsilon)$; a value s that is not a regular value is a *critical value*. If, in addition, for $\epsilon > 0$ there exists a homeomorphism $\omega : f^{-1}(s) \times (s - \epsilon, s + \epsilon) \to f^{-1}(s - \epsilon, s + \epsilon)$ such that $f \cdot \omega$ is the projection on the second factor, then the regular value is called a *topologically regular value*.

One denotes the set of critical values of f by $\mathcal{C}r(f)$.

(4) The map f is called a *tame map*, if in addition of being weakly tame it satisfies:

i) the set of critical values $\mathcal{C}_r(f) \subset \mathbb{R}$ is discrete,

ii) the number $\epsilon(f) := \inf\{|c - c'| \mid c, c' \in \mathcal{C}r(f), c \neq c'\}$ satisfies $\epsilon(f) > 0$. If X is compact, then (i) implies (ii).

The map f is called a *topologically tame map* if in addition of being weakly tame (i) and (ii) above are satisfied for topologically critical values instead of critical values. Clearly topologically tame implies tame.

(5) An ANR for which the set of tame resp. topologically tame maps is dense in the space of all maps with the compact-open topology is called a *good ANR* resp. *topologically good* ANR. Clearly topologically good ANR's are good ANR's.

The reader should be aware of the following facts:

Observation 2.6.

(i) If f is a weakly tame map, then the compact ANR $f^{-1}([a, b])$ has the homotopy type of a finite simplicial complex (cf. [Milnor, J. (1959)]), and therefore has finite-dimensional homology over any field κ.

(ii) If X is a locally finite simplicial complex and f is a proper continuous map linear on each simplex, then f is weakly tame, all regular and critical values are topologically regular and critical values and the set of critical values is discrete. If, in addition, X is compact then f is topologically tame.

(iii) If M is a smooth manifold and f is proper smooth map which is locally polynomial[10], in particular if f is a Morse map, then f is weakly tame and the conclusions in (ii) hold. If M is compact then

---

[10]That is, for any point of M there exist local coordinates around the point such f in these coordinates is a polynomial function.

f is topologically tame. The same holds for proper real analytic maps defined on real analytic or subanalytic spaces.

Composition of a tame or topologically tame map with a homeomorphism of the domain remains tame, respectively topologically tame.

(iv) If X is homeomorphic to a compact simplicial complex, in particular to a triangulable manifold, then the set of topologically tame and then of tame maps is dense in the set of all continuous maps equipped with the compact-open topology [11]. The same remains true if X is a compact Hilbert cube manifold, defined in the next section. In particular, these spaces are topologically good ANR's.

(v) If K is a compact ANR, then $f : X \to \mathbb{R}$ or $\mathbb{S}^1$ weakly tame (tame, topologically tame) makes the composition $\overline{f}_K : X \times K \to \mathbb{R}$ or $\mathbb{S}^1$ of f with the projection $X \times K \to X$ on the first factor weakly tame (respectively, tame, topologically tame) and one has $Cr(f) = Cr(\overline{f}_K)$.

The results of this book use only the homological implications of weakly tameness. Here we consider only the homology with coefficients in a fixed field κ. We list some of the homological implications of weakly tameness:

- for any compact interval I, $\dim H_*(f^{-1}(I))$ is finite since this space is a compact ANR;

- for any two closed intervals $I_1 \subset I_2$, $H_*(f^{-1}(I_2), f^{-1}(I_2 \setminus \overset{\circ}{I_1})) = H_*(f^{-1}(I_1), f^{-1}(\partial I_1))$, where $\overset{\circ}{I}$ denotes the interior of I and ∂I the boundary of I;

- for any two closed intervals the Mayer-Vietoris long exact sequence in homology associated with the spaces $f^{-1}(I_1)$, $f^{-1}(I_2)$, $f^{-1}(I_1 \cap I_2)$, and $f^{-1}(I_1 \cup I_1)$ is exact.

One can weaken the definition of *weakly tame* by requiring the above properties to hold for homology only, and refer to such maps as *homologically weakly tame* with respect to the field κ.

For a homologically weakly tame map, and in particular for a weakly tame map, a real number a or an angle θ is a *homologically regular value* with respect to the fixed field κ, if there exists $\epsilon > 0$ such that for t, t' with $a - \epsilon < t \leq t' < a + \epsilon$, respectively $\theta - \epsilon < t \leq t' < \theta + \epsilon$, the inclusions $f^{-1}(t) \subset f^{-1}([t, t'])$ and $f^{-1}(t') \subset f^{-1}([t, t'])$ induce isomorphisms in homology. A real number or angle which is not a homologically regular value

---

[11]This is because a continuous map on a simplicial complex is arbitrary closed to a simplicial map on a subdivision of the complex.

is called a *homologically critical value*. The set of a homological regular values, denoted by $h_\kappa \mathcal{R}(f)$, is obviously open, while the set of homological critical values, denoted by $h_\kappa \mathcal{C}r(f)$, is closed. Clearly $\mathcal{R}(f) \subseteq h_\kappa \mathcal{R}(f)$ and $\mathcal{C}r(f) \supseteq h_\kappa \mathcal{C}r(f)$.

The set $h_\kappa \mathcal{C}r(f)$ is always discrete, so $\epsilon'(f) = \inf d(c, c')$, c, c' different homologically critical values, is well defined and in case f is tame, hence also homologically weakly tame, $\epsilon'(f) \geq \epsilon(f)$.

If one is interested only in closed and open barcodes as considered in Chapter 5, then one can weaken further the notions of homologically regular values and homologically critical values for a real-valued map to *w-homologically regular values* and w-homologically critical values by requiring only the isomorphism of the inclusions $\mathbb{I}_t(r) \subseteq \mathbb{I}_{t'}(r)$ and $\mathbb{I}^{t'}(r) \subseteq \mathbb{I}^t(r)$ for any t, t' with $a - \epsilon < t \leq t' < a + \epsilon$ and any r. Here $\mathbb{I}_t(r) = \mathrm{img}(H_r(X_t) \to H_r(X))$ and $\mathbb{I}^t(r) = \mathrm{img}(H_r(X^t) \to H_r(X))$. One can extend these notions to angle-valued maps by using the infinite cyclic cover discussed in Subsection 2.3.4.[12]

These notions can be relevant for Chapter 5, whose statements can be all strengthened by replacing homologically tame, homologically regular values, and homologically critical values by w-*homologically tame*, w-*homologically regular value*, and w-*homologically critical value*. In this case the set of the w-homologically regular values is still open and larger than $h_\kappa \mathcal{R}(f)$, and the set of w-homologically critical values is contained in $h_\kappa \mathcal{C}r(f)$. The case of angle-valued maps can be derived from the case of real-valued ones by passing to the infinite cyclic cover discussed below.

2.3.2 *Compact Hilbert cube manifolds*

All the information about Hilbert cube manifolds presented in this subsection can be found in [Chapman, T.A. (1976)] and [Chapman, T.A. (1977)].

Recall that the Hilbert cube Q is the infinite product $Q = \prod_{i \in \mathbb{Z}_{\geq 0}} I_i = I^\infty$, with $I_i = I = [0, 1]$. The topology of Q is given by the metric $d(u, v) = \sum_i |u_i - v_i|/2^i$ with $u = \{u_i \in I, i \in \mathbb{Z}_{\geq 0}\}$ and $v = \{v_i \in I, i \in \mathbb{Z}_{\geq 0}\}$. A compact Hilbert cube manifold is a compact Hausdorff space locally homeomorphic to the Hilbert cube. The space Q is a compact ANR, and so is $X \times Q$ when X is a compact ANR.

---

[12]One uses the fact that the regular/critical angles for an angle-valued map $f : X \to \mathbb{S}^1$ are exactly the images of the regular/critical values of the infinite cyclic cover $\tilde{f} : \tilde{X} \to \mathbb{R}$ under the infinite cyclic cover $\mathbb{R} \to \mathbb{S}^1$.

For any positive integer n, write $Q = I^n \times Q'_n$, $Q'_n = \prod_{i \in \mathbb{Z}_{\geq n}} I_i$, and denote by $\pi_n : Q \to I^n$ the projection on the first factor and by $\pi_n^X : X \times Q \to X \times I^n$ the product $\pi_n^X = id_X \times \pi_n$.

For $F : X \times Q \to \mathbb{R}$, let F_n be the restriction of F to $X \times I^n = X \times (I^n \times 0)$ and $\overline{F}_n$ be the composition $\overline{F}_n := F_n \cdot \pi_n^X$.

For $f : X \to \mathbb{R}$ or $f : X \to \mathbb{S}^1$ put $\overline{f}_Q := f \cdot \pi_X$, where $\pi_X : X \times Q \to X$ is the projection on X.

Observation 2.7.

(i) If $f : X \to \mathbb{R}$ or $f : X \to \mathbb{S}^1$ is a weakly tame, resp. tame. resp. topologically tame map, then so is $\overline{f}_Q$.

(ii) The sequence of maps $\overline{F}_n$ converges uniformly to the map F.

Items (1) and (2) in Theorem 2.7 below are the main results about Hilbert cube manifolds. Their proof can be found in [Chapman, T.A. (1976)].

Theorem 2.7.

(i) (R. Edwards) *If X is a compact ANR then $X \times Q$ is a Hilbert cube manifold.*

(ii) (T. Chapman)

a) *If X is a compact Hilbert cube manifold, then X is homeomorphic to $X \times Q$.*

b) *If $\omega : X \to Y$ is a homotopy equivalence between two Hilbert cube manifolds with Whitehead torsion $\tau(\omega) = 0$, then ω is homotopic to a homeomorphism $\omega' : X \to Y$.*

c) *If $\omega : X \to Y$ is a homotopy equivalence between two Hilbert cube manifolds, then for any K compact ANR with Euler characteristic $\chi(K) = 0$ there exists a homeomorphism $\omega' : X \times K \to Y \times K$ such that ω' and $\omega \times id_K$ are homotopic.*

Recall that the Whitehead torsion was considered first by J. H. Whitehead for a homotopy equivalence $\omega : X \to Y$ between two finite simplicial complexes, and extended by Chapman for homotopy equivalences between arbitrary compact ANRs. This torsion is an element $\tau(\omega) \in \text{Wh}(\pi_1(X, x))$, where $\text{Wh}(\Gamma)$ denotes the Whitehead group of Γ, an abelian group associated with a discrete group Γ, introduced by J. H. Whitehead, cf. [Milnor, J. (1966)]. It is also known [Milnor, J. (1966)] that if K is a finite cell complex (actually, a compact ANR) with $\chi(K) = 0$, then $\tau(\omega \times Id_K) = 0$.

We also have the following observation, whose proof was provided to us by S. Ferry, cf. [Burghelea, D. (2015a)].

Proposition 2.8. *Any compact Hilbert cube manifold is a good* ANR.

Indeed, Theorem 2.7 implies that any Hilbert cube manifold is homeomorphic to $K \times Q$ with K a simplicial complex, and since K is good, in view of Observation 2.7, so is $K \times Q$.

2.3.3 *Infinite cyclic covers*

Suppose X is an ANR. The integral cohomology group $H^1(X; \mathbb{Z})$ can be identified with the set $[X, \mathbb{S}^1]$ of homotopy classes of continuous maps from X to $\mathbb{S}^1$, cf [Hatcher (2002)]. The map $f : X \to \mathbb{S}^1$ induces the group homomorphism $f^* : H^1(\mathbb{S}^1) \to H^1(X; \mathbb{Z})$. Once an orientation on $\mathbb{S}^1$ is chosen, a generator u of the cohomology group $H^1(\mathbb{S}^1; \mathbb{Z})$ is provided and the specified identification $f \rightsquigarrow \xi_f \in H^1(X; \mathbb{Z})$ with $\xi_f = f^*(u)$ is obtained.

Clearly, f homotopic to g implies $\xi_f = \xi_g$. One can also show that each cohomology class $\xi \in H^1(X; \mathbb{Z})$ is of this form, and that $\xi_f = \xi_g$ implies f homotopic to g. A good reference for these facts is [Hatcher (2002)].

The simplest example of infinite cyclic cover is the map $\pi : \mathbb{R} \to \mathbb{S}^1 = \{z \in \mathbb{C} | |z| = 1\}, \pi(t) = e^{it}$. The translations in $\mathbb{R}$ by multiples of 2π provide a free action of the group $\mathbb{Z}$ on $\mathbb{R}$, $\mu : \mathbb{Z} \times R \to \mathbb{R}$, $\mu(k, t) := t + 2\pi k$, whose quotient space is homeomorphic to $\mathbb{S}^1$.

An *infinite cyclic cover of* X associated to $\xi \in H^1(X : Z)$ is a continuous map $\widetilde{\pi} : \widetilde{X} \to X$ which satisfies the following property.

For any continuous map $f : X \to \mathbb{S}^1$ with $\xi_f = \xi$ there exists a map $\widetilde{f} : \widetilde{X} \to \mathbb{R}$ (a lift of f) such that the diagram

$$
\begin{array}{ccc}
R & \xrightarrow{\ \pi\ } & \mathbb{S}^1 \\
\widetilde{f} \big\uparrow & & \big\uparrow f \\
\widetilde{X} & \xrightarrow{\ \widetilde{\pi}\ } & X
\end{array}
\tag{2.29}
$$

is a pullback diagram. This means that the diagram is commutative and the map $\omega : \widetilde{X} \to \mathbb{R} \times X$ defined by $\omega = \widetilde{f} \times \widetilde{\pi}$ is a homeomorphism of $\widetilde{X}$ onto the image of ω. The map $\widetilde{f}$ is also called an *infinite cyclic cover of* f.

The following are important facts about $\widetilde{X}$ and $\widetilde{f}$ the reader should be aware of.

(i) *The free* $\mathbb{Z}$-*action:*

For any infinite cyclic cover $\widetilde{\pi} : \widetilde{X} \to X$ there exists a unique free action $\mu : \mathbb{Z} \times \widetilde{X} \to \widetilde{X}$ which makes $\widetilde{f}$ a $\mathbb{Z}$-equivariant map and makes $\widetilde{\pi}$ induce a homeomorphism from $\widetilde{X}/\mathbb{Z}$ onto X. Any homeomorphism $\mu(n, \cdot) : \widetilde{X} \to \widetilde{X}$ is called a *deck transformation*, and one denotes by τ the homeomorphism $\tau = \mu(1, \cdot)$, the generator of the group of deck transformations.

(ii) *Existence of infinite cyclic covers*:
Given X and $f : X \to \mathbb{R}$, there is a canonical infinite cyclic cover for ξ_f defined by f with $\widetilde{X} = \{(t, x) \in \mathbb{R} \times X \mid (\pi(t) = f(x)\}$, $\widetilde{\pi}(t, x) = x$, $\widetilde{f}(t, x) = t$, and $\mu(n, (t, x)) = (2\pi n + t, x)$.

(iii) *Uniqueness of infinite cyclic covers*:
If $\widetilde{\pi}_i : \widetilde{X}_i \to X_i$, $i = 1, 2$, are two infinite cyclic covers for $\xi \in H^1(X; \mathbb{Z})$, then there exists a homeomorphism $\theta : X_1 \to X_2$ which intertwines the free actions μ_1 and μ_2 and satisfies $\widetilde{\pi}_2 \cdot \theta = \widetilde{\pi}_1$.

(iv) *Uniqueness of the liftings*: Given an infinite cyclic cover $\widetilde{\pi} : \widetilde{X} \to X$ of (X, ξ) and a map $f : X \to \mathbb{S}^1$ with $\xi_f = \xi$, then for any two liftings $\widetilde{f}_1$ and $\widetilde{f}_2$ of f there exists a deck transformation τ^k such that $f_1 = f_2 \cdot \tau^k$.

All these statements can be recovered from any text book on algebraic topology which discusses coverings, for example [Hatcher (2002)] or [Munkres, J. (1984)].

2.3.4 Simplicial complexes, cell complexes and incidence matrices

An open k-simplex is a k-dimensional affine manifold (the set of common zeros of a finite collection of linear equalities and strict linear inequalities) diffeomorphic to the convex hull of $(k + 1)$-linearly independent vectors $v_1, v_2, \ldots, v_k \in V \simeq \mathbb{R}^N$, i.e., the set of vectors $v \in V$ of the form

$$v = t_0 v_0 + t_1 v_1 + t_2 v_2 + \cdots t_k v_k \text{ with } 0 < t_i < 1 \text{ and } \sum t_i = 1.$$

A k-simplex σ has $(k + 1)$-faces $\partial_i \sigma$, each a $(k - 1)$-simplex obtained by removing one of the vertices v_i. As any connected manifold, σ is equipped with two possible orientations defined by a total order of its vertices. Two total orders which differ by an even permutation define the same orientation.

For P a hyperplane in $V = \mathbb{R}^N$, $P := \{(x_1, \ldots, x_N) \in \mathbb{R}^N \mid \alpha_1 \cdot x_1 + \cdots + \alpha_N \cdot x_N = 0, \; \alpha_i \in \mathbb{R}\}$, one denotes by

$$P_\pm := \{(x_1, \ldots, x_N) \in \mathbb{R}^N \mid \pm(\alpha_1 \cdot x_1 + \cdots + \alpha_N \cdot x_N) > 0\}.$$

In this book a *cell* or a *generalized simplex* is an affine manifold diffeomorphic to either one of $\sigma \cap P$, $\sigma \cap P_+$, or $\sigma \cap P_-$, provided the hyperplane P does not contain any of the vertices of σ. A generalized simplex has faces, each of them a generalized simplex.

A **simplicial complex** X appears as the disjoint union of an at most countable family $\mathcal{X}$ of open simplices $\sigma \in \mathcal{X}$ in an $\mathbb{R}$-vector space V which satisfies the following property $\mathcal{P}$.

Property $\mathcal{P}$:

P1. If $\sigma \in \mathcal{X}$, then $\partial_i \sigma \in \mathcal{X}$.

P2. For each $\tau \in \mathcal{X}$, the set $\{\sigma \in \mathcal{X} \mid \tau \subset \overline{\sigma}\}$ is finite.

Here $\overline{\sigma}$ denoted the topological closure of σ in V.

The set $X = \bigcup_{\sigma \in \mathcal{X}} \sigma$ is equipped with an obvious topology, the coarser topology which insures that $\overline{\sigma}$ defined inductively on dimension of σ by

$$\overline{\sigma} = \sigma \bigcup_{0 \leq i \leq \dim \sigma} \overline{\partial_i \sigma}$$

is a closed set.

A **cell complex** Y in this book is a space which is a disjoint union of a collection $\mathcal{Y}$ of generalized simplices which satisfies the same properties P_1 and P_2 above. The topology is the coarser topology which ensures that the closure $\overline{\sigma}$ of each generalized simplex σ is the union of σ with the closure of all its faces. The definition of closure is understood inductively on dimension.

The notions of simplicial subcomplex and cell subcomplex are obvious.

To a finite simplicial or cell complex X or Y with $N = \sharp\mathcal{X}$ or $N = \sharp\mathcal{Y}$ equipped with

(i) a *good total order* "$\prec$" on the set $\mathcal{X}$, and

(ii) an orientation on each simplex or generalized simplex,

one associates an $N \times N$ upper triangular matrix $M(X)$ with zeros on the diagonal and entries $I(\tau, \sigma)$ equal to $0, 1, -1$, defined as follows.

The incidence number $I(\tau, \sigma) = 0$ if $\overline{\tau} \cap \overline{\sigma} = \emptyset$, and $I(\tau, \sigma) = +1/-1$ if τ is a face of σ and the orientation of σ induces on τ the orientation/reverse orientation of τ.

The total order on $\mathcal{X}$ or on $\mathcal{Y}$ is said to be *good* if τ being a face of σ implies $\tau \prec \sigma$. This indeed ensures that the matrix $M(X)$ is upper triangular with zeros on the diagonal.

The simplest way to provide data (i) and (ii) as above for a simplicial complex X is to choose a total order on the set of vertices of X. This order induces a canonical total order $\prec$ on the set of all simplices of X, the lexicographic order, as well as an orientation on each simplex. Indeed, the labeling of vertices $\{1, 2, 3, \ldots, K\}$ of X implies that any k-simplex σ corresponds to a unique string $(i_0 < i_1 < \cdots < i_k)$ which implicitly equips σ with the orientation and defines the following total order relation given by

$$(i_0 < i_1 < \cdots < i_{k_1}) \prec (i'_0 < i'_1 \cdots < i'_{k_2})$$

if $k_1 < k_2$, or $k_1 = k_2 = k$ and $i_0 = i'_0, \cdots i_{r-1} = i'_{r-1}$ and $i_r < i'_r$, $r \le k$.

Recall that a continuous map $f : X \to \mathbb{R}$, X a simplicial complex, is *simplicial* if restricted to any open simplex is linear. A continuous map $f : X \to \mathbb{S}^1$ is *simplicial* if a cyclic cover $\widetilde{f} : \widetilde{X} \to \mathbb{R}$ of f is simplicial. It is not hard to check that for X a finite simplicial complex:

  (i) any simplicial map is tame (actually topologically tame);
  (ii) the critical values of f are among the values of f on vertices;
  (iii) any t different from the values of f on vertices is a topologically regular value.

Let $f : X \to \mathbb{R}$ or $f : X \to \mathbb{S}^1$ and let t be a value in the target which is not among the values of f on vertices.

In case f is real-valued, for any simplex σ denote
$\sigma(t) := \sigma \cap f^{-1}(t),$
$\sigma(t)_- := \sigma \cap f^{-1}(-\infty, t),$
$\sigma(t)_+ := \sigma \cap f^{-1}(t, \infty).$
In case f is angle-valued one replaces f by $\widetilde{f}$ and use the same definitions.

Clearly, $\sigma(t)$ and $\sigma_\pm(t)$ are generalized simplices, and when nonempty $\dim \sigma(t) = \dim \sigma - 1$ and $\dim \sigma_\pm(t) = \dim \sigma$. An orientation on σ induces orientations on $\sigma(t)$ and $\sigma_\pm(t)$ with $I(\sigma(t), \sigma_+(t)) = 1$ and $I(\sigma(t), \sigma_-(t)) = -1$. If $\sigma' > \sigma$, then $I(\sigma(t), \sigma') = 0$ and $I(\sigma_\pm(t), \sigma') = I(\sigma, \sigma')$.

Also clearly, the collection $\mathcal{X}_t := \{\sigma(t) \mid \sigma \cap f^{-1}(t) \ne \emptyset\}$ defines a cell complex. If X is a simplicial complex whose set of simplicies is equipped with a good total order and each simplex with an orientation, then that order restricts to a good total order on the cells of $\mathcal{X}_t$, the orientations of simplices in $\mathcal{X}$ induce orientations on the generalized simplices in $\mathcal{X}_t$, and

$I(\tau(t), \sigma(t)) = I(\tau, \sigma)$. The incidence matrix $M(X_t)$ appears as a submatrix of $M(X)$.

Let t_1, t_2 be values $t_1 < t_2$ different from the values of the map f on vertices. Assume that there is no simplex σ which intersects nonempty both levels $f^{-1}(t_1)$ and $f^{-1}(t_2)$.

Under these assumptions, $X_{t_1} = f^{-1}(t_1)$, $X_{t_2} = f^{-1}(t_2)$, and $X_{[t_1,t_2]} = f^{-1}([t_1,t_2])$ have a canonical structure of cell complex, with X_{t_1} and X_{t_2} subcomplexes of $X_{[t_1,t_2]}$. The collection $\mathcal{X}_{[t_1,t_2]}$ of the cells of $X_{[t_1,t_2]}$ is

$$
\begin{cases}
\{\sigma \mid \sigma \subset f^{-1}(t_1, t_2)\sqcup \\
\{\sigma_+(t_1) \mid \sigma \cap f^{-1}(t_1) \neq \emptyset\}\sqcup \\
\{\sigma(t_1) \mid \sigma \cap f^{-1}(t_1) \neq \emptyset\}\sqcup \\
\{\sigma_-(t_2) \mid \sigma \cap f^{-1}(t_2) \neq \emptyset\}\sqcup \\
\{\sigma(t_2) \mid \sigma \cap f^{-1}(t_2) \neq \emptyset\}.
\end{cases}
$$

As long as the cell complex $X_{[t_1,t_2]}$ is concerned, one can use the total order on the simplices of X and derive a good total order on the cells of $X_{[t_1,t_2]}$ as follows: take the cells of X_{t_1} with the induced order, followed by the cells of X_{t_2} with the induced order, followed by the cells of $\{\sigma \in \mathcal{X} \mid \sigma \subset f^{-1}((t_1, t_2))\}$ with the induced order. Note that all subsets $\mathcal{X}_{t_1}, \mathcal{X}_{t_2}$, and $\mathcal{X}_{[t_1,t_2]} \setminus (\mathcal{X}_{t_1} \cup \mathcal{X}_{t_2})$ can be viewed as subsets of $\mathcal{X}$ and the incidence matrix is of the form

$$
M(X_{[t_1,t_2]}) = \begin{bmatrix} M(X_{t_1}) & 0 & A \\ 0 & M(X_{t_2}) & B \\ 0 & 0 & C \end{bmatrix}.
$$

The incidence matrices $M(X)$, $M(X_t)$, and $M(X_{[t_1,t_2]})$ are inputs for the calculation of the homology of X, X_t, and $X_{[t_1,t_2]}$, as well as of the inclusion-induced linear maps between the homologies of these spaces by effective algorithms, in particular the *persistence algorithm* cf. [Zomorodian, A., Carlsson, G. (2005)], [Cohen-Steiner, D., Edelsbrunner, H., Morozov, D. (2006)]. The discussion above for two values t_1 and t_2 can be extended to a finite collection of values $t_1 \neq t_2 \neq \cdots \neq t_r$ leading to the collection of cell complexes X, X_{t_i}, $X_{[t_i,t_j]}$ and their incidence matrices.

2.3.5 *Configurations*

Configurations of points with multiplicity

A configuration of points with multiplicity in the topological space X is a map with finite support

$$\delta : X \to \mathbb{Z}_{\geq 0}.$$

The support of δ, denoted by $\operatorname{supp} \delta$, is defined by

$$\operatorname{supp} \delta := \{x \in X \mid \delta(x) \neq 0\},$$

and the cardinality of δ, denoted by $\sharp \delta$, is defined by

$$\sharp \delta := \sum_{x \in \operatorname{supp} \delta} \delta(x).$$

One denotes by $\operatorname{Conf}(X)$ the set of all configuration and by $\operatorname{Conf}_N(X)$ the subset of configurations of cardinality N, Clearly

$$\operatorname{Conf}(X) = \bigsqcup_N \operatorname{Conf}_N(X).$$

The set $\operatorname{Conf}_N(X) = S^N(X)$ is actually the N-fold symmetric product of X, precisely the quotient of the cartesian product of N copies of X by the action of the group Σ_N of permutations of N elements. The set $\operatorname{Conf}(X)$, and then $\operatorname{Conf}_N(X)$ carry a natural topology, the *collision topology*, defined below, and when we regard $\operatorname{Conf}_N(X)$ and $\operatorname{Conf}(X)$ as topological spaces we mean exactly this topology.

When $K \subset X$ is a closed subset of X, another equally natural topology on the set $\operatorname{Conf}(X \setminus K)$ that accounts for the influence of K is the *bottleneck topology* [13]. One denotes by $\operatorname{Conf}_{bn}(X; K)$ the set $\operatorname{Conf}(X \setminus K)$ equipped with the bottleneck topology.

To describe the *collision topology* one introduces the following notation. For a system $\{U_1, U_2, \ldots, U_r; n_1, n_2, \ldots, n_r\}$ with U_i disjoint open sets of X and n_i positive integers one denotes by

$$\mathcal{U}(U_1, U_2, \ldots U_r; n_1, n_2, \ldots n_r) :=$$

$$\left\{ \delta \in \operatorname{Conf}(X) \mid \operatorname{supp} \delta \subset \bigcup_i U_i \text{ and } \sum_{x \in U_i \cap \operatorname{supp} \delta} \delta(x) = n_i \right\}.$$

All configurations $\delta \in \mathcal{U}(U_1, U_2, \ldots U_r; n_1, n_2, \ldots, n_r)$ have the same cardinality, the integer $\sum_{i=1}^r n_i$. The sets of configurations $\mathcal{U}(\cdots)$ provide what

---

[13] A name we learned from [Cohen-Steiner, D., Edelsbrunner, H., Harer, J. (2007)].

is usually called a base for the collision topology. A set $\mathcal{U} \subset \mathrm{Conf}(X)$ is open in the collision topology if for any $\delta \in \mathcal{U}$ with support $\{x_1, \ldots, x_r\}$ and $\delta(x_r) = n_r$ one can find a system $(U_1, U_2, \ldots, U_r; n_1, n_2, \ldots, n_r,)$ $x_i \in U_i$ such that $\mathcal{U}(U_1, U_2, \ldots, U_r; n_1, n_2, \ldots, n_r) \subseteq \mathcal{U}$. With this topology $\mathrm{Conf}_N(X)$ are components[14] of $\mathrm{Conf}(X)$. It is immediate that $\mathrm{Conf}_N(X)$ equipped with the collision topology is exactly $S^N(X) = X^N / \Sigma_N$ equipped with the quotient topology derived from the product topology on X^N.

To describe the bottleneck topology one considers systems $\{U_1, U_2, \ldots, U_r, V; n_1, n_2, \ldots, n_r\}$ with U_i and V disjoint open sets of X, $V \supset K$, and n_i positive integers, and one denotes by

$$\mathcal{U}(U_1, U_2, \ldots, U_r, V; n_1, n_2, \ldots, n_r) :=$$

$$\left\{ \delta \in \mathrm{Conf}(X) \mid \mathrm{supp}\, \delta \subset \bigcup_i U_i \cup V \text{ and } \sum_{x \in U_i \cap \mathrm{supp}\, \delta} \delta(x) = n_i \right\}.$$

Note that the configurations in $\mathcal{U}(U_1, U_2, \ldots, U_r, V; n_1, n_2, \ldots, n_r)$ might not all have the same cardinality. Their support has exactly $\sum_i n_i$ points counted with multiplicity in $\bigcup_i U_i$, and all other points of the support are included in V. A subset O of $\mathrm{Conf}(X \setminus K)$ is open in the bottleneck topology if for any $\delta \in O$ one can find $\{U_1, U_2, \ldots, U_r, V; n_1, n_2, \ldots, n_r\}$ s.t. $\delta \in \mathcal{U}(U_1, U_2, \ldots, U_r, V; n_1, n_2, \ldots, n_r) \subseteq O$.

If the topological space X is a metric space, i.e., the topology is derived from a metric d, then $\mathrm{Conf}_N(X)$ carries a metric $\underline{d}$ which induces the collision topology. The metric $\underline{d}$ is defined by

$$\underline{d}(\langle x_1, \ldots, x_N \rangle, \langle y_1, \ldots, y_N \rangle) = \inf_f (\sup_i d(x_i, f(x_i)),$$

where the infimum is taken over all bijective maps $f : \{x_1, \ldots, x_N\} \to \{y_1, \ldots, y_N\}$, and $\langle x_1, \ldots, x_N \rangle \langle y_1, \ldots, y_N \rangle$ are the elements of $S^N(X)$ represented by the N-tuples $\langle x_1, \ldots, x_N \rangle$ and $\langle y_1, \ldots y_N, \rangle$, respectively. If d is a complete metric, so is $\underline{d}$.

Similarly, a metric on X gives rise to a metric $\underline{d}$ on $\mathrm{Conf}(X \setminus K)$, referred to as the *bottleneck metric*, first proposed in [Cohen-Steiner, D., Edelsbrunner, H., Harer, J. (2007)] for the case $X = \mathbb{R}^2$, $K = \Delta$, which induces the bottleneck topology. Since the formula proposed in [Cohen-Steiner, D., Edelsbrunner, H., Harer, J. (2007)] requires explanations and this metric in not used in this book, we do not describe it. The bottleneck metric is never complete unless $K = \emptyset$, in which case $\mathrm{Conf}(X) = \mathrm{Conf}_{\mathrm{bn}}(X; \emptyset)$.

Finally, note the following:

---

[14]When X connected, then connected components

(i) If $X = \mathbb{C}$, then $\mathrm{Conf}_N(X)$ can be identified with the degree-N monic polynomials with complex coefficients, and if $X = \mathbb{C} \setminus 0$, to the degree-$N$ monic polynomials with non-zero free coefficient. To the configuration δ whose support consists of the points $z_1, z_2, \ldots, z_k$ with $\delta(z_i) = n_i$ one associates the monic polynomial $P^\delta(z) = \prod_i (z - z_i)^{n_i}$. Then as topological spaces $\mathrm{Conf}_N(\mathbb{C})$ and $\mathrm{Conf}_N(\mathbb{C} \setminus 0)$ can be identified with $\mathbb{C}^N$ and $\mathbb{C}^{N-1} \times (\mathbb{C} \setminus 0)$, respectively.

(ii) If $X = \mathbb{T} := \mathbb{R}^2/\mathbb{Z}$ is the quotient of $\mathbb{R}^2$ by the action $\mu(n, (a, b)) = (a + 2\pi n, b + 2\pi n)$, then the space $\mathbb{T}$ can be identified to $\mathbb{C} \setminus 0$ by the homeomorphism $\langle a, b \rangle \mapsto e^{ia+(b-a)}$, and the spaces $\mathrm{Conf}_N(\mathbb{T})$ and $\mathrm{Conf}_N(\mathbb{C} \setminus 0)$ are homeomorphic. Here $\langle a, b \rangle \in \mathbb{T}$ denotes the μ-orbit of (a, b).

(iii) The *canonical metric* $\underline{D}$ on $\mathrm{Conf}_N(\mathbb{R}^2)$ or $\mathrm{Conf}_N(\mathbb{T})$ refers to the metrics derived from the complete Euclidean metric D on $\mathbb{R}^2$ or on $\mathbb{T} = \mathbb{R}^2/\mathbb{Z}$. Both these metrics are complete.

Configurations of subspaces (split submodules)

Let A be a unital ring. In this book A is either a field κ, or the field $\kappa[t^{-1}, t]]$ the field of Laurent power series with coefficients in κ, or the ring $\kappa[t^{-1}, t]$ of Laurent polynomials.

For V a f.g. free A-module, we denote by $\mathcal{P}(V)$ the collection of split submodules[15] of V.

A configuration of split submodules of V is a map $\widehat{\delta} : X \to \mathcal{P}(V)$ with finite support[16] such that the linear map induced by inclusions $i :$ $\bigoplus_{x \in \mathrm{supp}\,\widehat{\delta}} \widehat{\delta}(x) \to V$ is an isomorphism. The set of these configurations is denoted by $\mathrm{CONF}_V(X)$ and viewed as a refinement of $\mathrm{Conf}_N(X)$, $N = \mathrm{rank}\,V$.

$\mathrm{CONF}_V(X)$ comes equipped with a natural topology, the *collision topology* defined, as in the case of configurations of points, with the help of a base given by systems $(U_1, U_2, \ldots, U_k; V_1, V_2, \ldots, V_k)$ with $U_1, U_2, \ldots, U_k$ disjoint open subsets of X and free submodules $V_1, V_2, \ldots, V_k$ satisfying $V_i \cap V_j = 0$ and $\bigoplus_{1 \le i \le k} V_i = V$, and consisting of the configurations $\widehat{\delta}$ with

(1) $\mathrm{supp}\,\widehat{\delta} \subset \bigsqcup_i U_i$,
(2) $\bigoplus_{x \in U_i} \widehat{\delta}(x) = V_i$.

---

[15]A submodule V' of V is *split* if the inclusion $V' \subseteq V$ has a left inverse.
[16]As before, $\mathrm{supp}\,\widehat{\delta} := \{x \in X \mid \mathrm{rank}(\widehat{\delta}(x)) \ne 0\}$.

The assignment $\widehat{\delta}(x) \leadsto \dim(\widehat{\delta}(x))$ provides a continuous map

$$\dim : \mathrm{CONF}_V(X) \to \mathrm{Conf}_N(X).$$

A configuration $\widehat{\delta}$ with $\dim \widehat{\delta} = \delta$ is called a *refinement* of δ.

In case $A = \mathbb{R}$ or $A = \mathbb{C}$, instead of V_i one can take open sets $O_i \subset G_{k_i}(V)$, where $G_r(V)$ is the Grassmanian of r-dimensional subspaces of V (which has the obvious topology) and instead of (2) above one considers

(2') $\bigoplus_{x \in U_i} \widehat{\delta}(x) \in O_i$.

This provides a coarser topology still making $\widehat{\delta}(x) \leadsto \dim(\widehat{\delta}(x))$ a continuos map. If in addition V is equipped with a Hermitian scalar product then one denotes by $\mathrm{CONF}_V^{\perp}(X)$ the subset of configurations in $\mathrm{CONF}_V(X)$ which in addition satisfy

(3) $\widehat{\delta}(x) \perp \widehat{\delta}(y)$ for $x \neq y$,

equipped with this coarser topology.

Of course this coarser topology can be considered for other field whenever a natural topology can be provided on $G_r(V)$.

Example. The simplest example of configurations of points and refinement to a configuration of subspaces is provided by a linear map $T : V \to V$, with V a complex vector space of finite dimension N.

In this case $X = \mathbb{C}$, $\delta \in \mathrm{Conf}_N(\mathbb{C})$ and $\widehat{\delta} \in \mathrm{CONF}_V(\mathbb{C})$, with $\delta(z)$ the multiplicity of the eigenvalue z, i.e., z not an eigenvalue is viewed as z being an eigenvalue of multiplicity zero, and $\widehat{\delta}(z)$ is the generalized eigenspace[17] of the eigenvalue z.

Configurations of subquotients

Let A be a unital (commutative) ring and V be a free f.g. A-module. A *subquotient* of V is a pair $\omega = (W, W')$ with $W' \subset W$ split submodules of V. For each such subquotient ω one considers the free module $\widehat{\omega} = W/W'$.

The subquotient is called *virtually trivial* if $W' = W$.

Call *splitting* any linear map $i_\omega : W/W' \to W$ which is a right inverse of the canonical projection $\pi_\omega : W \to W/W'$, i.e., $\pi_\omega \cdot i_\omega = id$. A splitting of ω realizes the quotient W/W' as a split submodule of W, and then of V.

Denote by $\widetilde{\mathcal{P}}(V)$ the collection of subquotients of V.

A *configuration of subquotients of V indexed by points in the space X* is given by a map $\widetilde{\omega} : X \to \widetilde{\mathcal{P}}(V)$ with finite support, i.e., $\widetilde{\omega}(x)$ is virtually trivial for all but finitely many points $x \in X$, which satisfies properties P1,

---

[17]The generalized eigenspace of the eigenvalue λ is $\bigcup_r \ker(T - \lambda E)^r$.

P2, and P3 below. Configurations of subquotients are called in [Burghelea, D. (2016a)] *special configurations*.

To formulate these three properties we need some notation.

Denote by $\alpha = (W_\alpha, W'_\alpha)$, $\beta = (W_\beta, W'_\beta)$, $\gamma = (W_\gamma, W'_\gamma), \ldots$ elements of $\widetilde{\mathcal{P}}(V)$, by $\mathcal{A} := \mathcal{A}(\widetilde{\omega})$ the image of $\widetilde{\omega}$ in $\widetilde{\mathcal{P}}(V)$, and for any $\alpha \in \mathcal{A}$ denote

$$\mathcal{A}_\alpha := \{\beta \in \mathcal{A} \mid W_\beta \subseteq W_\alpha\}.$$

P1. The set $\mathcal{A}(\widetilde{\omega}) = \mathcal{A}$ is finite.

P2. If $\beta = (W_\beta, W'_\beta) \in \mathcal{A}_\alpha$, $\beta \neq \alpha$, then $W_\beta \subseteq W'_\alpha$.

P3. For any $\alpha \in \mathcal{A}$, one has

$$\sum_{\beta \in \mathcal{A}_\alpha} \operatorname{rank} \left(W_\beta / W'_\beta \right) = \operatorname{rank} W_\alpha,$$

$$\sum_{\alpha \in \mathcal{A}} \operatorname{rank} \left(W_\alpha / W'_\alpha \right) = \operatorname{rank} V.$$

Once a collection of splittings $i_\alpha : W_\alpha/W'_\alpha \to W_\alpha$ indexed by $\alpha \in \mathcal{A}$ is given, a configuration $\widetilde{\omega}$ of subquotients provides a configuration $\widehat{\widetilde{\omega}} \in \operatorname{CONF}_V(X)$ of split submodules, $\widehat{\widetilde{\omega}}(x) = i_\alpha(W_{\widetilde{\omega}(x)}/W'_{\widetilde{\omega}(x)})$.

In the case where A is the field $\mathbb{C}$ or $\mathbb{R}$ and V is a Hilbert space, each subquotient $W_{\widetilde{\omega}(x)}/W'_{\widetilde{\omega}(x)}$ has a *canonical splitting*, defined by the orthogonal complement of $W'(x)$ in $W(x)$, hence a canonical way to convert a configuration $\widetilde{\omega}$ of subquotients into an element $\widehat{\widetilde{\omega}} \in \operatorname{CONF}_V(X)$. In case V is equipped with a Hilbert space structure and $\widetilde{\omega}$ is a configuration of subquotients, we will write $\widehat{\omega} = \widehat{\widetilde{\omega}}$ for the configuration of subspaces of V derived from $\widetilde{\omega}$.

2.3.6 *Algebraic topology of a pair $(X, \xi \in H^1(X; \mathbb{Z}))$*

Let κ be a fixed field. For a space X denote by $H_r(X)$ the (singular) homology in degree r with coefficients in κ. Recall that $H_*(X)$ is obtained as the homology of the singular chain complex $(C_*(X), \partial_*)$ whose component $C_k(X)$ is the κ-vector space generated by all singular simplices in dimension k, cf. [Munkres, J. (1984)] or [Hatcher (2002)]. In case X is equipped with a triangulation, hence identified to a simplicial complex or cell complex as described in the previous subsection, the homology vector spaces can be calculated by using a considerably smaller subcomplex, the one whose components are generated only by the simplices of the triangulation. When the simplicial complex is finite these data can be fed in a computer (via the incidence matrix) and the dimensions $\dim H_r(X)$ can be effectively calculated. Because of that we regard these dimensions as *computer friendly invariants*. Denote by $\beta_r(X)$ the dimension of $H_r(X)$. These numbers are known as the Betti numbers of X over the field κ.

Denote by $H^1(X;\mathbb{Z})$ the (singular) cohomology group in degree one with integer coefficients and recall from Subsection 2.3.3 that for X an ANR the set $H^1(X;\mathbb{Z})$ parametrizes the homotopy classes of continuous maps $f : X \to \mathbb{S}^1$ via $f \rightsquigarrow \xi_f$.

For $\xi \in H^1(X;\mathbb{Z})$, let $\pi : \widetilde{X} \to X$ be an infinite cyclic cover and $\tau : \widetilde{X} \to \widetilde{X}$ the generator of the group of deck transformations, which induces the isomorphism $\tau_r : H_r(\widetilde{X}) \to H_r(\widetilde{X})$ of the κ-vector space $H_r(\widetilde{X})$, and hence a structure of $\kappa[t^{-1}, t]$-module. Note that if X is a finite simplicial complex, then in view of the description of the homology vector space in terms of the chain complex generated by simplices, $H_r(\widetilde{X})$ is in general an infinite-dimensional κ-vector space, but always a f.g. $\kappa[t^{-1}, t]$-module. Since $\kappa[t^{-1}, t]$ is a principal ideal domain, cf. [Lang, S. (2002)], the $\kappa[t^{-1}, t]$-torsion submodule, $\mathrm{Tor}(H_r(\widetilde{X}))$, when regarded as a κ-vector space, is of finite dimension. Let us write $V_r(X;\xi) := \mathrm{Tor}(H_r(\widetilde{X}))$ for this vector space and $T_r = T_r(X;\xi) : V_r(X;\xi) \to V_r(X;\xi)$ for the linear isomorphism provided by the multiplication by t, in our case induced by τ_r.

Let $H_r^N(X;\xi) := H_r(\widetilde{X})/\mathrm{Tor}(H_r(\widetilde{X}))$, which is a free f.g $\kappa[t^{-1}, t]$-module. Since $\kappa[t^{-1}, t]$ is a principal ideal domain, $H_r(\widetilde{X})$ is isomorphic to $H_r^N(X;\xi) \oplus \mathrm{Tor}(H_r(\widetilde{X}))$. One denotes by $\beta_r^N(X;\xi)$ the rank of $H^N(X;\xi)$.

1. Novikov homology and the monodromy of (X, ξ)

We refer to $H_r^N(X;\xi)$ as the Novikov homology, to $\beta_r^N(X;\xi)$ as the Novikov-Betti number in degree r, and to the similarity class of the linear isomorphism $T_r(X;\xi)$ as the *homological r-monodromy* over the field κ.

Most experts call Novikov homology the $\kappa[t^{-1}, t]]$-vector space

$$H_r(\widetilde{X}) \otimes_{\kappa[t^{-1}, t]} \kappa[t^{-1}, t]] = (H_r(\widetilde{X})/\mathrm{Tor}H_r(\widetilde{X})) \otimes_{\kappa[t^{-1}, t]} \kappa[t^{-1}, t]],$$

in which case $\beta_r^N(X;\xi)$ is equal to the dimension of this vector space.

We also refer to the characteristic polynomial of $T_r(X;\xi)$, as the r-th Alexander polynomial and denote it by $A_r(X;\xi)(z)$, and to the rational function

$$A(X;\xi)(z) := \prod_r A_r(X;\xi)(z)^{(-1)^r}$$

as the **Alexander function**.

2. Alexander polynomial of a knot

A knot $K \subset S^3$ is an embedded smooth (or more general locally flat) simple curve (i.e., homeomorphic to the oriented circle $\mathbb{S}^1$) in the three-dimensional sphere S^3. Consider $X = S^3 \setminus N$, where N is an open tubular neighborhood of K in S^3. Clearly X is homotopy equivalent to $S^3 \setminus K$. The Alexander dual of the generator $u \in H_1(K; \mathbb{Z})$ is an integral cohomology class $\xi \in H^1(X; \mathbb{Z})$. The Alexander polynomial of the knot, a fundamental invariant of the knot, is a polynomial with integer coefficients

$$a_r z^r + \cdots + a_1 z + a_0,$$

with $a_0 \neq 0$ and $a_r \geq 0$ defined as the only generator of the principal ideal $(P(t))$ defined by the isomorphism $H_r(\widetilde{X}; \mathbb{Z}) \equiv \mathbb{Z}[t^{-1}, t]/(P(t))$, where $\mathbb{Z}[t^{-1}, t]$ denotes the ring of Laurent polynomial with coefficients in $\mathbb{Z}$. The monic polynomial $(1/a_n) \cdot P(t)$ can be calculated as the characteristic polynomial of $T_1(X; \xi)$, the 1-monodromy of (X, ξ), with coefficients in $\mathbb{Q}$ cf. [Rolfsen, D. (1976)] and [Milnor, J. (1968)].

3. Homology with coefficients in a rank-one representation

A rank-one representation of the fundamental group $\pi_1(X, x)$ is provided by a group homomorphism $\pi_1(X, x) \to GL_1(\kappa)$, and since $GL_1(\kappa)$ is abelian, is equivalent to a homomorphism $H_1(X; \mathbb{Z}) \to GL_1(\kappa)$. An element $u \in \kappa \setminus 0$ defines a homomorphism $\hat{u} : \mathbb{Z} \to GL_1(\kappa) = \kappa \setminus 0$. An element $\xi \in H^1(X; \mathbb{Z})$ can be viewed as a homomorphism $\xi : H_1(X; \mathbb{Z}) \to \mathbb{Z}$ and then, together with $u \in \kappa \setminus 0$, provides by composition with $\hat{u}$ a rank-one representation (ξ, u) of $\pi_1(X, x)$. The homology $H_r(X; (\xi, u))$ is the κ-vector space which can be described as follows.

Let $\widetilde{X} \to X$ be the infinite cyclic cover defined by ξ. Consider $(C_*(\widetilde{X}), \partial_*)$ as a complex of left $\kappa[t^{-1}, t]$-modules, and using $u \in \kappa \setminus 0$ regard κ as a right $\kappa[t^{-1}, t]$-module with right multiplication by t given by $\lambda \cdot t := \lambda u$. Consider the chain complex of κ-vector spaces $\kappa \otimes_{\kappa[t^{-1}, t]} (C_*(\widetilde{X}), \partial_*)$. The homology of this complex is denoted by $H_r(X; (\xi, u))$, and if X is a compact ANR, the dimension of these vector spaces, denoted by $\beta_r(X; (\xi, u))$, is finite and is referred to as the u-*twisted Betti number*.

The following well-known fact can be derived by the reader as an exercise and can be also recovered from any textbook on homology with local coefficients.

Observation 2.8.
1. If $u = 1$, then $\beta_r(X, (\xi, 1)) = \beta_r(X)$.
2. For all but finitely many $u \in \kappa \setminus 0$, one has $\beta_r^N(X; \xi) = \beta_r(X; (\xi, u))$.

4. Poincaré Duality and related diagrams

For Y an n-dimensional closed topological κ-orientable manifold, $f : Y \to \mathbb{R}$ a real-valued topologically tame map, and a a topologically regular value of f let $Y_a := f^{-1}((-\infty, a])$ and $Y^a = f^{-1}([a, \infty))$. This implies that Y_a and Y^a are topologically submanifolds with boundary. Poincaré Duality provides the commutative diagrams

$$
\begin{array}{ccccc}
H_r(Y_a) & \xrightarrow{i_a(r)} & H_r(Y) & \xrightarrow{j_a(r)} & H_r(Y, Y_a) \\
\downarrow & & \downarrow & & \downarrow \\
(H_{n-r}(Y, Y^a))^* & \xrightarrow{j^a(n-r)^*} & (H_{n-r}(Y))^* & \xrightarrow{i^a(n-r)^*} & (H_{n-r}(Y^a))^*
\end{array}
$$

$$(2.30)$$

$$
\begin{array}{ccccc}
H_r(Y^a) & \xrightarrow{i^a(r)} & H_r(Y) & \xrightarrow{j^a(r)} & H_r(Y, Y^a) \\
\downarrow & & \downarrow & & \downarrow \\
(H_{n-r}(Y, Y_a))^* & \xrightarrow{j_a(n-r)^*} & (H_{n-r}(Y))^* & \xrightarrow{i_a(n-r)^*} & (H_{n-r}(Y_a))^*
\end{array}
$$

$$(2.31)$$

The vertical arrows, which are all isomorphisms in each diagram, are called the Poincaré Duality isomorphisms. The horizontal arrows are induced by the inclusions of $i_a : Y_a \to Y$ or $i^a : Y^a \to Y$ in Y and the inclusion of pairs $j_a : (Y, \emptyset) \to (Y, Y_a)$ or $j^a : (Y, \emptyset) \to (Y, Y^a)$.

The above diagrams are natural with respect to $a < b$, in the sense that there exists a collection of linear maps in homology, induced by the obvious inclusions of spaces and pairs which appear in diagrams (2.30) and (2.31) defining a morphism from the diagrams corresponding to a to the diagram corresponding to b and making all squares commutative.

5. Borel-Moore homology and Poincaré Duality

For an n-dimensional manifold Y, not necessarily compact, the Poincaré Duality can be better formulated using Borel-Moore homology, cf. [Borel,

A., Moore, J.C. (1960)], especially tailored for locally compact spaces Y and pairs (Y, K), with K closed subset of Y. Borel-Moore homology coincides with the standard homology when Y is compact. In general, for a locally compact space Y, it can be described as the inverse limit of the homology $H_r(Y, Y \setminus U)$ for all U that are open sets with compact closure. One denotes the Borel-Moore homology in dimension r by H_r^{BM}. A similar "inverse limit" description can be provided for the relative Borel-Moore homology, $H_r^{BM}(Y, K)$, cf. [Borel, A., Moore, J.C. (1960)].

For Y a n-dimensional topological κ-orientable manifold, $g : Y \to \mathbb{R}$ a proper topologically tame map, and a a regular value of g, Poincaré Duality provides the commutative diagrams

$$
\begin{array}{ccc}
H_r^{BM}(Y_a) \longrightarrow H_r^{BM}(Y) \longrightarrow H_r^{BM}(Y, Y_a) \\
\downarrow \qquad\qquad \downarrow \qquad\qquad \downarrow \\
(H_{n-r}(Y, Y^a))^* \longrightarrow (H_{n-r}(Y))^* \longrightarrow (H_{n-r}(Y^a))^*
\end{array}
$$

$$(2.32)$$

$$
\begin{array}{ccc}
H_r^{BM}(Y^a) \longrightarrow H_r^{BM}(Y) \longrightarrow H_r^{BM}(Y, Y^a) \\
\downarrow \qquad\qquad \downarrow \qquad\qquad \downarrow \\
(H_{n-r}(Y, Y_a))^* \longrightarrow (H_{n-r}(Y))^* \longrightarrow (H_{n-r}(Y_a))^*
\end{array}
$$

$$(2.33)$$

The vertical arrows in each diagram represent the Poincaré Duality isomorphisms. The horizontal arrows are induced by the inclusions of Y_a or Y^a in Y and the inclusion of pairs $(Y, \emptyset)$ in (Y, Y_a) or (Y, Y^a). In this book this homology will be considered only for locally compact ANRs X which admit a proper weakly tame maps $f : X \to \mathbb{R}$, in which case the Borel-Moore homology satisfies the equalities

$$
H_r^{BM}(X) = \varprojlim_{0 < l \to \infty} H_r(X, X_{-l} \sqcup X^l),
$$

$$
H_r^{BM}(X_a) = \varprojlim_{0 < l \to \infty} H_r(X_a, X_{a-l}),
$$

$$
H_r^{BM}(X^a) = \varprojlim_{0 < l \to \infty} H_r(X, X^{a+l}),
$$

$$
H_r^{BM}(X, X_a) = \varprojlim_{0 < l \to \infty} H_r(X, X_a \sqcup X^{a+l}),
$$

$$
H_r^{BM}(X, X^a) = \varprojlim_{0 < l \to \infty} H_r(X, X^a \sqcup X_{a-l}).
$$

$$(2.34)$$

The reader can use the formulae (2.34) as definitions without the need of substantial knowledge of this concept.

We apply diagrams (2.32) and (2.33) to $X = \widetilde{M}^n$ and $g = \widetilde{f}$, with $\widetilde{f} : \widetilde{M} \to \mathbb{R}$ the infinite cyclic cover of $f : M^n \to \mathbb{S}^1$, a tame map defined on a closed κ-orientable topological manifold with a, b regular values, and obtain

$$
\begin{array}{ccc}
H_r^{BM}(\widetilde{M}_a) & \xrightarrow{\;{}^{BM}i_a(r)\;} H_r^{BM}(\widetilde{M}) \xrightarrow{\;{}^{BM}j_a(r)\;} H_r^{BM}(\widetilde{M}, \widetilde{M}_a) \\
\downarrow & \downarrow \qquad\qquad \downarrow \\
(H_{n-r}(\widetilde{M}, \widetilde{M}^a))^* \xrightarrow{(j^a(n-r))^*} (H_{n-r}(\widetilde{M}))^* \xrightarrow{(i^a(n-r))^*} (H_{n-r}(\widetilde{M}^a))^*
\end{array}
$$
$$(2.35)$$

$$
\begin{array}{ccc}
H_r^{BM}(\widetilde{M}^b) & \xrightarrow{\;{}^{BM}i^b(r)\;} H_r^{BM}(\widetilde{M}) \xrightarrow{\;{}^{BM}j^b(r)\;} H_r^{BM}(\widetilde{M}, \widetilde{M}^b) \\
\downarrow & \downarrow \qquad\qquad \downarrow \\
(H_{n-r}(\widetilde{M}, \widetilde{M}_b))^* \xrightarrow{(j_b(n-r))^*} (H_{n-r}(\widetilde{M}))^* \xrightarrow{(i_b(n-r))^*} (H_{n-r}(\widetilde{M}_b))^* \; .
\end{array}
$$
$$(2.36)$$

The inclusions of pairs $(\widetilde{M}, \widetilde{M}_{-l'} \sqcup \widetilde{M}^{l'}) \subseteq (\widetilde{M}, \widetilde{M}_{-l} \sqcup \widetilde{M}^l)$ for $l' > l$ induce in homology an inverse system whose limit is $H_r^{BM}(\widetilde{M})$. Similar inclusions of pairs associated with $l' > l$ induce inverse systems whose limits are the remaining Borel-Moore homology vector spaces considered above.

Chapter 3

Graph Representations

3.1 Generalities on graph representations

An oriented graph Γ consists of:

(i) a set of vertices $\mathcal{V}$ denoted by $x, y, z, \ldots$;

(ii) a set of oriented edges $\mathcal{E}$ denoted by $a, b, c, \ldots$, each edge with a *tail* and a *head*, hence two maps $t : \mathcal{E} \to \mathcal{V}$, $h : \mathcal{E} \to \mathcal{V}$, such that

(iii) the map $\mathcal{E} \xrightarrow{t \times h} \mathcal{V} \times \mathcal{V}$ is finite to one.

In this book we will be concerned only with the following graphs, denoted by $\mathcal{Z}$ and G_{2m}, m a positive integer.

The *graph $\mathcal{Z}$* has the set of vertices $\mathcal{V} := \{x_i, i \in \mathbb{Z}\}$ and the set of oriented edges $\mathcal{E} := \{a_i : x_{2i-1} \to x_{2i}, b_i : x_{2i+1} \to x_{2i}\}$; hence the maps $t : \mathcal{E} \to \mathcal{V}$ and $h : \mathcal{E} \to \mathcal{V}$ are defined by

$$t(a_i) = x_{2i-1}, \quad t(b_i) = x_{2i+1},$$
$$h(a_i) = x_{2i}, \quad h(b_i) = x_{2i}.$$

The graph $\mathcal{Z}$ is depicted in Figure 3.1 below.

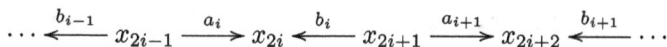

$$\cdots \xleftarrow{\ b_{i-1}\ } x_{2i-1} \xrightarrow{\ a_i\ } x_{2i} \xleftarrow{\ b_i\ } x_{2i+1} \xrightarrow{\ a_{i+1}\ } x_{2i+2} \xleftarrow{\ b_{i+1}\ } \cdots$$

Fig. 3.1   The graph $\mathcal{Z}$

The *graph G_{2m}* has the set of vertices $\mathcal{V} = \{x_i, i = 1, 2, \ldots, 2m\}$ and the set of oriented edges $\mathcal{E} = \{a_i : x_{2i-1} \to x_{2i}, b_i : x_{2i+1} \to x_{2i}, 1 \leq i \leq m\}$ with $x_{2m+1} = x_1$. The maps t and h are defined similarly. The graph G_{2m} is depicted in Figure 3.2 below.

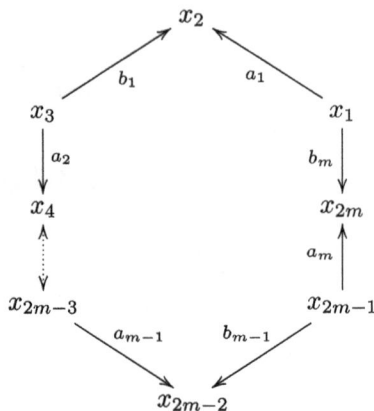

Fig. 3.2   The graph G_{2m}

In the literature the graph G_2 is known as the Kronecker graph and is graphically represented by

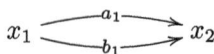

Fig. 3.3   The graph G_2

Fix a field κ. Given an oriented graph Γ, a Γ-representation ρ^Γ is an assignment which to each vertex x of Γ assigns a finite-dimensional vector space V_x, and to each oriented arrow $a : x \to y$ a linear map $\alpha(a) : V_x \to V_y$. When the graph is obvious from the context, the superscript "Γ" will be sometimes suppressed from the notation.

In the case of the graph $\mathcal{Z}$ one denotes explicitly a representation by

$$\rho^{\mathcal{Z}} := \{V_i, \alpha_i : V_{2i-1} \to V_{2i}, \beta_i : V_{2i+1} \to V_{2i}, \quad i \in \mathbb{Z}\},$$

and in case of the graph G_{2m} by

$$\rho^{G_{2m}} := \begin{cases} V_r, & 1 \le r \le 2m, \\ \alpha_i : V_{2i-1} \to V_{2i}, \; \beta_i : V_{2i+1} \to V_{2i}, & 1 \le i \le m, \\ V_{2m+1} = V_1. \end{cases}$$

The Γ-representation ρ^Γ is called

(i) *trivial*, if all $V_x = 0$,

(ii) *with finite support*, if $V_x = 0$ for all but finitely many vertices x,

(iii) *regular*, if all the linear maps corresponding to edges are isomorphisms.

The $\mathcal{Z}$-representation $\rho^{\mathcal{Z}}$ is *periodic* if there exists an integer N such that $V_i = V_{i+2N}$, $\alpha_i = \alpha_{i+N}$, $\beta_i = \beta_{i+2N}$. The integer N is called the *period of* ρ. If the representation $\rho^{\mathcal{Z}}$ has period N, then it also has kN as a period for any $k \in \mathbb{Z}_{\geq 1}$.

For a fixed graph Γ and two linear representations $\rho = \{V_x, \alpha(a)\}$, $\rho' = \{V_x', \alpha'(a)\}$, a *morphism* $\omega : \rho \to \rho'$ assigns to each vertex x a linear map $\omega_x : V_x \to V_x'$ so that for any arrow $a : x \to y$ one has $\alpha'(a) \cdot \omega_x = \omega_y \cdot \alpha(a)$.

The notions of *monomorphism, epimorphism,* and *isomorphism* are morphisms with all linear maps ω_x injective, surjective, and bijective, respectively. The notions of subrepresentation, quotient representation, (direct) sum are the obvious extensions of the same notions for vector spaces.

One says that the representation ρ^{Γ} is *indecomposable* if it is not isomorphic to the sum of two nontrivial representations. Such a representation will be often referred to simply as an "indecomposable".

For a fixed graph Γ one can consider the category whose objects are finite-dimensional Γ-representations with finite support, and whose morphisms are the *morphisms of representations*. This category is abelian. For the definition of abelian *categories* the reader can consult [Bucur, I., Deleanu, A. (1968)], or any book with *abelian categories* in the title.

As a consequence one has the following Krull-Remak-Schmidt decomposition theorem, cf. Chapter 6, page 154 in [Bucur, I., Deleanu, A. (1968)].

Theorem 3.1. *Any representation with finite support is isomorphic to a finite sum of indecomposable finite-dimensional representations with finite support which are unique up to isomorphism. More precisely, if $\rho = \rho_1 \oplus \rho_2 \oplus \cdots \oplus \rho_n$ and $\rho = \rho_1' \oplus \rho_2' \oplus \cdots \oplus \rho_m'$, then $n = m$ and there exists a permutation σ of the set $\{1, 2, \ldots, n\}$ such that $\rho_i' \simeq \rho_{\sigma(i)}$ for all i.*

In case of the graph $\mathcal{Z}$ a representation which is either periodic or with finite support is called a *good $\mathcal{Z}$-representation*. These are the only $\mathcal{Z}$-representations of interest in this book.

The Krull-Remak-Schmidt decomposition stated above continues to hold for good $\mathcal{Z}$-representations; however, for periodic representations infinitely many copies of indecomposables with finite support appear as *translations* of each indecomposable with finite support, cf. [Burghelea, D.,

Haller, S. (2015)] or Exercise E.1 in Section 3.4.

3.2   The indecomposable representations

We begin with a few examples of $\mathcal{Z}$- and G_{2m}-representations which turn out to be all indecomposables for these graphs.

Examples of $\mathcal{Z}$-representations with finite support

1. **Barcodes:** For any pair of integers r, s with $r \leq s$ one considers the representation

$$\rho_{r,s}^{\mathcal{Z}} := \begin{cases} V_l = \begin{cases} \kappa, & \text{if } r \leq l \leq s, \\ 0, & \text{otherwise}, \end{cases} \\ \alpha_i = \begin{cases} Id, & \text{if } r+1 \leq 2i \leq s, \\ 0, & \text{otherwise}, \end{cases} \\ \beta_i = \begin{cases} Id, & \text{if } r \leq 2i \leq s-1, \\ 0, & \text{otherwise}. \end{cases} \end{cases}$$

Such representations will be referred to as $\mathcal{Z}$-*barcodes* or simply *bar codes*. For reasons that will be made clear in next section, we will use the following (perhaps odd) notation that divides the set of barcodes in four different types: closed, open, closed-open, and open-closed:

(1) $\rho^{\mathcal{Z}}([i,j]) = \rho^{\mathcal{Z}}_{2i,2j}$, $i \leq j$, closed barcode,
(2) $\rho^{\mathcal{Z}}((i,j)) = \rho^{\mathcal{Z}}_{2i+1,2j-1}$, $i < j$, open barcode,
(3) $\rho^{\mathcal{Z}}([i,j)) = \rho^{\mathcal{Z}}_{2i,2j-1}$, $i < j$, closed-open barcode,
(4) $\rho^{\mathcal{Z}}((i,j]) = \rho^{\mathcal{Z}}_{2i+1,2j}$, $i < j$, open-closed barcode[1].

2. **The regular representation $\rho_\infty^{\mathcal{Z}}$:** Define
$\rho_\infty^{\mathcal{Z}} := \{V_i = \kappa, \alpha_i = Id, \beta_i = Id, i \in \mathbb{Z}\}$.

Observation 3.1.

(1) Any finite-dimensional regular $\mathcal{Z}$-representation ρ is isomorphic to the sum of finitely many copies of $\rho_\infty^{\mathcal{Z}}$. More precisely, $\rho \simeq \bigoplus_{\alpha \in \mathcal{S}}(\rho_\infty^{\mathcal{Z}})_\alpha$. One can interpret the finite set $\mathcal{S}$ as a basis $\mathcal{S} = \{v_1, v_2, \ldots, v_r\}$ for the vector space V_0 or for any other space V_k

---

[1] The geometric reason invoked above is explained by the fact that only the even integers will index critical values of real or angle-valued maps — the odd integers will index chosen regular values.

and, if we do this, we obtain a canonical isomorphism from ρ to $\bigoplus_{\alpha \in \mathcal{S}} (\rho_\infty^{\mathcal{Z}})_\alpha$, with $(\rho_\infty^{\mathcal{Z}})_\alpha$ a copy of $\rho_\infty^{\mathcal{Z}}$.

(2) A nontrivial morphism $\theta = \{\theta_i\} : \rho = \rho_{r,s}^{\mathcal{Z}} \to \rho' = \rho_{r',s'}^{\mathcal{Z}}$ implies that $\theta_i : V_i \to V_i'$ are isomorphisms for $\sup\{r, r'\} \leq i \leq \inf\{s, s'\}$, which also means that such a nontrivial morphism exists only when $\sup\{r, r'\} \leq \inf\{s, s'\}$. It is not hard to derive the additional relations between the integers r, s, r' and s' when such a morphism exists. See Exercises E2 in Section 3.4.

Examples of G_{2m}-representations

Type I (G_{2m}-barcodes):

As in the case of barcodes for $\mathcal{Z}$-representations, these representations are labeled by four types of intervals I of the form

$[i, l]$ with $1 \leq i \leq m$, $i \leq l$, called closed barcode,
(i, l) with $1 \leq i \leq m$, $i < l$, called open barcode,
$[i, l)$ with $1 \leq i \leq m$, $i < l$, called closed-open barcode,
$(i, l]$ with $1 \leq i \leq m$, $i < l$, called open-closed barcode,

and denoted by $\rho^{G_{2m}}([i, j])$, $\rho^{G_{2m}}((i, j))$, $\rho^{G_{2m}}([i, j))$, and $\rho^{G_{2m}}((i, j])$, respectively. Precisely, for $1 \leq i \leq j \leq 2m$, we have

$$\begin{cases} \rho^{G_{2m}}([i, j + km]) \\ \rho^{G_{2m}}((i, j + km)) \\ \rho^{G_m}([i, j + km)) \\ \rho^{G_{2m}}((i, j + km]) \end{cases} \text{ has } V_r = \begin{cases} \kappa^{k+1}, \text{ if } 2i \leq r \leq 2j, \ \kappa^k, \text{ otherwise} \\ \kappa^{k+1}, \text{ if } 2i < r < 2j, \ \kappa^k, \text{ otherwise} \\ \kappa^{k+1}, \text{ if } 2i \leq r < 2j, \ \kappa^k, \text{ otherwise} \\ \kappa^{k+1}, \text{ if } 2i < r \leq 2j, \ \kappa^k, \text{ otherwise} \end{cases}$$

and for $1 \leq j \leq i \leq 2m$ we have

$$\begin{cases} \rho^{G_{2m}}([i, j + km]) \\ \rho^{G_{2m}}((i, j + km)) \\ \rho^{G_{2m}}([i, j + km]) \\ \rho^{G_{2m}}((i, j + km]) \end{cases} \text{ has } V_r = \begin{cases} \kappa^{k-1} \text{ if } j < r < i, \ \kappa^k \text{ otherwise} \\ \kappa^{k-1} \text{ if } j \leq r \leq i, \ \kappa^k \text{ otherwise} \\ \kappa^{k-1} \text{ if } j < r \leq i, \ \kappa^k \text{ otherwise} \\ \kappa^{k-1} \text{ if } j \leq r < i, \ \kappa^k \text{otherwise} \end{cases} .$$

In each case, if α_i or β_i have the source and the target isomorphic (to κ^k or κ^{k-1}), then they are representable by the identity matrix, and in the remaining cases by diagonal matrices of maximal rank.

An easy graphical way to visualize the representation $\rho(\{i, j\})$, $1 \leq i, j \leq m$ is the following.

Suppose the vertices $x_1, x_2, \ldots, x_{2m-1}, x_{2m}$ are located counterclockwise on the unit circle (not visible in Figure 3.1) with center $O = (0, 0)$, say at the angles $0 < t_1 < \theta_1 < t_2 < \theta_2 < \cdots < \theta_m \leq 2\pi$, with $\theta_r = 2r\pi/m$ and $t_i = (2r - 1)\pi/m$. The angles t_r correspond to odd labeled vertices x_{2r-1} and the angles θ_r to even labeled vertices x_{2r}. The radii from 0 to x_i are shown in Figure 3.4 by the dotted lines Ox_i.

Draw the counterclockwise spiral curve [2] in the (x, y)-plane from $a = \theta_i$ to $b = \theta_j + 2\pi k$ with the ends a black or an empty circle to indicate "closed" or "open". A black circle indicates that the end is on this spiral, an empty circle that is not.

The vector space V_i is generated by the intersection points of the spiral with the radius Ox_i corresponding to the vertex x_i and α_i and β_i are defined on generators as follows: a generator e of V_{2i+1} is sent to the generator e' of V_{2i} if connected by a piece of spiral, and to 0 if not. The spiral in Figure 3.4 below corresponds to $k = 2$, and defines the representation $\rho^{G2m}([i, j+2m])$.

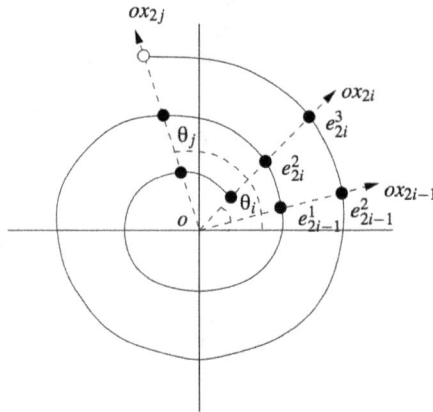

Fig. 3.4   The spiral for $[i, j + 2m)$

Exercise E.4 in Section 3.4 asks the reader to describe for all integers $i, j \leq m$ and k the representation $\rho^{G2m}([i, j + mk])$, $\rho^{G2m}([i, j + mk))$, $\rho^{G2m}([i, j + mk))$, and $\rho^{G2m}((i, j + mk])$ i.e. the vector spaces V_r and the matrices representing the α_i and β_i.

Type II (Jordan blocks/cells):

These representations are labeled by Jordan blocks $J = (V, T)$, i.e.,

---

[2] Given by: $\begin{cases} x(t) = r(t)\cos(tb + (1-t)a), \\ y(t) = r(t)\sin(tb + (1-t)a), \end{cases}$ $0 \leq t \leq 1,\ r(t) = (1 + t)$

pairs $T : V \to V$, with V a finite-dimensional vector space and T a linear isomorphism, which cannot be written as $(V_1, T_1) \oplus (V_2, T_2)$ with $\dim V_i > 0$, and are denoted by $\rho^{G_{2m}}(J)$. We usually denote by $\rho^{G_{2m}}(V, T)$ any regular G_{2m}-representation with all vector spaces $V_i = V$ and

$$\alpha_i = \begin{cases} T, & i = 1, \\ Id, & 2 \leq i \leq m, \end{cases} \qquad \beta_i = Id, \ 1 \leq i \leq m.$$

If $J = (\kappa^k, T(\lambda, k))$ we also write $\rho^G(J) := \rho^G(\lambda, k)$. Recall that $T(\lambda, k)$ is the $k \times k$ matrix

$$T(\lambda, k) = \begin{pmatrix} \lambda & 1 & 0 & \cdots & 0 & 0 \\ 0 & \lambda & 1 & \cdots & 0 & 0 \\ 0 & 0 & \lambda & \cdots & 0 & 0 \\ \vdots & \vdots & \vdots & \ddots & \vdots & \vdots \\ 0 & 0 & 0 & \cdots & \lambda & 1 \\ 0 & 0 & 0 & \cdots & 0 & \lambda \end{pmatrix}$$

for $k > 1$ and $T(\lambda, 1) = (\lambda)$.

Observation 3.2.

(1) Any regular G_{2m}-representation is equivalent to the representation $\rho^{G_{2m}}(V, T)$ with $V = V_1$ and $T = \beta_m^{-1} \cdot \alpha_m \cdot \beta_{m-1}^{-1} \cdots \beta_1^{-1} \cdot \alpha_1$.
(2) A regular representation $\rho^{G_{2m}}(V, T)$ is indecomposable iff the pair (V, T) is indecomposable.

One refers to both, the labeling interval $\{i, j\}$ and the representation $\rho^G(\{i, j\})$,[3] as a *barcode*, and to both the indecomposable pair $J = (V, T)$ and the representation $\rho^G(J)$ as a *Jordan block*.

As pointed out in Section 2.1 for arbitrary κ, if a Jordan block has an eigenvalue in κ, then this is the only eigenvalue and the matrix is conjugate to a Jordan cell. The matrix T of a Jordan block (V, T) representing T, when considered over the algebraic closure of κ, decomposes (up to similarity) into a direct sum of Jordan cells corresponding to eigenvalues in $\overline{\kappa} \setminus \kappa$, all conjugate [4].

The main result about $\mathcal{Z}$- and G_{2m}-representation is contained in the following theorem.

---

[3] One writes "{" and "}" as a single notation for [and (and for] and), respectively.
[4] All images by all Galois automorphisms of the field extension $\kappa \subset \overline{\kappa}$ of an element in $\overline{\kappa} \setminus \kappa$.

Theorem 3.2.

(1) *The $\mathcal{Z}$-representation $\rho_\infty^{\mathcal{Z}}$ and the G_{2m}-representations which correspond to Jordan blocks are indecomposable.*

(2) *The representations defined by $\mathcal{Z}$-barcodes and by G_{2m}-barcodes are indecomposable.*

(3) *All indecomposable G_{2m}-representations correspond to barcodes and to Jordan blocks, and are those described above. All indecomposable $\mathcal{Z}$-representations with finite support correspond to barcodes, and the only periodic indecomposable $\mathcal{Z}$-representation is $\rho_\infty^{\mathcal{Z}}$.*

Items (1) and (2) are more or less straightforward consequences of Observations 3.1 and 3.2 (see Exercise E4 in Section 3.4).

Item (3) is implicit in the work of Nazarova and of Donovan and Freislich [Nazarova, L.A. (1973)] [Donovan, P., Freislich, M.R. (1973)], who describe the indecomposable representations for many graphs, including G_{2m}. An alternative proof can be also provided using the properties of the elementary transformations described for the algorithm presented in Section 3.3, cf. Exercise E.10 in Section 3.4.

To summarize one has:

(1) For any G_{2m}-representation $\rho = \rho^{G_{2m}}$ one denotes by:

    – $\mathcal{B}(\rho)$ the collection of all bar codes (with proper multiplicity when appear multiple times as independent summands of ρ),

    – $\mathcal{B}^c(\rho)$, $\mathcal{B}^o(\rho)$, $\mathcal{B}^{c,o}(\rho)$ and $\mathcal{B}^{o,c}(\rho)$ the subcollections of barcodes with both ends closed, open, closed-open, and open-closed, respectively.

    – $\mathcal{J}(\rho)$ the collection of all Jordan blocks (with proper multiplicity when appear multiple times as independent summands of ρ).

Clearly, $\mathcal{B}_r(\rho) = \mathcal{B}^c(\rho) \sqcup \mathcal{B}^o(\rho) \sqcup \mathcal{B}^{c,o}(\rho) \sqcup \mathcal{B}^{o,c}(\rho)$, and $\mathcal{J}(\rho)$ are finite sets. We write $\rho(I)$ and $\rho(J)$ for the indecomposable representations indexed by the barcode I and by the Jordan block J, respectively.

For $\lambda \in \kappa \setminus 0$ one denotes by

    – $\mathcal{J}_\lambda(\rho)$ the collection of Jordan blocks $J = (V, T)$ with T having λ as an eigenvalue. In this case each such Jordan block is of the form $J = (\kappa^k, T(\lambda, k))$, cf. Proposition 2.2.

By the Krull-Remak-Schmidt decomposition theorem one has

$$\rho \simeq \bigoplus_{I \in \mathcal{B}(\rho)} \rho^{G_{2m}}(I) \oplus \bigoplus_{J \in \mathcal{J}(\rho)} \rho^{G_{2m}}(J) \tag{3.1}$$

(2) For any good $\mathcal{Z}$-representation $\rho = \rho^{\mathcal{Z}}$ one denotes by:
- $\mathcal{B}(\rho)$ the collection of all barcodes (with multiplicity),
- $\mathcal{B}^c(\rho)$, $\mathcal{B}^o(\rho)$, $\mathcal{B}^{c,o}(\rho)$ and $\mathcal{B}^{o,c}(\rho)$ the subcollections of closed, open, closed-open, and open-closed bar codes,
- $\mathcal{J}(\rho)$ the collection of all copies of $\rho_\infty^{\mathcal{Z}}$ which appear as independent direct summands in ρ.[5]

Again $\mathcal{B}_r(\rho) = \mathcal{B}^c(\rho) \sqcup \mathcal{B}^o(\rho) \sqcup \mathcal{B}^{c,o}(\rho) \sqcup \mathcal{B}^{o,c}(\rho)$ and one has

$$\rho \simeq \bigoplus_{I \in \mathcal{B}(\rho)} \rho^{\mathcal{Z}}(I) \oplus \bigoplus_{\alpha \in \mathcal{J}(\rho)} (\rho_\infty^{\mathcal{Z}})_\alpha. \tag{3.2}$$

In view of (3.2) and Observation 3.1, any periodic $\mathcal{Z}$-representation contains regular subrepresentation ρ_{reg}, unique up to isomorphisms, which splits off ρ, such that:

$$\rho \simeq \rho_{\text{reg}} \oplus \rho/\rho_{\text{reg}},$$
$$\mathcal{J}(\rho) = \mathcal{J}(\rho_{\text{reg}}), \quad \mathcal{J}(\rho/\rho_{\text{reg}}) = \emptyset, \tag{3.3}$$
$$\mathcal{B}(\rho) = \mathcal{B}(\rho/\rho_{\text{reg}}), \quad \mathcal{B}(\rho_{\text{reg}}) = \emptyset.$$

Any $\mathcal{Z}$-representation with finite support has $\mathcal{J}(\rho) = \emptyset$ and $\mathcal{B}_r(\rho)$ finite. In view of the Krull-Remak-Schmidt theorem, any statement about G_{2m}-representations or about good $\mathcal{Z}$-representations formulated in this book will be verified for the indecomposable representations described above, and if it holds true, it remains true for arbitrary representations.

G_{2m}-representations and the associated linear relations

Given a G_{2m}-representation

$$\rho = \begin{cases} V_r, \ 1 \le r \le 2m, \\ \alpha_i : V_{2i-1} \to V_{2i}, \ \beta_i : V_{2i+1} \to V_{2i}, \ 1 \le i \le m, \\ V_{2m+1} = V_1 \end{cases}$$

we introduce the relations $R(\rho)_i : V_{2i-1} \rightsquigarrow V_{2i+1}$ and $R^i(\rho) : V_{2i-1} \rightsquigarrow V_{2i-1}$ defined by

$$R(\rho)_i := R(\alpha_i, \beta_i),$$
$$R(\rho)^i := R(\rho)_{i+m} \cdots R(\rho)_{i+1} \cdot R(\rho)_i,$$

---

[5]There is no danger of confusion in using the same notations $\mathcal{B}$ and $\mathcal{J}$ because they appear in connection with different types of graphs, $\mathcal{Z}$ and G_{2m}.

and consider the linear isomorphism $R^i(\rho)_{\mathrm{reg}}$.

Given a linear isomorphism $T : V \to V$, with V a finite-dimensional vector space over κ, denote by $\mathcal{J}(T)$ the set of Jordan blocks of T.

Proposition 3.1. *If ρ is a G_{2m}-representation, then the set $\mathcal{J}(\rho^{G_{2m}})$ of Jordan blocks of the representation $\rho^{G_{2m}}$ and the set $\mathcal{J}(R^i(\rho)_{\mathrm{reg}})$ of Jordan blocks of the linear isomorphism $R^i(\rho)_{\mathrm{reg}}$ coincide.*

Proof. The statement is obviously true for indecomposable representations and it follows in general by observing that $\mathcal{J}(\rho_1 \oplus \rho_2) = \mathcal{J}(\rho_1) \sqcup \mathcal{J}(\rho_1)$ and $\mathcal{J}(\rho_1 \oplus \rho_2) = \mathcal{J}(\rho_1) \sqcup \mathcal{J}(\rho_1)$, which are immediate consequences of the definitions.   □

3.2.1   Two basic constructions

The infinite cyclic cover

The infinite cyclic cover of a G_{2m}-representation $\rho = \{V_r, a_i, b_i, 1 \le r \le 2m, 1 \le i \le m\}$ is the periodic $\mathcal{Z}$-representation $\tilde{\rho} := \{\tilde{V}_r, \tilde{a}_i, \tilde{b}_i, r, i \in \mathbb{Z}\}$ defined by $\tilde{V}_{r+2mk} = V_r$, $\tilde{a}_{i+km} = a_i$, and $\tilde{b}_{i+km} = b_i$, with period m. Conversely, any periodic $\mathcal{Z}$-representation with period m is the infinite cyclic cover of a G_{2m}-representation. When applied to the indecomposables $\rho^{G_{2m}}(I)$ or $\rho^{G_{2m}}(J)$ one obtains:

$$\widetilde{\rho^{G_{2m}}(I)} = \bigoplus_{k \in \mathbb{Z}} \rho^{\mathcal{Z}}(I + mk)$$

$$\widetilde{\rho^{G_{2m}}(J)} = \bigoplus_{1 \le i \le \dim V_J} (\rho^{\mathcal{Z}}_\infty)_i, \quad J = (V_J, T_J), \tag{3.4}$$

where $I + r$, $r \in \mathbb{Z}$ denotes the translate of the interval I by r units.

The truncation $T_{k,l}$

The truncation of a $\mathcal{Z}$-representation (respectively G_{2m}-representation) is defined for any pair of integers k, l with $k \le l$ (respectively, k, l with $1 \le k \le l \le m$).

If $\rho = \{V_r, \alpha_i, \beta_i\}$ and $T_{k,l}(\rho) = \{V'_r, \alpha'_i, \beta'_i\}$, then we put

$$V'_r = \begin{cases} V_r, & \text{if } 2k \leq r \leq 2l, \\ 0, & \text{otherwise,} \end{cases}$$

$$\alpha'_r = \begin{cases} \alpha_r, & \text{if } k+1 \leq r \leq l, \\ 0, & \text{otherwise,} \end{cases} \tag{3.5}$$

$$\beta'_r = \begin{cases} \beta_r, & \text{if } k \leq r \leq l-1, \\ 0, & \text{otherwise.} \end{cases}$$

The truncation is functorial and commutes with taking direct sums.

Applying this to indecomposable $\mathcal{Z}$-representations one obtains

$$\begin{aligned} T_{k,l}(\rho^{\mathcal{Z}}_\infty) &= \rho^{\mathcal{Z}}([k,l]), \\ T_{k,l}(\rho^{\mathcal{Z}}(I)) &= \rho^{\mathcal{Z}}(I \cap [k,l]). \end{aligned} \tag{3.6}$$

Applying to indecomposable G_{2m}-representations one obtains

$$\begin{aligned} T_{k,l}(\rho^{G_{2m}}(I)) &= \bigoplus_{r \in \mathbb{Z}} \rho^G(I_r), \qquad I_r = (I + rm) \cap [k,l], \\ T_{k,l}(\rho^{G_{2m}}(J)) &= \bigoplus_{n} \rho^G([k,l]), \quad n = \dim V. \end{aligned} \tag{3.7}$$

Observation 3.3.

(a) If ρ is a G_{2m}-representation, then

$$\mathcal{B}^c(T_{k,l}(\rho)) =$$
$$\{I \in \mathcal{B}(\tilde{\rho}) : I \cap [k,l] \neq \emptyset \text{ and closed}\} \sqcup \{\sharp \mathcal{J}(\tilde{\rho})\text{-copies of } [k,l]\},$$
$$\mathcal{B}^o(T_{k,l}(\rho)) = \{I \in \mathcal{B}^o(\tilde{\rho}) : I \subset [k,l]\}.$$

(b) If ρ is a good $\mathcal{Z}$-representation, then

$$\mathcal{B}^c(T_{k,l}(\rho)) =$$
$$\{I \in \mathcal{B}(\rho) : I \cap [k,l] \neq \emptyset \text{ and closed}\} \sqcup \{\sharp \mathcal{J}(\tilde{\rho})\text{-copies of } [k,l]\},$$
$$\mathcal{B}^o(T_{k,l}(\rho)) = \{I \in \mathcal{B}^o(\rho) : I \subset [k,l]\}.$$

3.2.2 The $\kappa[t^{-1}, t]$-module associated to a G_{2m}-representation

For ρ a G_{2m}-representation, choose

(i) a Krull-Remak-Schmidt decomposition (3.1) into barcode and Jordan block representations.

(ii) For any $J = (V_J, T_J) \in \mathcal{J}(\rho)$ choose a basis $S_J := \{v_1, v_2, \ldots, v_{\dim V_J}\}$ of V_J, denote by $S(\rho) := \bigsqcup S_J$, and identify $\bigoplus V_J$ with $\kappa[S(\rho)]$ [6], and correspondingly $T = \bigoplus T_J$ with a linear isomorphism $T : \kappa[S(\rho)] \to \kappa[S(\rho)]$.

The above choices provide a decomposition of the infinite cyclic cover $\widetilde{\rho}$,

$$\widetilde{\rho} \simeq \bigoplus_{I \in \mathcal{B}(\widetilde{\rho})} \rho^{\mathcal{Z}}(I) \oplus \bigoplus_{i \in S(\rho)} (\rho_\infty^{\mathcal{Z}})_i$$

with $\mathcal{B}(\widetilde{\rho}) = \{(I + mk) \mid I \in \mathcal{B}(\rho), k \in \mathbb{Z}\}$.

Consider the κ-vector space $\kappa[\mathcal{B}(\widetilde{\rho}) \sqcup S(\rho)]$ the direct sum of the $\kappa[\mathcal{B}(\widetilde{\rho})]$ and $\kappa[S(\rho)]$ and equip each of these vector spaces with a structure of f.g. $\kappa[t^{-1}, t]$-module as follows. The translation by m-units, $I \to I + m$, of each barcode I defines a free $\mathbb{Z}$-action on the set $\mathcal{B}(\widetilde{\rho})$, hence a free $\kappa[t^{-1}, t]$-module structure on $\kappa[\mathcal{B}(\widetilde{\rho})]$. The linear isomorphism T defines a $\kappa[t^{-1}, t]$-module structure on $\kappa[S(\rho)]$. The first vector space is the free part of our f.g. $\kappa[t^{-1}, t]$-module, while the second is the torsion part of this module. From this perspective the decomposition of the infinite cyclic representation $\widetilde{\rho}$ into two components, the first component equal to the sum of barcode type indecomposables and the second component, the regular part, corresponds exactly to the decomposition of any f.g. module over a principal ideal domain as a sum of a free module and a torsion module.

3.2.3   *The matrix $M(\rho)$ and the representation ρ_u*

To the G_{2m}-representation $\rho = \{V_r, \alpha_i, \beta_i\}$, $1 \le r \le 2m$, $1 \le i \le m$, one associates the linear map

$$M(\rho): \bigoplus_{1 \le i \le m} V_{2i-1} \to \bigoplus_{1 \le i \le m} V_{2i}$$

defined by the block matrix

$$\begin{pmatrix} \alpha_1 & -\beta_1 & 0 & \cdots & & 0 \\ 0 & \alpha_2 & -\beta_2 & \ddots & & \vdots \\ \vdots & \ddots & \ddots & \ddots & & 0 \\ 0 & \cdots & 0 & \alpha_{m-1} & -\beta_{m-1} \\ -\beta_m & 0 & \cdots & & 0 & \alpha_m \end{pmatrix}.$$

---

[6] see Definition (3.1) below

To the $\mathcal{Z}$-representation $\rho = \{V_r, \alpha_i, \beta_i\}$ one associates the linear map

$$M(\rho): \bigoplus_{i \in \mathbb{Z}} V_{2i-1} \to \bigoplus_{i \in \mathbb{Z}} V_{2i}$$

defined by the infinite block matrix with the entries

$$M(\rho)_{2r-1,2s} = \begin{cases} \alpha_r, & \text{if } s = r, \\ \beta_{r-1}, & \text{if } s = r - 1, \\ 0, & \text{otherwise.} \end{cases}$$

The G_{2m}-representation $\rho_u = \{V'_r, \alpha'_i, \beta'_i\}$ is given by $V'_r = V_r$, $\alpha'_1 = u\alpha_1$, $\alpha'_i = \alpha_i$ for $i \neq 1$, and $\beta'_i = \beta_i$.

For a representation ρ define

$$\dim \ker(\rho) := \dim \ker M(\rho), \quad \dim \operatorname{coker}(\rho) := \dim \operatorname{coker} M(\rho). \qquad (3.8)$$

With the definitions above one has the following:

Observation 3.4.

(a) $(\rho_1 \oplus \rho_2)_u = (\rho_1)_u \oplus (\rho_2)_u$,
(b) $\rho^G(\lambda, k)_u = \rho^G(u\lambda, k)$,
(c) $\rho^G(\{i, j + km\})_u \equiv \rho^G(\{i, j + km\})$,
(d) $\dim \ker(\rho_1 \oplus \rho_2) = \dim \ker(\rho_1) + \dim \ker(\rho_2)$,
(e) $\dim \operatorname{coker}(\rho_1 \oplus \rho_2) = \dim \operatorname{coker}(\rho_1) + \dim \operatorname{coker}(\rho_2)$.

and

Proposition 3.2. *[Burghelea, D., Dey, T (2013)]*

(a) *For indecomposable G_{2m}-representations of barcode type one has*

    (a1) $\dim \ker \rho^G([i, l]) = 0$, $\dim \operatorname{coker} \rho^G([i, l]) = 1$,
    (a2) $\dim \ker \rho^G([i, l)) = 0$, $\dim \operatorname{coker} \rho^I([i, l)) = 0$,
    (a3) $\dim \ker \rho^G((i, l]) = 0$, $\dim \operatorname{coker} \rho^G((i, l]) = 0$,
    (a4) $\dim \ker \rho^G((i, l)) = 1$, $\dim \operatorname{coker} \rho^G((i, l)) = 0$,

    and for indecomposable $\mathcal{Z}$-representations with finite support one has

    (a5) $\dim \ker \rho^{\mathcal{Z}}([i, l]) = 0$, $\dim \operatorname{coker} \rho^{\mathcal{Z}}([i, l]) = 1$,
    (a6) $\dim \ker \rho^{\mathcal{Z}}([i, l)) = 0$, $\dim \operatorname{coker} \rho^{\mathcal{Z}}([i, l)) = 0$,
    (a7) $\dim \ker \rho^{\mathcal{Z}}((i, l]) = 0$, $\dim \operatorname{coker} \rho^{\mathcal{Z}}((i, l]) = 0$,
    (a8) $\dim \ker \rho^{\mathcal{Z}}((i, l)) = 1$, $\dim \operatorname{coker} \rho^{\mathcal{Z}}((i, l)) = 0$.

(b) *For indecomposable G_{2m}-representations of Jordan cell type one has*

    (b1) $\dim \ker \rho^G(J) = 0$, *if* $J \neq (\kappa^k, T(1, k))$, *and*
        $\dim \ker \rho^G(\kappa^k, T(1, k)) = 1$,

(b2) dim coker $\rho^G(J) = 0$, *if* $J \neq (\kappa^k, T(1, k))$, *and*
     dim coker $\rho^G(\kappa^k, T(1, k)) = 1$,

and for the $\mathcal{Z}$-representation $\rho_\infty^{\mathcal{Z}}$ one has

(b3) dim ker$(\rho_\infty^{\mathcal{Z}}) = 0$,
(b4) dim coker$(\rho_\infty^{\mathcal{Z}}) = 1$.

(c) *If* $\rho = \{V_i, \alpha_i, \beta_i\}$ *is a regular $\mathcal{Z}$-representation, i.e., all α_i and β_i are isomorphisms, then* ker $M(\rho) = 0$, *and for every i the canonical inclusion $V_{2i} \to \bigoplus_{r \in \mathbb{Z}} V_{2r}$ followed by the projection onto* coker $M(\rho)$ *provides an isomorphism $V_{2i} \cong$ coker $M(\rho)$.*

To prove Proposition 3.2 one notices that the calculation of the kernel of $M(\rho)$ reduces to the description of the space of solutions of the linear system

$$\alpha_1(v_1) = \beta_1(v_3),$$
$$\alpha_2(v_3) = \beta_2(v_5),$$
$$\vdots \tag{3.9}$$
$$\alpha_m(v_{2m-1}) = \beta_m(v_1).$$

We leave this proof as an exercise to the reader (see Exercise E.7 in Section 3.4).

Proposition 3.2 can be refined. To do so we need some additional notation.

Definition 3.1. For a set S denote by $\kappa[S]$ the vector space generated by S, i.e. the vector space of κ-valued maps on S with finite support, and by $\kappa[[S]]$ the vector space of all κ-valued maps on S. If S is finite, $\kappa[S] = \kappa[[S]]$.

For S_1, S_2 subsets of S the canonical linear maps $\kappa[S_1] \to \kappa[S_2]$, $\kappa[S_1] \to \kappa[[S_2]]$, or $\kappa[[S_1]] \to \kappa[[S_2]]$ are the unique linear maps which restrict to the identity on $S_1 \cap S_2$ and are zero on $S_1 \setminus S_2$. More precisely, if $f \in \kappa[[S_1]]$ then it can be uniquely decomposed as $f = f_1 + f_2$ with supp$f_1 \subseteq S_1 \cap S_2$ and supp$f_2 \subseteq S_1 \setminus S_2$. The canonical linear map sends f_1 to itself and f_2 to zero.

There is here a conflict of notation: $\kappa[S]$ for the vector space generated by the set S and $\kappa[t^{-1}, t]$ for the ring of Laurent polynomials in one variable with coefficients in κ. Both notations are used in this book, but fortunately in contexts which prevent ambiguity.

Proposition 3.3.

(a) *Let ρ be a G_{2m}-representation. Then every decomposition $\rho = \bigoplus_{I \in \mathcal{B}(\rho)} \rho^G(I) \oplus \bigoplus_{J \in \mathcal{J}(\rho)} \rho^G(J)$ induces isomorphisms*

$$\Psi^c \colon \kappa[\mathcal{B}^c(\rho) \sqcup \mathcal{J}_1(\rho)] \to \operatorname{coker} M(\rho),$$

$$\Psi^o \colon \kappa[\mathcal{B}^o(\rho) \sqcup \mathcal{J}_1(\rho)] \to \ker M(\rho),$$

compatible with truncations.

(b) *Let ρ be a good $\mathcal{Z}$-representation. Then every decomposition $\rho^{\mathcal{Z}} = \bigoplus_{I \in \mathcal{B}(\rho)} \rho^{\mathcal{Z}}(I) \oplus \bigoplus_{\alpha \in \mathcal{J}(\rho)} (\rho_\infty^{\mathcal{Z}})_\alpha$, induces isomorphisms*

$$\Psi^c \colon \kappa[\mathcal{B}^c(\rho) \sqcup \mathcal{J}(\rho)] \to \operatorname{coker} M(\rho),$$

$$\Psi^o \colon \kappa[\mathcal{B}^o(\rho)] \to \ker M(\rho),$$

that are compatible with truncations. If $\rho^{\mathcal{Z}}$ has finite support, then $\mathcal{J}(\rho) = \emptyset$, and if $\rho^{\mathcal{Z}}$ is periodic, then $\mathcal{J}(\rho)$ is the finite set described above.

Recall that for a G_{2m}-representation $\mathcal{J}_1(\rho)$ denotes the set of Jordan blocks with eigenvalue 1. All $J \in \mathcal{J}_1(\rho)$ are actually Jordan cells. The construction of Ψ^c and Ψ^o is tautological for indecomposable representations. For an arbitrary representation ρ, the decompositions (3.1) and (3.2) permit to assemble the Ψ^cs and Ψ^os into the linear isomorphisms Ψ^c and Ψ^o, as stated. Note that a specified decomposition (3.1) of ρ provides a decomposition (3.2) of $\tilde{\rho}$ with

$$\mathcal{B}(\tilde{\rho}) = \{I + mk \mid I \in \mathcal{B}(\rho), k \in \mathbb{Z}\},$$

and correspondingly of the truncations $T_{k,l}(\tilde{\rho})$ and $T_{k,l}(\rho)$ cf. Observation 3.3.

Let us explain in more detail what "compatible with the truncations" means. The inclusions of sets $\{i \mid k \leq i \leq l\} \subseteq \{i \mid k' \leq i \leq l'\} \subset \mathbb{Z}$ for $i' \leq i$ and $l' \geq l$ induce the commutative diagram

$$
\begin{array}{ccccc}
\bigoplus_{k \leq i \leq l} V_{2i-1} & \longrightarrow & \bigoplus_{k' \leq i \leq l'} V_{2i-1} & \longrightarrow & \bigoplus_i V_{2i-1} \\
\Big\downarrow {\scriptstyle M(T_{k,l}(\rho))} & & \Big\downarrow {\scriptstyle M(T_{k',l'}(\rho))} & & \Big\downarrow {\scriptstyle M(\rho)} \\
\bigoplus_{k \leq i \leq l} V_{2i} & \longrightarrow & \bigoplus_{k' \leq i \leq l'} V_{2i} & \longrightarrow & \bigoplus_i V_{2i}
\end{array}
\tag{3.10}
$$

and then the linear maps

$$\ker M(T_{k,l}(\rho)) \xrightarrow{\ i\ } \ker M(T_{k',l'}(\rho)) \xrightarrow{\ i'\ } \ker M(\rho) \tag{3.11}$$

and

$$\operatorname{coker} M(T_{k,l}(\rho)) \xrightarrow{\ j\ } \operatorname{coker} M(T_{k',l'}(\rho)) \xrightarrow{\ j'\ } \operatorname{coker} M(\rho). \tag{3.12}$$

The linear maps i and i' are injective, since by Observation 3.4(a) we have the inclusions $\mathcal{B}(T_{k,l}(\rho))^\circ \subseteq \mathcal{B}(T_{k',l'}(\rho))^\circ \subseteq \mathcal{B}(\rho)^\circ \subseteq \mathcal{B}(\rho)^\circ \sqcup \mathcal{L}$, with the set $\mathcal{L}$ specified below, which make the linear maps

$$\kappa[\mathcal{B}^\circ(T_{k,l}(\rho))] \longrightarrow \kappa[\mathcal{B}^\circ(T_{k',l'}(\rho))] \longrightarrow \kappa[\mathcal{B}^\circ(\rho) \sqcup \mathcal{L}] \qquad (3.13)$$

injective. We also have the linear maps

$$\kappa[\mathcal{B}^c(T_{k,l}(\rho))] \longrightarrow \kappa[\mathcal{B}^c(T_{k',l'}(\rho))] \longrightarrow \kappa[\mathcal{B}^c(\rho) \sqcup \mathcal{L}] , \qquad (3.14)$$

not necessarily injective, described below. If $I \in \mathcal{B}^c(T_{k,l}(\rho))$ is also an element of $\mathcal{B}^c(T_{k',l'}(\rho))$, cf. Observation 3.3 (a) for the description of $\mathcal{B}^c(T_{k,l}(\rho))$, then the linear map $\kappa[\mathcal{B}^c(T_{k,l}(\rho))] \to \kappa[\mathcal{B}^c(T_{k',l'}(\rho))]$ in the sequence (3.14) sends the barcode I into itself; otherwise to zero. Similarly for the linear map $\kappa[\mathcal{B}^c(T_{k',l'}(\rho))] \to \kappa[\mathcal{B}^c(\rho) \sqcup \mathcal{L}]$. The compatibility with truncation means the commutativity of the diagrams (3.15) and (3.16) below:

$$\begin{array}{ccccc}
\ker M(T_{k,l}(\rho)) & \xrightarrow{\ i\ } & \ker M(T_{k',l'}(\rho)) & \xrightarrow{\ i'\ } & \ker M(\rho) \\
\Psi^\circ \uparrow & & \Psi^\circ \uparrow & & \Psi^\circ \uparrow \\
\kappa[\mathcal{B}^\circ(T_{k,l}(\rho))] & \longrightarrow & \kappa[\mathcal{B}^\circ(T_{k',l'}(\rho))] & \longrightarrow & \kappa[\mathcal{B}^\circ(\rho) \sqcup \mathcal{L}]
\end{array} \qquad (3.15)$$

and

$$\begin{array}{ccccc}
\operatorname{coker} M(T_{k,l}(\rho)) & \xrightarrow{\ j\ } & \operatorname{coker} M(T_{k',l'}(\rho)) & \xrightarrow{\ j'\ } & \operatorname{coker} M(\rho) \\
\Psi^c \uparrow & & \Psi^c \uparrow & & \Psi^c \uparrow \\
\kappa[\mathcal{B}^c(T_{k,l}(\rho))] & \longrightarrow & \kappa[\mathcal{B}^c(T_{k',l'}(\rho))] & \longrightarrow & \kappa[\mathcal{B}^c(\rho) \sqcup \mathcal{L}].
\end{array} \qquad (3.16)$$

In the diagram (3.15) the bottom sequence is the sequence (3.13), $\mathcal{L} = \emptyset$ if ρ is a good $\mathcal{Z}$-representation and $\mathcal{L} = \mathcal{J}_1(\rho)$ if ρ is a G_{2m}-representation. In the diagram (3.16) the bottom sequence is the sequence (3.14), $\mathcal{L} = \mathcal{J}(\rho)$ if ρ is a good $\mathcal{Z}$-representation and $\mathcal{L} = \mathcal{J}_1(\rho)$ if ρ is a G_{2m}-representation.

We close this section with an observation about the $\mathcal{Z}$-representation $\tilde{\rho}$, $\tilde{\rho} = \{\tilde{V}_r, \tilde{\alpha}_i, \tilde{\beta}_i\}$ associated with a G_{2m}-representation $\rho = \{V_r, \alpha_i, \beta_i\}$.

Observation 3.5.

In case $\tilde{\rho} = (\widetilde{\rho^{G_{2m}}})$, the shift in indices $r \mapsto r + 2m$ for the vector spaces V_r's, and the shift in indices $i \mapsto i + m$, for the linear maps α_i, β_i, induce the linear endomorphism τ_m on the kernel and on the co-kernel of the associated matrices $M(\tilde{\rho})$, making $\ker M(\tilde{\rho})$ and $\operatorname{coker} M(\tilde{\rho})$ $\kappa[t^{-1}, t]$-modules.

The translation of intervals $I \to I + m$ defines free $\mathbb{Z}$-actions on $\mathcal{B}^c(\tilde{\rho})$ and $\mathcal{B}^\circ(\tilde{\rho})$, and hence $\kappa[t^{-1}, t]$−module structures on $\kappa[\mathcal{B}^c(\tilde{\rho})]$ and $\kappa[\mathcal{B}^\circ(\tilde{\rho})]$, which coupled with the isomorphism $T : \kappa[\mathcal{J}(\tilde{\rho})] \rightarrow \kappa[\mathcal{J}(\tilde{\rho})]$ define the $\kappa[t^{-1}, t]$-module $\kappa[\mathcal{B}^c(\tilde{\rho}) \sqcup \mathcal{J}(\tilde{\rho})]$ and $\kappa[\mathcal{B}^o(\tilde{\rho})]$.

Clearly the linear isomorphisms Ψ^o and Ψ^c become isomorphisms of $\kappa[t^{-1}, t]$-modules.

3.3 Calculation of indecomposables (an algorithm)

It is convenient to visualize a $\mathcal{Z}$- or G_{2m}-representation ρ as a diagram

$$
\begin{array}{ccccccc}
V_1 & V_3 & \cdots & V_{2m-3} & V_{2m-1} & V_{2m+1} \, , & (3.17) \\
\downarrow\searrow & \downarrow\searrow & & \downarrow & \downarrow\searrow & \downarrow\searrow \\
\cdots & V_2 & V_4 & \cdots & V_{2m-2} & V_{2m}
\end{array}
$$

which in case of a G_{2m}-representation has $V_{2m+1} = V_1$ and $V_r = 0$ for $r \geq 2m + 1$.

Note that every $\mathcal{Z}$-representations with finite support can be regarded as a G_{2m}-representation for m large enough so it will be enough to treat the case of G_{2m}-representations only.

In this section we describe a way to calculate the decomposition of a G_{2m}-representation into barcodes and Jordan blocks representations. The presentation essentially reproduces Section 6 in [Burghelea, D., Dey, T (2013)].

The algorithm we propose takes the matrix $M(\rho)$ constructed from α_i, β_i and uses the transformations $T_1(i), T_2(i), T_3(i), T_4(i)$ defined below to transform $M(\rho)$ into the matrix $M(\rho') = T...(\cdots)M(\rho)$, in which the total number of rows and columns is strictly smaller. Each such transformation modifies the representation ρ into the representation ρ', while keeping the indecomposable Jordan block representations unaffected, but possibly changing the barcode representations. Some of these barcodes remain the same, some are eliminated, and some are shortened by one unit, as described below. For each transformation one records the changes to reconstruct the original bar codes. The elementary transformations are applied as long as the linear maps α_i and β_i display some lack of injectivity and surjectivity. When no such transformation is applicable, the algorithm terminates with all α_i and β_i invertible. The barcodes can be reconstructed reading backwards the eliminations/modifications performed.

3.3.1   *Elementary transformations*

The following diagrams[7] depict the transformations. The indices increase from right to left, signifying that the vector spaces are placed counterclockwise with increasing indices around a graph G_{2m}.

Transformation $T_1(i)(\rho)$:

$$\cdots \xleftarrow{\alpha_{i+1}} V_{2i+1} \xrightarrow{\beta_i} V_{2i} \xleftarrow{\alpha_i} V_{2i-1} \xrightarrow{\beta_{i-1}} V_{2i-2} \longleftarrow \cdots$$

$$V'_{2i-1} = V_{2i-1}/\ker(\beta_{i-1}),\ V'_{2i} = V_{2i}/\alpha_i(\ker(\beta_{i-1})),\ V'_k = V_k \text{ for } k \neq 2i-1, 2i.$$

Transformation $T_2(i)(\rho)$:

$$\cdots \xleftarrow{\alpha_{i+1}} V_{2i+1} \xrightarrow{\beta_i} V_{2i} \xleftarrow{\alpha_i} V_{2i-1} \xrightarrow{\beta_{i-1}} V_{2i-2} \longleftarrow \cdots$$

$$V'_{2i} = \beta_i(V_{2i+1}) \quad V'_{2i-1} = \alpha_i^{-1}(\beta_i(V_{2i+1})),\ V'_k = V_k \text{ for } k \neq 2i-1, 2i.$$

Transformation $T_3(i)(\rho)$:

$$\cdots \xrightarrow{\beta_{i+1}} V_{2i+2} \xleftarrow{\alpha_{i+1}} V_{2i+1} \xrightarrow{\beta_i} V_{2i} \xleftarrow{\alpha_i} V_{2i-1} \xrightarrow{\beta_i} \cdots$$

$$V'_{2i} = \alpha_i(V_{2i-1}),\ V'_{2i+1} = \beta_i^{-1}(\alpha_i(V_{2i-1})),\ V'_k = V_k \text{ for } k \neq 2i+1, 2i.$$

---

[7]In the diagrams $V_0 = V_{2m}$ and $\beta_0 = \beta_m$.

Transformation $T_4(i)(\rho)$:

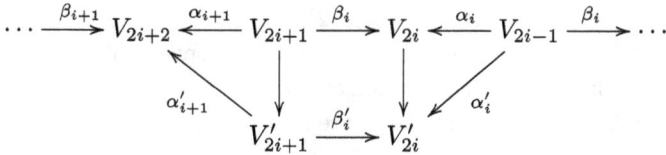

$$V'_{2i+1} = V_{2i+1}/\ker(\alpha_{i+1}), V'_{2i} = V_{2i}/\beta_i(\ker(\alpha_{i+1})) \; V'_k = V_k \text{ for } k \neq 2i+1, 2i.$$

Note that T_2 and T_3 provide subrepresentations while T_1 and T_4 provide quotient representations.

Since each of the above transformation commutes with the direct sums, it follows in a straightforward manner that the elementary transformations modify the barcodes as follows.

- $T_1(i)$ eliminates the barcodes $(i-1, i), (i-1, i]$ and shortens the bar codes $(i-1, k]$ to $(i, k]$ if $i \leq 2$ and shortens the bar codes $(m, k]$, $m < k$, to $(1, k-m]$.
- $T_2(i)$ eliminates the barcodes $[i, i], (i-1, i]$ and shortens the barcodes $\{l, i + km]$ to $\{l, i - 1 + km]$ for $k \geq 0$.
- $T_3(i)$ eliminates the barcodes $[i, i+1), [i, i]$ and shortens the barcodes $[i, k\}$ to $[i+1, k\}$ for $i < m$ and the barcodes $(m, k\}$, $m < k$, to $(1, k-m\}$.
- $T_4(i)$ eliminates the barcodes $(i, i+1), [i, i+1)$ and shortens the barcodes $\{l, (i+1) + km)$ to $\{l, i + km)$ for $k \geq 0$.

To decide how many barcodes are eliminated one uses Proposition 3.4 below. Let $\sharp\{\cdot, \cdot\}$ denote the number of type $\{\cdot, \cdot\}$ barcodes in the representation ρ.

Proposition 3.4.

(i) $\sharp(i, i+1) = \dim(\ker \beta_i \cap \ker \alpha_{i+1})$.

(ii) $\sharp[i, i] = \dim(V_{2i}/((\beta_i(V_{2i+1}) + \alpha_i(V_{2i-1})))$.

(iii) $\sharp(i, i+1] = \dim(\beta_i(V_{2i+1}) + \alpha_i(\ker \beta_{i-1})) - \dim(\beta_i(V_{2i+1}))$.

(iv) $\sharp[i, i+1) = \dim(\alpha_i(V_{2i-1}) + \beta_i(\ker \alpha_{i+1})) - \dim(\alpha_i(V_{2i-1}))$.

The proof is straightforward.

As one can see from the diagrams above, when β_{i-1} is injective the representations ρ and ρ' coincide, and we say that $T_1(i)$ is not applicable. Similarly, when β_i is surjective $T_2(i)$ is not applicable, when α_i is surjective $T_3(i)$ is not applicable, and when α_{i+1} is injective $T_4(i)$ is not applicable.

When all α_i, β_i are invertible, no elementary transformation is applicable and at this stage the algorithm (Step 2) terminates.

To explain how the algorithm works, it is convenient to consider the following block matrices B_{2i-1} and B_{2i}, $i = 1, \ldots, m$, which become submatrices of M_ρ when the entries β_i are replaced by $-\beta_i$. Let

$$B_{2i-1} = \begin{pmatrix} \alpha_i & \beta_i \\ 0 & \alpha_{i+1} \end{pmatrix}, \quad B_{2i} = \begin{pmatrix} \beta_i & 0 \\ \alpha_{i+1} & \beta_{i+1} \end{pmatrix} \tag{3.18}$$

for $i = 1, 2, \ldots, m-1$ and

$$B_{2m-1} = \begin{pmatrix} \alpha_m & \beta_m \\ 0 & \alpha_1 \end{pmatrix}, \quad B_{2m} = \begin{pmatrix} \beta_m & 0 \\ \alpha_1 & \beta_1 \end{pmatrix}. \tag{3.19}$$

We modify M_ρ by modifying successively each block B_k. When $m > 1$ the algorithm iterates over the blocks in multiple passes. In a single pass, it processes the blocks $B_1, B_2, \ldots, B_{2m}$ in this order.

When $B_{2(i-1)} = \begin{pmatrix} \beta_{i-1} & 0 \\ \alpha_i, & \beta_i \end{pmatrix}$ is processed, then:

1. If β_{i-1} is not injective, we apply $T_1(i)$. This boils down to changing the bases of V_{2i-1} and V_{2i} so that the matrix $B_{2(i-1)}$ becomes

$$\left(\begin{array}{c|c||c} \beta_{i-1,1} & 0 & 0 \\ \hline \alpha_{i,1}^1 & \alpha_{i,2}^1 & \beta_i^1 \\ \hline \alpha_{i,1}^2 & 0 & \beta_i^2 \end{array} \right),$$

with $\left(\beta_{i-1,1} \ 0 \right)$ in column echelon form and $\begin{pmatrix} \alpha_{i,2}^1 \\ 0 \end{pmatrix}$ in row echelon form.

In this block matrix the first and third columns correspond to V'_{2i-1} and V_{2i+1}, respectively, and the first and third rows to $V_{2(i-1)}$ and V'_{2i}, respectively. The second column and row become "irrelevant", as a result of which the modified block matrix $B_{2(i-1)}$ becomes

$$\begin{pmatrix} \beta'_{i-1} & 0 \\ \alpha'_i & \beta'_i \end{pmatrix} = \begin{pmatrix} \beta_{i-1,1} & 0 \\ \alpha_{i,1}^2 & \beta_i^2 \end{pmatrix}.$$

2. If β_i is not surjective, we apply $T_2(i)$. This boils down to changing the bases of V_{2i-1} and V_{2i} so that the matrix $B_{2(i-1)}$ becomes

$$\left(\begin{array}{c|c||c} \beta_{i-1,1} & \beta_{i-1,2} & 0 \\ \hline \alpha_{i,1}^1 & \alpha_{i,2}^1 & \beta_i^1 \\ \hline \alpha_{i,1}^2 & 0 & 0 \end{array} \right),$$

with $\begin{pmatrix} \beta_i^1 \\ 0 \end{pmatrix}$ in row echelon form and $\begin{pmatrix} \alpha_{i,1}^2 & 0 \end{pmatrix}$ in column echelon form.

In this block matrix the second and third columns correspond to V_{2i-1}' and V_{2i+1}, respectively, and the first and second rows to $V_{2(i-1)}'$ and V_{2i}', respectively. We make the first column and third row "irrelevant", as a result of which the modified block matrix $B_{2(i-1)}$ becomes $\begin{pmatrix} \beta_{i-1}' & 0 \\ \alpha_i' & \beta_i' \end{pmatrix} = \begin{pmatrix} \beta_{i-1,2} & 0 \\ \alpha_{i,2}^1 & \beta_i^1 \end{pmatrix}$.

When B_{2i-1} is processed, then:

3. If α_i is not surjective, we apply $T_3(i)$. This boils down to changing the bases of V_{2i+1} and V_{2i} so that the matrix B_{2i-1} becomes

$$\left(\begin{array}{c||c|c} \alpha_i^1 & \beta_{i,1}^1 & \beta_{i,2}^1 \\ \hline 0 & \beta_{i,1}^2 & 0 \\ \hline\hline 0 & \alpha_{i+1,1} & \alpha_{i+1,2} \end{array} \right),$$

with $\begin{pmatrix} \alpha_i^1 \\ 0 \end{pmatrix}$ in row echelon form and $\begin{pmatrix} \beta_{i,1}^2 & 0 \end{pmatrix}$ in column echelon form.

In this block matrix the first and third columns correspond to V_{2i-1} and V_{2i+1}', respectively, and the first and third rows to V_{2i}' and V_{2i+2}, respectively. We make the second column and second row "irrelevant", as a result of which the modified block matrix B_{2i-1} becomes $\begin{pmatrix} \alpha_i' & \beta_i' \\ 0 & \alpha_{i+1}' \end{pmatrix} = \begin{pmatrix} \alpha_i^1 & \beta_{i,2}^1 \\ 0 & \alpha_{i+1,2} \end{pmatrix}$.

4. If α_{i+1} is not injective, we apply $T_4(i)$. This boils down to changing the bases of V_{2i+1} and V_{2i}, so that the matrix B_{2i-1} becomes

$$\left(\begin{array}{c||c|c} \alpha_i^1 & \beta_{i,1}^1 & \beta_{i,2}^1 \\ \hline \alpha_i^2 & \beta_{i,1}^2 & 0 \\ \hline\hline 0 & \alpha_{i+1,1} & 0 \end{array} \right),$$

with $\begin{pmatrix} \alpha_{i+1,1} & 0 \end{pmatrix}$ in column echelon form and $\begin{pmatrix} \beta_{i,2}^1 \\ 0 \end{pmatrix}$ in row echelon form.

In this block matrix the first and second columns correspond to V_{2i-1} and V_{2i+1}', respectively, and the second and third rows to V_{2i}' and $V_{2(i+1)}$, respectively. We make the third column and first row "irrelevant", as a result of which the modified block matrix B_{2i-1} becomes $\begin{pmatrix} \alpha'_i & \beta'_i \\ 0 & \alpha'_{i+1} \end{pmatrix} = \begin{pmatrix} \alpha_i^2 & \beta_{i,1}^2 \\ 0 & \alpha_{i+1,1} \end{pmatrix}$.

Explicit formulae for α's and β's are given at the end of this section. At each pass the algorithm may eliminate or change barcodes, and if this happens, the matrix has less columns or rows. If this does not happen, the algorithm terminates, and indicates that there is no barcode left. At termination, all α_i and β_i become isomorphisms. The bar codes can be recovered by keeping track of all eliminations of the barcodes after each elementary transformation. A barcode which is not eliminated in a pass gets shrunk by exactly two units during that pass, that is, a barcode $\{i, j\}$ shrinks to $\{i+1, j-1\}$ by exactly two distinct elementary transformations. For example, if $m = 5$ the barcode $(1, 5]$ during the pass became $(2, 4]$ as a result of applying $T_1(1)$ when inspecting B_1 and $T_2(5)$ when inspecting B_9.

When a bar code $[i, i]$ is eliminated, say, in the k-th pass, we know that it corresponds to a barcode $[i - k + 1, i + k - 1]$ in the original representation. Similarly, other barcodes of type $\{i, i + 1\}$ eliminated at the k-th pass correspond to the barcode $\{i - k + 1, i + k\}$. In both cases, the multiplicity of the barcodes can be determined from the multiplicity of the eliminated barcodes thanks to Proposition 3.3.

When $m = 1$, the operations on the minors above are not well defined. In this case we extend the quiver G_2 to G_4 ($m = 2$) by adding fake levels t_2, s_2 where $H_r(X_{t_2}) = H_r(X_{s_2}) = H_r(X_{s_1})$ and α_2, β_2 are identities[8].

A high-level pseudocode for the Step 2 can be written as follows:

3.3.2 *Algorithm for deriving barcodes from $M(\rho)$*

Consider the block submatrices $B_1, \ldots, B_m$ of $M(\rho)$.
Repeat

    for $j := 1$ to $2m$ do

        1. if $j = 2i - 1$ is odd

            A. if α_{i+1} is not injective, update $B_{2i-1} := T_4(i)(B_{2i-1})$.
            B. if α_i is not surjective, update $B_{2i-1} := T_3(i)(B_{2i-1})$.
            C. delete any rows and columns rendered irrelevant.

        2. if $j = 2i$ is even

            A. if β_{i+1} is not surjective, update $B_{2i} := T_2(i+1)(B_{2i})$.

---

[8]Other, easier methods can also be used in this case.

B. if β_i is not injective, update $B_{2i} := T_1(i+1)(B_{2i})$.

C. delete any rows and columns rendered irrelevant.

endfor

until M_ρ is not empty or has not been updated.

Output M_ρ.

Example 1

Suppose κ has characteristic different from 2 and 3. To illustrate how Step 2 works, we consider the representation given by

$$\alpha_1 = \begin{pmatrix} 1 & 1 & 2 \\ -3 & 4 & 2 \\ -2 & 1 & 2 \end{pmatrix}, \quad \alpha_2 = \begin{pmatrix} 1 & 0 \\ 0 & 1 \end{pmatrix}, \quad \alpha_3 = \begin{pmatrix} 1 & 0 & 0 \\ 0 & 1 & 0 \end{pmatrix}, \quad \alpha_4 = \begin{pmatrix} 1 & 0 \\ 0 & 1 \end{pmatrix},$$

$$\beta_1 = \begin{pmatrix} 1 & 0 \\ 0 & 1 \\ 0 & 0 \end{pmatrix}, \quad \beta_2 = \begin{pmatrix} 1 & 0 & 0 \\ 0 & 1 & 0 \end{pmatrix}, \quad \beta_3 = \begin{pmatrix} 1 & 0 \\ 0 & 1 \end{pmatrix}, \quad \beta_4 = \begin{pmatrix} 1 & 0 & 0 \\ 0 & 1 & 0 \end{pmatrix}. \tag{3.20}$$

- Inspect B_1 and B_2. No changes are necessary.
- Inspect B_3. Since α_3 is not injective, one modifies the block by applying $T_4(2)$, which makes both α_3 and β_2 equal to $\begin{pmatrix} 1 & 0 \\ 0 & 1 \end{pmatrix}$.
- Inspect the blocks B_4, B_5, B_6, B_7. No changes are necessary.
- Inspect B_8. Since β_4 is not injective, one modifies the block by applying $T_1(1)$, which leads to $\alpha_1 = \begin{pmatrix} -4 & 3 \\ -3 & 0 \end{pmatrix}$ and $\beta_1 = \begin{pmatrix} -1 & 1 \\ -1 & 0 \end{pmatrix}$.

Indeed, the block B_8 is given by

$$\left(\frac{\beta_4 \,\|\, 0}{\alpha_1 \,\|\, \beta_1} \right) = \begin{pmatrix} 1 & 0 & 0 \,\|\, 0 & 0 \\ 0 & 1 & 0 \,\|\, 0 & 0 \\ \hline 1 & 1 & 2 \,\|\, 1 & 0 \\ -3 & 4 & 2 \,\|\, 0 & 1 \\ -2 & 1 & 2 \,\|\, 0 & 0 \end{pmatrix}.$$

Since β_4 is already in column echelon form one only has to change the base of V_2 to bring the last column of α_1 in row echelon form which ends up with

$$\begin{pmatrix} 1 & 0 & 0 & \| & 0 & 0 \\ 0 & 1 & 0 & \| & 0 & 0 \\ \hline 1 & 1 & 2 & \| & 1 & 0 \\ -4 & 3 & 0 & \| & -1 & 1 \\ -3 & 0 & 0 & \| & -1 & 0 \end{pmatrix}.$$

Therefore, $\alpha'_1 = \begin{pmatrix} -4 & 3 \\ -3 & 0 \end{pmatrix}$, $\beta'_1 = \begin{pmatrix} 1 & 0 \\ 0 & 1 \end{pmatrix}$, $\beta'_4 = \begin{pmatrix} -1 & 1 \\ -1 & 0 \end{pmatrix}$.

The algorithm stops as all α'_is and β'_is are at this time invertible. The last transformation $T_1(1)$ has eliminated only the barcode $(4, 5]$ and the previous, which was the first transformation, $T_4(2)$, has eliminated only the barcode $(2, 3)$. This can be concluded from Proposition 4.4. In view of the properties of these two transformations, one concludes that these were the only two barcodes. So the barcodes are $(4, 5]$, $(2, 3)$.

At termination, all α_i and β_i, become isomorphisms because otherwise one of the transformations would be applicable. The Jordan cells can be recovered from the Jordan decomposition of the matrix

$$T = \beta_{i-1}^{-1} \cdot \alpha_{i-1} \cdot \beta_{i-2}^{-1} \cdots \beta_1^{-1} \cdot \alpha_1 \cdot \beta_m^{-1} \cdot \alpha_m \cdots \beta_{i+1}^{-1} \cdot \alpha_{i+1} \cdot \beta_i^{-1} \cdot \alpha_i \quad \text{for any } i.$$

In our example $T = \begin{pmatrix} 0 & -1 \\ 1 & -1 \end{pmatrix} \cdot \begin{pmatrix} -4 & 3 \\ -3 & 0 \end{pmatrix}$, similar to the Jordan cell $J = (3, 2)$.

Standard linear algebra routines permit the calculation of the Jordan cells for familiar algebraically closed fields. Note that if κ is not algebraically closed it may not be possible to decompose the matrix T into Jordan cells unless we consider the algebraic closure of κ. It is however possible to decompose the matrix T up to similarity as a sum of indecomposable invertible matrices while remaining in the class of matrices with entries in the field κ.

In the example above T is similar to $\begin{pmatrix} 3 & 1 \\ 0 & 3 \end{pmatrix}$ the Jordan cell $(\lambda = 3, k = 2)$.

We leave the reader to work out the case $\kappa = \mathbb{Z}_2$ and $\kappa = \mathbb{Z}_3$.

3.3.3   Implementation of $T_1(i), T_2(i), T_3(i), T_4(i)$

Recall from Subsection 2.1.1 that $R(M)$ and $L(M)$ are the invertible matrices which make $M \cdot R(M)$ and $L(M) \cdot M$ matrices in column echelon form and row echelon form respectively.

1. $T_1(i)$ acts on the block matrix $B_{2(i-1)} = \begin{pmatrix} \beta_{i-1} & 0 \\ \alpha_i & \beta_i \end{pmatrix}$. First one modifies $B_{2(i-1)}$ to the block matrix $\begin{pmatrix} \beta_{i-1,1} & 0 & 0 \\ \alpha_{i,1} & \alpha_{i,2} & \beta_i \end{pmatrix}$, where $(\beta_{i-1,1}\ 0) = \beta_{i-1} \cdot R(\beta_{i-1})$ and $(\alpha_{i,1}\ \alpha_{i,2}) = \alpha_i \cdot R(\beta_{i-1})$. Then, one passes to the block matrix $\begin{pmatrix} \beta_{i-1,1} & 0 & 0 \\ \alpha_{i,2}^1 & \alpha_{i,2}^1 & \beta_i^1 \\ \alpha_{i,2}^2 & 0 & \beta_i^2 \end{pmatrix}$ where $\begin{pmatrix} \alpha_{i,2}^1 \\ 0 \end{pmatrix} = L(\alpha_{i,2}) \cdot \alpha_{i,2}$, $\begin{pmatrix} \alpha_{i,1}^1 \\ \alpha_{i,2}^2 \end{pmatrix} = L(\alpha_{i,2}) \cdot \alpha_{i,1}$ and $\begin{pmatrix} \beta_i^1 \\ \beta_i^2 \end{pmatrix} = L(\alpha_{i,2}) \cdot \beta_i$. The modified block matrix is $\begin{pmatrix} \beta_{i-1,1} & 0 \\ \alpha_{i,2}^2 & \beta_i^2 \end{pmatrix}$.

2. $T_2(i)$ acts on the block matrix $B_{2(i-1)} = \begin{pmatrix} \beta_{i-1} & 0 \\ \alpha_i & \beta_i \end{pmatrix}$. First one modifies $B_{2(i-1)}$ to the block matrix $\begin{pmatrix} \beta_{i-1} & 0 \\ \alpha_i^1 & \beta_i^1 \\ \alpha_i^2 & 0 \end{pmatrix}$, where $\begin{pmatrix} \beta_i^1 \\ 0 \end{pmatrix} = L(\beta_i) \cdot \beta_i$ and $\begin{pmatrix} \alpha_i^1 \\ \alpha_i^2 \end{pmatrix} = L(\beta_i) \cdot \alpha_i$. Then, one passes to the block matrix $\begin{pmatrix} \beta_{i-1,1} & \beta_{i-1,2} & 0 \\ \alpha_{i,1}^1 & \alpha_{i,2}^1 & \beta_i^1 \\ \alpha_{i,1}^2 & 0 & 0 \end{pmatrix}$ with $\begin{pmatrix} \alpha_{i,1}^2 \\ 0 \end{pmatrix} = \alpha_{i,1}^2 \cdot R(\alpha_{i,1}^2)$, $\begin{pmatrix} \alpha_{i,1}^1 \\ \alpha_{i,2}^1 \end{pmatrix} = \alpha_{i,1} \cdot R(\alpha_{i,1}^2)$, and $\begin{pmatrix} \beta_{i-1,1} \\ \beta_{i-1,2} \end{pmatrix} = \beta_{i-1} \cdot R(\alpha_{i,1})$. The modified block matrix is $\begin{pmatrix} \beta_{i-1,2} & 0 \\ \alpha_{i,2}^1 & \beta_i^1 \end{pmatrix}$.

3. $T_3(i)$ acts on the block matrix $B_{2i-1} = \begin{pmatrix} \alpha_i & \beta_i \\ 0 & \alpha_{i+1} \end{pmatrix}$. First one modifies B_{2i-1} to the block matrix $\begin{pmatrix} \alpha_i^1 & \beta_i^1 \\ 0 & \beta_i^2 \\ 0 & \alpha_{i+1} \end{pmatrix}$, where $\begin{pmatrix} \alpha_i^1 \\ 0 \end{pmatrix} = \alpha_i \cdot R(\alpha_i)$ and $\begin{pmatrix} \beta_i^1 \\ \beta_i^2 \end{pmatrix} = \beta_i \cdot R(\alpha_i)$. Then, one passes to the block matrix $\begin{pmatrix} \alpha_i^1 & \beta_{i,1}^1 & \beta_{i,2}^1 \\ 0 & \beta_{i,1}^2 & 0 \\ 0 & \alpha_{i+1,1} & \alpha_{i+1,2} \end{pmatrix}$ with $(\beta_{i,1}^2\ 0) = \beta_i^2 \cdot R(\beta_i^2)$, $(\beta_{i,1}^1\ \beta_{i,2}^1) = \beta_i^1 \cdot R(\beta_i^2)$ and $(\alpha_{i+1,1}\ \alpha_{i+1,2}) = \alpha_{i+1} \cdot R(\beta_i^2)$. The modified block matrix

is $\begin{pmatrix} \alpha_i^1 & \beta_{i,2}^1 \\ 0 & \alpha_{i+1,2} \end{pmatrix}$.

4. $T_4(i)$ acts on the block matrix $B_{2i-1} = \begin{pmatrix} \alpha_i & \beta_i \\ 0 & \alpha_{i+1} \end{pmatrix}$. First one modi-

fies B_{2i-1} to the block matrix $\begin{pmatrix} \alpha_i & \beta_{i,1} & \beta_{i,2} \\ 0 & \alpha_{i+1,1} & 0 \end{pmatrix}$, where $\begin{pmatrix} \alpha_{i+1,1} & 0 \end{pmatrix} = \alpha_{i+1} \cdot R(\alpha_{i+1})$ and $\begin{pmatrix} \beta_{i,1} & \beta_{i,2} \end{pmatrix} = \beta_i \cdot R(\alpha_{i+1})$.

Then, one passes to the block matrix $\begin{pmatrix} \alpha_i^1 & \beta_{i,1}^1 & \beta_{i,2}^1 \\ \alpha_i^2 & \beta_{i,1}^2 & 0 \\ 0 & \alpha_{i+1,1}^2 & 0 \end{pmatrix}$ with

$\begin{pmatrix} \beta_{i,2}^1 \\ 0 \end{pmatrix} = L(\beta_{i,2}) \cdot \beta_{i,2}$, $\begin{pmatrix} \beta_{i,1}^1 \\ \beta_{i,1}^2 \end{pmatrix} = L(\beta_{i,2}) \cdot \beta_{i,1}$, and $\begin{pmatrix} \alpha_i^1 \\ \alpha_i^2 \end{pmatrix} =$

$L(\beta_{i,2}) \cdot \alpha_i$. The modified block matrix is $\begin{pmatrix} \alpha_i^2 & \beta_{i,1}^2 \\ 0 & \alpha_{i+1,1} \end{pmatrix}$.

Example 2

Suppose that κ is a field of characteristic different from 2 and 3. In the example of the angle-valued map described in Chapter 4, Figure 4.7 $m = 7$ and the only interesting representation is $\rho_1{}^9$. One begins with the representation ρ_1 given by

$$\alpha_1 = \begin{pmatrix} 1 & 0 & 0 \\ 0 & 1 & 0 \\ 0 & 0 & 1 \end{pmatrix}, \ \alpha_2 = \begin{pmatrix} 0 & 0 \\ 1 & 0 \\ 0 & 1 \end{pmatrix}, \ \alpha_3 = \begin{pmatrix} 1 & 0 & 0 \\ 0 & 1 & 0 \\ 0 & 0 & 1 \end{pmatrix}; \alpha_4 = \begin{pmatrix} 1 & 0 \\ 0 & 1 \end{pmatrix},$$

$$\alpha_5 = \begin{pmatrix} 0 & 1 & 0 \\ 0 & 0 & 1 \end{pmatrix}, \ \alpha_6 = \begin{pmatrix} 1 & 0 \\ 0 & 1 \end{pmatrix}, \ \alpha_7 = \begin{pmatrix} 1 & 0 & 0 \\ 0 & 2 & 0 \\ 0 & -3 & 2 \end{pmatrix},$$

$$\beta_1 = \begin{pmatrix} 0 & 0 \\ 1 & 0 \\ 0 & 1 \end{pmatrix} \ \beta_2 = \begin{pmatrix} 1 & 0 & 0 \\ 0 & 1 & 0 \\ 0 & 0 & 1 \end{pmatrix} \ \beta_3 = \begin{pmatrix} 0 & 0 \\ 1 & 0 \\ 0 & 1 \end{pmatrix}, \ \beta_4 = \begin{pmatrix} 0 & 1 & 0 \\ 0 & 0 & 1 \end{pmatrix},$$

$$\beta_5 = \begin{pmatrix} 1 & 0 \\ 0 & 1 \end{pmatrix}, \ \beta_6 = \begin{pmatrix} 0 & 1 & 0 \\ 0 & 0 & 1 \end{pmatrix}, \ \beta_7 = \begin{pmatrix} 1 & 0 & 0 \\ 0 & 1 & 0 \\ 0 & 0 & 1 \end{pmatrix}.$$

(3.21)

and one ends up with the representation $\rho(5)$ described below.

The representation ρ_1 has all $\alpha_i = \beta_i = Id_1$, $i = 1,\ldots,7$ and the

---

[9]In dimension 0 the representation ρ_0 has all $\alpha_i = Id_1$, and in dimension 2 the representation ρ_2 is the trivial representation.

representation ρ_2 is the trivial representation. There are 14 blocks defined by the formulae (3.22) and (3.23).

Recall that

$$B_{2i-1} = \begin{pmatrix} \alpha_i & \beta_i \\ 0 & \alpha_{i+1} \end{pmatrix}, \quad B_{2i} = \begin{pmatrix} \beta_i & 0 \\ \alpha_{i+1} & \beta_{i+1} \end{pmatrix} \tag{3.22}$$

for $i = 1, 2, \ldots, m-1$, and

$$B_{2m-1} = \begin{pmatrix} \alpha_m & \beta_m \\ 0 & \alpha_1 \end{pmatrix}, \quad B_{2m} = \begin{pmatrix} \beta_m & 0 \\ \alpha_1 & \beta_1 \end{pmatrix}. \tag{3.23}$$

Further,

- Inspecting $B_{2i-1} = \begin{pmatrix} \alpha_i & \beta_i \\ 0 & \alpha_{i+1} \end{pmatrix}$ no change is performed if α_i is surjective and α_{i+1} is injective, but
  if α_i is not surjective one applies $T_3(i)$,
  if α_{i+1} is not injective one applies $T_4(i)$.
- Inspecting $B_{2i} = \begin{pmatrix} \beta_i & 0 \\ \alpha_{i+1} & \beta_{I+1} \end{pmatrix}$ no change is performed if β_i is injective and β_{i+1} is surjective, but
  if β_i is not injective one applies $T_1(i+1)$,
  if β_{i+1} is not surjective one applies $T_2(i+1)$.

With this in mind one inspects the blocks $B_1, B_2, \ldots, B_{13}, B_{14}$, and when needed one applies the appropriate changes and moves to the next box to be inspected.

(1) No changes are needed inspecting B_1 and B_2.

Inspecting $B_3 = \begin{pmatrix} \alpha_3 & \beta_3 \\ 0 & \alpha_4 \end{pmatrix}$ observe that α_2 is not surjective hence one applies $T_3(2)$ which modifies α_2 into $\alpha'_2 = \begin{pmatrix} 1 & 0 \\ 0 & 1 \end{pmatrix}$, α_3 into

$\alpha'_3 = \begin{pmatrix} 0 & 0 \\ 1 & 0, \\ 0 & 1 \end{pmatrix}$, and β_2 into $\beta'_2 = \begin{pmatrix} 1 & 0 \\ 0 & 1 \end{pmatrix}$.

Update ρ to the representation $\rho(1)$.

(2) For the updated representation $\rho(1)$ inspect $B_4 = \begin{pmatrix} \beta_2 & 0 \\ \alpha_3 & \beta_3 \end{pmatrix}$. Observe β_3 is not surjective, hence one applies $T_2(3)$ which modifies α_3 into $\alpha'_3 = \begin{pmatrix} 1 & 0 \\ 0 & 1 \end{pmatrix}$, and β_3 into $\beta'_3 = \begin{pmatrix} 1 & 0, \\ 0 & 1 \end{pmatrix}$.

With the representation updated to $\rho(2)$, inspect the next blocks; no changes are needed inspecting B_5 and B_6.

(3) Inspecting B_7 note that α_5 is not injective, hence one applies $T_4(4)$ which modifies α_5 into $\alpha'5 = \begin{pmatrix} 1 & 0 \\ 0 & 1 \end{pmatrix}$, β_4 into $\beta'_4 = \begin{pmatrix} 1 & 0 \\ 0 & 1 \end{pmatrix}$.

Again with the representation updated to $\rho(3)$ no changes are needed inspecting B_8, B_9, B_{10} and B_{11}, provided that for the field κ the matrix $M = \alpha_7$ is invertible, which is indeed the case for M as defined in Chapter 4 for κ of characteristic different from 2.

(4) Inspecting B_{12} note that β_6 is not injective and one has to apply $T_1(7)$ and modify

α_7 into $\alpha' = \begin{pmatrix} 2 & 0 \\ -3 & 2 \end{pmatrix}$,

β_7 into $\beta'_7 = \begin{pmatrix} 0 & 1 & 0 \\ 0 & 0 & 1 \end{pmatrix}$,

β_6 into $\beta'_6 = \begin{pmatrix} 1 & 0 \\ 0 & 1 \end{pmatrix}$.

Update the representation to $\rho(4)$.

Inspect B_{13}, no change.

(5) Inspect $B_{14} = \begin{pmatrix} \beta_7 & 0 \\ \alpha_1 & \beta_1 \end{pmatrix}$, note that β_7 is not injective, and then apply $T_1(8)$ and modify

β_7 into $\beta'_7 = \begin{pmatrix} 1 & 0 \\ 0 & 1 \end{pmatrix}$,

α_1 into $\alpha'_1 = \begin{pmatrix} 1 & 0 \\ 0 & 1 \end{pmatrix}$,

β_1 into $\beta'_1 = \begin{pmatrix} 1 & 0 \\ 0 & 1 \end{pmatrix}$, and update to a representation $\rho(5)$, which at this time has all linear maps α_i, β_i are invertible.

The algorithm stops. We derive $T = \begin{pmatrix} 2 & 0 \\ -3 & 2 \end{pmatrix}$ hence Jordan cell $(2; 2)$

The chain of modifications was

$$\rho \xrightarrow{T_3(2)} \rho(1) \xrightarrow{T_2(3)} \rho(2) \xrightarrow{T_4(4)} \rho(3) \xrightarrow{T_1(7)} \rho(4) \xrightarrow{T_1(8)} \rho(5)$$

Backwards calculations:

(i) $T_1(8)$: in view of this action and of Proposition 3.4, $\rho(4)$ has only one barcode, $(7, 8]$.

(ii) $T_1(7))$: in view of this action and of Proposition 3.4, $\rho(3)$ has only one barcode, $(6, 8]$.

(iii) $T_4(4)$: in view of this action and of Proposition 3.4 $\rho(2)$ has two barcodes, $(6, 8]$ and $(4, 5]$.

(iv) $T_2(3)$: in view of this action and of Proposition 3.4, $\rho(1)$ has three barcodes, $(6, 8]$, $(4, 5]$, and $[3, 3]$.

(v) $T_3(2)$: in view of this action and of Proposition 3.4, ρ has three barcodes, $(6, 8]$, $(4, 5]$, and $[2, 3]$.

Clearly then, the collection of bar codes and of Jordan cells is: bar codes $[2, 3], (4, 5), (6, 8]$ and Jordan cell $(2; 2)$ for a field κ of characteristic $\neq 2$. We leave to the reader to complete the calculations in the case of the fields with two and three elements, $\kappa = \mathbb{Z}_2$ and $\kappa = \mathbb{Z}_3$.

3.4   Exercises

E.1 Describe all indecomposable $\mathcal{Z}$-representations and formulate and prove the Krull-Remak-Schmidt decomposition theorem for arbitrary $\mathcal{Z}$-representations (compare with [Warfield, R.B. (1969)]).

E.2 Prove that any indecomposable $\mathcal{Z}$-representation with finite support is a barcode and any regular indecomposable $\mathcal{Z}$-representation is isomorphic to $\rho_\infty^\mathcal{Z}$.

E.3 Make explicit the relations between the integers r, s, r', s' which ensure the existence of a nontrivial morphism from the barcode $\rho_{r,s}^\mathcal{Z}$ to the barcode $\rho_{r',s'}^\mathcal{Z}$.

E.4 Describe the vector spaces V_r and the linear maps α_i and β_i for all representations $\rho^{G_{2m}}([i, j + mk])$, $\rho^{G_{2m}}([i, j + mk))$, $\rho^{G_{2m}}((i, j + mk))$, and $\rho^{G_{2m}}((i, j + mk])$.

E.5 Verify Proposition 3.2; in particular, calculate ker $\rho_\infty^\mathcal{Z}$ and coker $\rho_\infty^\mathcal{Z}$.

E.6 Calculate the barcodes and the Jordan cells for the representation ρ_1 in Section 3.3.3 for the field $\kappa = \mathbb{Z}_2$.

E.7 Calculate the barcodes and the Jordan cells for the representation ρ_1 in Section 3.3.3 for the field $\kappa = \mathbb{Z}_3$.

E.8 Use elementary transformations to verify that any indecomposable G_{2m} representation which is not regular is a barcode.

Chapter 4

Barcodes and Jordan Blocks via Graph Representations

4.1 The graph representations associated to a map

Consistent with the notations in the previous chapters, denote by:

$\mathbb{Z}$ the set of integers, $\mathbb{R}$ the set of real numbers, $\mathbb{C}$ the set of complex numbers, often identified with the plane $\mathbb{R}^2$ by $\mathbb{R}^2 \ni (a, b) \mapsto z + a + ib \in \mathbb{C}$,

$\Delta \subset \mathbb{R}^2$ the *diagonal* $\Delta := \{(x, x) \in \mathbb{R}^2 \mid x \in \mathbb{R}\}$,

$\mathbb{T} := \mathbb{R}^2 / \mathbb{Z}$ the quotient space for the action $\mu : \mathbb{Z} \times \mathbb{R}^2 \to \mathbb{R}^2$ given by $\mu(n, (a, b)) = (a + 2\pi n, b + 2\pi n)$,

$\mathbb{S}^1 = \Delta / \mathbb{Z} = \mathbb{R} / \mathbb{Z}$ the quotient space for the action $\mu : \mathbb{Z} \times \mathbb{R} \to \mathbb{R}$ given by $\mu(n, x) = (x + 2\pi n)$, also regarded as $\mathbb{S}^1 = \{z = e^{i\theta} \in \mathbb{C}\}$.

The set $\mathbb{T}$ can be identified with $\mathbb{C} \setminus 0$ by the assignment $\mathbb{T} \ni \langle a, b \rangle \mapsto e^{ia+(b-a)}$, where $\langle a, b \rangle$ denotes the orbit of $(a, b) \in \mathbb{R}^2$ for the action μ of $\mathbb{Z}$ on $\mathbb{R}^2$.

Even if not mentioned, all spaces X are ANR's, and all maps $f : X \to \mathbb{R}$ and $f : X \to \mathbb{S}^1$ are proper continuous maps. In the second case this implies that X is compact. To simultaneously consider real- and angle-valued maps we sometimes use $\mathbb{U}$ for either $\mathbb{R}$ or $\mathbb{S}^1$.

In this book $\mathbb{S}^1$ is viewed either as the subset of complex numbers $\{z \in \mathbb{C} \mid |z| = 1\}$, or as the set of angles $\{\theta \in \mathbb{R} / 2\pi\mathbb{Z}\}$, often identified to $[0, 2\pi)$, and the map $\pi : \mathbb{R} \to \mathbb{S}^1$ is given by $\pi(t) = e^{\sqrt{-1}t}$ or $\pi(t) = \theta \in [0, 2\pi)$, with $t - \theta$ an integer multiple of 2π.

Recall from Chapter 2 Subsection 2.2.1:

Observation 4.1.

(i) The map $f : X \to \mathbb{U}$ is *weakly tame* if for any $u \in \mathbb{U}$, $f^{-1}(u)$ is an ANR. Then for any closed interval $I \subset \mathbb{U}$, $f^{-1}(I)$ is an ANR.

(ii) The value $u \in \mathbb{U}$ is a regular value for f if there exists an open neighborhood U of u in $\mathbb{U}$ such for any $u' \in U$, $f^{-1}(U)$ retracts by deformation to $f^{-1}(u')$, and is a critical value if not a regular value. Denote by $Cr(f)$ the set of critical values.

(iii) The map $f : X \to \mathbb{U}$ is *tame* if the following two conditions hold:

    (i) the set $Cr(f)$ is discrete,

    (ii) $\epsilon(f) := \inf_{c,c' \in Cr(f), c \neq c'} d(c, c') > 0$.

(iv) If f is tame, $s_1 < s_2 < s_3$ are three successive critical values[1], and α, β are such that $s_1 < \alpha < s_2 < \beta < s_3$, then the inclusions

$$f^{-1}(s_2) \subseteq f^{-1}([s_2, \beta)) \subseteq f^{-1}((\alpha, \beta)),$$
$$f^{-1}(s_2) \subseteq f^{-1}((\alpha, s_2]) \subseteq f^{-1}((\alpha, \beta)),$$
$$f^{-1}(s_1, s_2) \supseteq f^{-1}((s_1 \alpha]) \supseteq f^{-1}(\alpha),$$
$$f^{-1}(s_1 s_2) \supseteq f^{-1}(([\alpha, s_2)) \supseteq f^{-1}(\alpha)$$

are homotopy equivalences.

Let $f_1 : X_1 \to \mathbb{U}$ and $f_2 : X_1 \to \mathbb{U}$ be two tame maps. A continuous map $\omega : X_1 \to X_2$ which satisfies $f_2 \cdot \omega = f_1$ and restricts to a homotopy equivalence between fibers above each $u \in \mathbb{U}$ is called a *fiberwise weak homotopy equivalence*[2].

To a tame map $f : X \to \mathbb{U}$ one associates the following data set, denoted by $\mathbf{D(f)}$:

1. Data $\mathbf{D(f)}$ for a tame map f
consist of:

(1) The collection of critical values $\cdots s_{i-1} < s_i < s_{i+1} \cdots$, with $\epsilon(f) = \inf_{i \in \mathbb{Z}}(s_{i+1} - s_i)$. Note that when $\mathbb{U} = \mathbb{S}^1$ and the map f has critical angles $0 \le \theta_1 < \theta_2 < \cdots < \theta_m < 2\pi$, then $s_{i+mk} = \theta_i + 2\pi k$ are the critical values for any infinite cyclic cover $\tilde{f} : \tilde{X} \to \mathbb{R}$ of f.

(2) The compact spaces $X_{2i} := f^{-1}(s_i)$ and the open sets $X_{2i-1} := f^{-1}(s_{i-1}, s_i)$. If f is angle-valued with m critical angles $0 \le \theta_1 < \theta_2 < \cdots < \theta_m < 2\pi$, in order to make sense of X_1 one introduces $\theta_0 = \theta_m - 2\pi$.

(3) The homotopy class of maps $\widehat{a}_i : X_{2i-1} \to X_{2i}$ and $\widehat{b}_i : X_{2i+1} \to X_{2i}$ described below.

---

[1]Two critical values are *successive* if all the values between them are regular.

[2]Actually a fiberwise weak homotopy equivalence from X_1 to X_2 is given by a sequence of such maps, $\omega(1), \omega(2), \ldots, \omega(2k+1), \omega(2i+1) : X(2i+1) \to X(2i), \omega(2i) : X(2i) \to X(2i-1)$, with $X_1 = X(1)$ and $X_2 = X(2k+1)$.

Note that if $\widetilde{f} : \widetilde{X} \to \mathbb{R}$ is an infinite cyclic cover of an angle-valued map $f : X \to \mathbb{S}^1$ and $\pi(t) = \theta$ then $(\widetilde{f})^{-1}(t)$ and $f^{-1}(\theta)$ are homeomorphic by the restriction of $\widetilde{\pi} : \widetilde{X} \to X$, the infinite cyclic cover defined by ξ_f. Note also that if in addition f has m critical values then $\widetilde{X}_i$ is canonically homeomorphic and $\widetilde{X}_{2mk+i}$ by the restriction of τ^k, where τ is the generator of the group of deck transformations of $\widetilde{X} \to X$, cf. Subsection 2.3.3 in Chapter 2.

In order to define the homotopy classes $\widehat{a}_i$ and $\widehat{b}_i$, consider

(i) $X_{2i-1}^l := f^{-1}([s_{i-1}, s_i))$, $X_{2i-1}^r := f^{-1}((s_{i-1}, s_i])$,[3] $X_{2i-1} = X_{2i-1}^l \cap X_{2i-1}^r$,
and the inclusions

(ii) $\mathrm{in}_{i-1}^l : X_{2i-1} := f^{-1}((s_{i-1}, s_i)) \to f^{-1}([s_{i-1}, s_i)) = X_{2i-1}^l$,

(iii) $\mathrm{in}_i^r : X_{2i-1} := f^{-1}((s_{i-1}, s_i)) \to f^{-1}((s_{i-1}, s_i]) = X_{2i}^r$,

(iv) $\iota_{i-1}^r : X_{2i-2} = f^{-1}(s_{i-1}) \to f^{-1}([s_{i-1}, s_i)) = X_{2i-1}^l$,

(v) $\iota_i^l : X_{2i} = f^{-1}(s_i) \to f^{-1}((s_{i-1}, s_i]) = X_{2i-1}^r$.

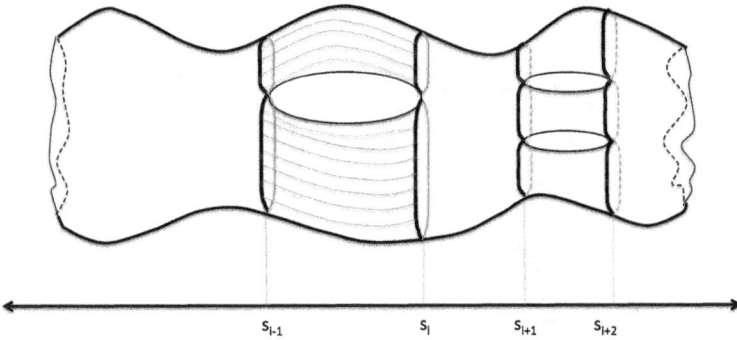

Fig. 4.1   (critical) Levels

Since by Observation 4.1 (iv) the tameness of f implies that ι_{i-1}^r and ι_i^l are homotopy equivalences, consider the homotopy inverses

(vi) $\pi_{i-1}^r : X_{2i-1}^l \to X_{2i-2}$, and

(vii) $\pi_i^l : X_{2i-1}^r \to X_{2i}$,

---

[3] "l" and "r" stand for *left* and *right*.

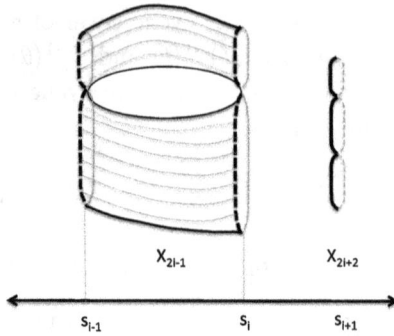

Fig. 4.2   Interlevel and level

and define

$$a_i = \pi_i^r \cdot \text{in}_i^r \quad \text{and} \quad b_i = \pi_{i-1}^l \cdot \text{in}_{i-1}^l .$$

Clearly the homotopy classes of a_i and b_i, denoted by $\widehat{a}_i$ and $\widehat{b}_i$, are well defined, i.e., independent on the choices of π_i^r and π_i^l.

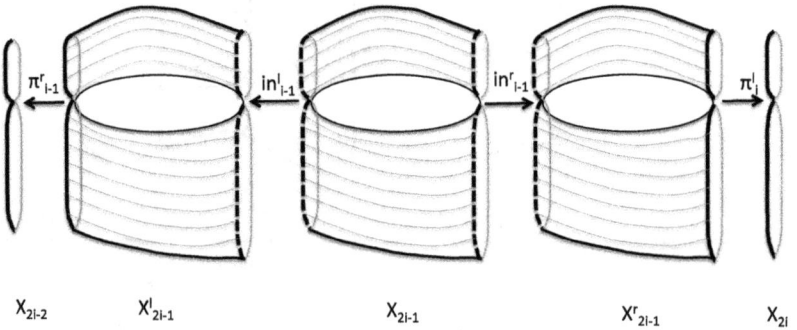

Fig. 4.3   Relevant pieces

Note that if $t_i's$ are any values in the intervals (s_{i-1}, s_i), hence regular values, it follows from the tameness of f, by Observation 4.1 (iv) above, that the inclusions $f^{-1}(t_i) \subset f^{-1}(s_{i-1}, s_i)$ and $f^{-1}(s_i) \subset f^{-1}([t_i, t_{i+1}])$ are homotopy equivalences.

2. The graph representations $\rho_r(f)$ for a tame map f

For a field κ and a nonnegative integer r, the data $\mathbf{D(f)}$ provide a Γ-representation $\rho_r(f)$ of κ-vector spaces with $V_r = H_r(X_r)$ and α_i, β_i the linear maps induced in homology by the homotopy classes $\hat{a}_i$ and $\hat{b}_i$, respectively.

If f is real-valued, then the graph Γ is $\mathcal{Z}$, and if f is angle-valued with m critical values, then the graph Γ is G_{2m}. If f is real-valued and X is compact then, the $\mathcal{Z}$-representation $\rho_r(f)$ has finite support.

If $\widetilde{f}$ is the infinite cyclic cover of an angle-valued map f with m critical angles, then the $\mathcal{Z}$-representation $\rho_r(\widetilde{f})$ is the same as $\widetilde{\rho}_r(f)$, hence periodic with period m.

Observation 4.2. The representation $\rho_r(f)$ is equivalent with the representation defined by $V_{2i-1} = H_r(f^{-1}(t_i))$, $V_{2i} = H_r(f^{-1}([t_i, t_{i+1}]))$, with α_i induced by the inclusion $f^{-1}(t_i) \subset f^{-1}([t_i, t_{i+1}])$ and β_i by the inclusion $f^{-1}(t_{i+1}) \subset f^{-1}([t_i, t_{i+1}])$

This description of $\rho_r(f)$ uses only homology of compact ANR's, convenient for algorithms, but involves the choice of regular values t_i, which makes the description of $\rho_r(f)$ less canonical. This will be used in Section 4.5, where an algorithm for the calculation of barcodes and Jordan cell for simplicial maps is provided. The algorithm involves incidence matrices of finite cell complexes which are of the form $f^{-1}(t_i)$ and $f^{-1}([t_i, t_{i+1}])$.

4.2 Barcodes and Jordan blocks of a tame map

For f a real-valued tame map with critical values

$$\cdots < c_i < c_{i+1} < c_{i+2} \cdots$$

the sets of barcodes $\mathcal{B}_r^c(f)$, $\mathcal{B}_r^o(f)$, $\mathcal{B}_r^{c,o}(f)$, $\mathcal{B}_r^{o,c}(f)$ are obtained from the intervals of $\mathcal{B}_r^c(\rho_r(f))$, $\mathcal{B}_r^o(\rho_r(f))$, $\mathcal{B}_r^{c,o}(\rho_r(f))$, $\mathcal{B}_r^{o,c}(\rho_r(f))$, respectively, by replacing each pair of ends i, j by c_i, c_j.

For f an angle-valued tame map with critical angles

$$0 \leq \theta_1 < \theta_2 \cdots < \theta_m < 2\pi$$

the sets of barcodes $\mathcal{B}_r^c(f)$, $\mathcal{B}_r^o(f)$, $\mathcal{B}_r^{c,o}(f)$, $\mathcal{B}_r^{o,c}(f)$ are obtained from the intervals of $\mathcal{B}_r^c(\rho_r(f))$, $\mathcal{B}_r^o(\rho_r(f))$, $\mathcal{B}_r^{c,o}(\rho_r(f))$, $\mathcal{B}_r^{o,c}(\rho_r(f))$, respectively, by replacing the ends $i, j + mk$ by $\theta_i, \theta_j + 2\pi k$, and the sets of Jordan blocks

$\mathcal{J}_r(f)$ are the same as the sets $\mathcal{J}_r(\rho_r(f))$. Passing to the algebraic closure $\overline{\kappa}$ if κ is not algebraically closed, the Jordan blocks can be decomposed into Jordan cells which correspond to pairs (λ, k), $\lambda \in \overline{\kappa} \setminus 0$. We will use the same notation $\mathcal{J}_r(f)$ for the sets of Jordan cells, since they provide the same information.

The sets $\mathcal{J}_r(f)$ of Jordan blocks consists of indecomposable pairs $J = (V_J, T_J)$ of finite-dimensional κ-vector space V_J and linear isomorphism $T_J : V_J \to V_J$. For each r one denotes $N_r(f) := \sum_{J \in \mathcal{J}_r(f)} \dim V_J$, $V_r(f) := \bigoplus_{J \in \mathcal{J}_r(f)} V_J$ and $T_r(f) := \bigoplus_{J \in \mathcal{J}_r(f)} T_J$.

If f is a tame real-valued map and X is compact, then the sets of critical values and the sets $\mathcal{B}_r^{\cdots}(f)$ are finite. If X is not compact, hence only locally compact, this is not necessarily the case; the representation $\rho_r(f)$ might contain other type of indecomposable representations than just barcodes. We are interested only in the case of $\widetilde{f} : \widetilde{X} \to \mathbb{R}$, the infinite cyclic cover of a map $f : X \to \mathbb{S}^1$ (and this only to better understand the case of tame angle-valued maps). In this case the representation $\rho_r(\widetilde{f})$ contains, in addition to the collection of barcodes $\mathcal{B}_r^{\cdots}(\widetilde{f})$, $N_r(f)$ copies of $\rho_\infty^{\mathbb{Z}}$. The sets of barcodes $\mathcal{B}_r(\widetilde{f})$, if not empty, are infinite and consists of all $2\pi k$-translates of the barcodes in $\mathcal{B}_r^{\cdots}(f)$ when viewed as intervals with the left end in $[0, 2\pi)$. The collection of $N_r(f)$ copies of $\rho_\infty^{\mathbb{Z}}$ is denoted by $\mathcal{J}_r(\widetilde{f})$; its members can be regarded as *infinite barcodes*, since they generate barcodes for truncations of $T_{i,j}(\rho_r(\widetilde{f}))$; however, we will avoid this name.

To summarize, for X a compact ANR

- the invariants we associate to a tame real-valued map, a field κ and an integer r are:
  - (a) the critical values,
  - (b) the four collections of intervals referred to as barcodes;

- the invariants we associate to a tame angle-valued map, a field κ, and an integer r are:
  - (a) the critical angles,
  - (b) the four collections of intervals, barcodes,
  - (c) the collection of Jordan cells (pairs $(\lambda \in \overline{\kappa}, k \in \mathbb{Z}_{\geq 1})$, equivalently, the characteristic polynomial of $[T_r(f)]$ with its characteristic divisors.

Clearly, in view of the definition proposed the above elements remain the same for fiberwise weak homotopy equivalent tame maps.

There are two convenient ways to organize the barcodes:

(i) as a configuration of points in the complex plane,
(ii) as a chain complex of vector spaces, referred below as the AM-*complex* (*alternative to the Morse complex*) for f a real-valued map, and as the AN-*complex* (*alternative to the Novikov complex*) for f an angle-valued map.

4.2.1 The configurations δ_r^f

It will be convenient to collect the closed r-barcodes and the open $(r-1)$-barcodes as configurations of points in $\mathbb{R}^2 = \mathbb{C}$ for $f : X \to \mathbb{R}$, X compact, and in $\mathbb{T} = \mathbb{C} \backslash 0$ for $f : X \to \mathbb{S}^1$. The "convenience" is justified by Theorem 4.1 below and Theorems 5.1–5.6 in Chapter 5.

The **configuration δ_r^f for a f real-valued map** is the map with finite support $\delta_r^f : \mathbb{R}^2 = \mathbb{C} \to \mathbb{Z}_{\geq 0}$ which assigns to $z = a + ib$ the number of the closed r-barcodes $[a,b] \in \mathcal{B}_r^c(f)$ if $a \leq b$, and the number of open $(r-1)$-barcodes $(b,a) \in \mathcal{B}_{r-1}^o(f)$ if $a > b$. Equivalently, one can regard δ_r^f as a monic polynomial

$$P_r^f(z) := \prod_i (z - z_i)^{\delta_r^f(z_i)}$$

with the roots z_i the complex numbers in the support of δ_r^f.

Figure 4.4 depicts such a configuration by means of *black bullet* points located above or on the diagonal, representing the closed r-barcodes, and *empty bullet* points located below the diagonal, representing the open $(r-1)$-barcodes.

The **configuration δ_r^f for f an angle-valued map** is the map with finite support $\delta_r^f : \mathbb{T} = \mathbb{R}^2/\mathbb{Z} \to \mathbb{Z}_{\geq 0}$ which assigns to the orbit $\langle a, b \rangle$ the number of closed r-barcodes $[a,b] \in \mathcal{B}_r^c$ if $a \leq b$ and the number of open $(r-1)$-barcodes $(b,a) \in \mathcal{B}_{r-1}^o$ if $a > b$. Recall that one can identify $\mathbb{T}$ with $\mathbb{C} \backslash 0$ by sending $\langle a, b \rangle$ to $e^{ia+(b-a)}$. Then the configuration described above becomes the configuration $\delta_r^f : \mathbb{C} \backslash 0 \to \mathbb{Z}_{\geq 0}$. Equivalently, one can regard it as a monic polynomial with nonzero free coefficient

$$P_r^f(z) := \prod_i (z - z_i)^{\delta_r^f(z_i)}$$

with the roots z_i the complex numbers in the support of δ_r^f.

Figure 4.5 depicts $\delta_r(f)$ as a configuration of points in $\mathbb{C} \backslash 0$; the *black bullet* points outside or on the unit circle correspond to closed r-barcodes,

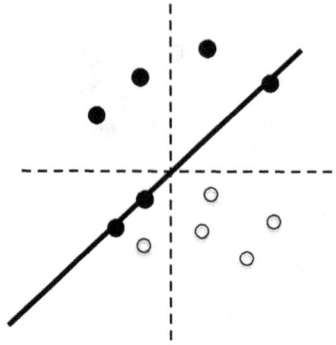

Fig. 4.4 δ_r^f for real-valued map

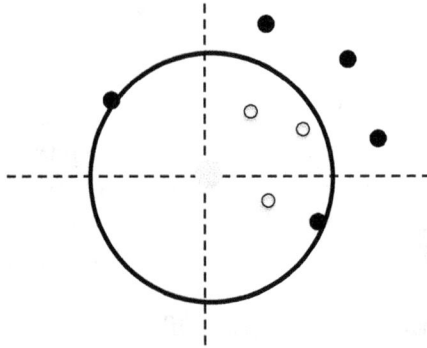

Fig. 4.5 δ_r^f for angle-valued map

while the *empty bullet* points inside the unit circle correspond to open $(r - 1)$-barcodes.

Clearly, given the polynomials $P_r^f(z)$ one can reconstruct the closed r and the open $(r - 1)$ barcodes. cf Exercise E4 in Section 4.4.

4.2.2 The AM *and* AN *complexes*

For a tame map $f : X \to \mathbb{R}$ with critical values

$$\cdots c_i < c_{i+1} < c_{i+2} \cdots ,$$

the barcodes $\mathcal{B}_r^c(f)$, $\mathcal{B}_r^o(f)$, and $\mathcal{B}_r^{c,o}(f)$ define a chain complex $\mathcal{C}_*^M(f) := (C_*(f), \partial_*(f))$ equipped with a filtration

$$\cdots \subseteq \mathcal{C}_*^M(f)(i) \subseteq \mathcal{C}_*^M(f)(i+1) \subseteq \cdots \subseteq \mathcal{C}_*^M(f).$$

Recall that for a set S, $\kappa[S]$ denotes the κ-vector space generated by S. The vector space $C_r(f)$ is given by

$$\boxed{\begin{aligned} C_r(f) &:= \kappa[\mathcal{B}_r^c(f) \sqcup \mathcal{B}_{r-1}^o(f) \sqcup \mathcal{B}_r^{c,o}(f) \sqcup \mathcal{B}_{r-1}^{c,o}(f)] \\ &= \kappa[\mathcal{B}_r^c(f)] \oplus \kappa[\mathcal{B}_{r-1}^o(f)] \oplus \kappa[\mathcal{B}_r^{c,o}(f)] \oplus \kappa[\mathcal{B}_{r-1}^{c,o}(f)] \end{aligned}}. \quad (4.1)$$

The boundary map $\partial_r(f) : C_r(f) \to C_{r-1}(f)$ is given by the block matrix

$$\partial_r(f) = \begin{Vmatrix} 0 & 0 & 0 & 0 \\ 0 & 0 & 0 & 0 \\ 0 & 0 & 0 & Id \\ 0 & 0 & 0 & 0 \end{Vmatrix}.$$

The subspaces $C_r(f)(i) \subseteq C_r(f)$ are defined by the same formulae (4.1) with $\mathcal{B}_r^{\cdots}(f)$ replaced by $\mathcal{B}_r^{\cdots}(f)(i)$, given by

$$\begin{aligned} \mathcal{B}_r^c(f)(i) &= \{[c', c''] \in \mathcal{B}_r^c \mid c' \le c_i\}, \\ \mathcal{B}_r^{c,o}(f)(i) &= \{[c', c'') \in \mathcal{B}_r^{c,o} \mid c' \le c_i\}, \\ \mathcal{B}_{r-1}^o(f)(i) &= \{(c', c'') \in \mathcal{B}_{r-1}^o \mid c'' \le c_i\}, \\ \mathcal{B}_{r-1}^{c,o}(f)(i) &= \{[c', c'') \in \mathcal{B}_{r-1}^{c,o} \mid c'' \le c_i\}. \end{aligned}$$

Obviously, $\partial_r(f)(C_r(f)(i)) \subset C_{r-1}(f)(i)$.

In case X is compact one refers to this complex as the AM complex (alternative Morse complex) since, when X is a closed smooth manifold and f is a Morse function, it is isomorphic to the Morse complex associated to a Morse-Smale vector field which has f as Lyapunov function tensored by κ, as described in Chapter 8 Section 8.1.

For a tame angle-valued map $f : X \to \mathbb{S}^1$ consider the field $\kappa[t^{-1}, t]]$ of Laurent power series with coefficients in κ and use the same formulae (4.1) to define for $C_r(f)$ and $\partial_r(f)$ but with $\kappa[t^{-1}, t]]$ instead of κ. One obtains a chain complex of $\kappa[t^{-1}, t]]$-vector spaces referred to as the AN complex and denoted by $C_*^N(f)$.

Again, when f is a Morse map, this complex of $\kappa[t^{-1}, t]]$-vector spaces is isomorphic to the Novikov complex associated to a Morse-Smale vector field which has f as a Lyapunov map tensored by κ, as described in Chapter 8 Section 8.1.

The (non-canonical) isomorphisms alluded to above also establishes the independence of the vector field for the Morse complex or for the Novikov complex when tensored by κ, a result known from [Cornea, O., Ranicki, A. (2003)].

The AM and AN complexes determines completely the number of closed plus the number of open barcodes and the number of closed-open barcodes for f. If one replaces $\mathcal{B}_r^{c,o}$ by $\mathcal{B}_r^{o,c}$ one obtains the AM or AN complex for the map $-f$ or f^{-1}, respectively.

4.2.3 *The relevant exact sequences*

The sets $\mathcal{B}_r(f)$ for f a tame real-valued map allow one to calculate the homology vector spaces $H_r(X)$ and $H_r(f^{-1}[a, b])$ with coefficients in the field κ. The sets $\mathcal{B}_r(f)$ and $\mathcal{J}_r(f)$ for f a tame angle-valued map allow one to calculate the Novikov homology $H_r^N(X; \xi)$ and the homology with coefficients in the representation (ξ, u), cf. Subsection 2.3.6, $H_r(X; (\xi, u))$, as well as other related homologies and for a pair $(X; \xi \in H^1(X; \mathbb{Z}))$, with ξ the cohomology class determined by the angle-valued map $f : X \to \mathbb{S}^1$. The definitions of these homologies were reviewed in Chapter 2 Subsection 2.3.6 and will be reminded whenever needed.

The tools which permit these calculations are provided by Proposition 4.1 below, complemented by Observation 4.3. This Observation follows from the tameness of f and the properties of truncation of the $\mathcal{Z}$- and G_{2m}-representations discussed in the Chapter 3.

Observation 4.3. If $f : X \to \mathbb{U}$ is a tame map, $f_{[a,b]} : f^{-1}([a, b]) \to \mathbb{U}$ the restriction of f, and $s_{i-1} < s_i \le s_j < s_{j+1}$ critical values and α, β such that $s_{i-1} < \alpha \le s_i \le s_j \le \beta < s_{j+1}$, then

$$\rho_r(f_{[s_i, s_j]}) = T_{i,j}(\rho_r(f))$$

and

$$\rho_r(f_{[\alpha, \beta]}) = \rho_r(f_{[s_i, s_j]}).$$

It is understood that s_{i-1}, s_i and s_j, s_{j+1} are consecutive critical values.

Proposition 4.1. *Let $f: X \to \mathbb{S}^1$ be a tame map and $\widetilde{f}: \widetilde{X} \to \mathbb{R}$ its infinite cyclic cover. Let $\rho_r = \rho_r(f)$ and $\widetilde{\rho}_r = \rho_r(\widetilde{f}) = \widetilde{\rho}_r(f)$ be the representations associated with f and $\widetilde{f}$. One has the following short exact sequences:*

$$0 \to \operatorname{coker} M((\rho_r)_u) \to H_r(X; (\xi_f, u)) \to \ker M((\rho_{r-1})_u) \to 0, \qquad (4.2)$$

which for $u = 1$ becomes

$$0 \to \operatorname{coker} M(\rho_r) \to H_r(X) \to \ker M(\rho_{r-1}) \to 0, \qquad (4.3)$$

and the short exact sequence

$$0 \to \operatorname{coker} M(\widetilde{\rho}_r) \to H_r(\widetilde{X}) \to \ker M(\widetilde{\rho}_{r-1}) \to 0. \qquad (4.4)$$

These sequences are compatible with the truncations.

In the case of the G_{2m}-representation $\rho_r(f)$ "compatibility with truncation" means that for any pairs of critical angles (θ_i, θ_j) and $(\theta_{i'}, \theta_{j'})$ with $0 < \theta_i \leq \theta_{i'} \leq \theta_{j'} \leq \theta_j \leq 2\pi$, the diagram (4.5) is commutative. In the case of the $\mathcal{Z}$-representation $\widetilde{\rho}_r(f) = \rho(\widetilde{f})$, this means that for any pair of critical values (c_i, c_j) and $(c_{i'}, c_{j'})$ with $c_i \leq c_{i'} \leq c_{j'} \leq c_j$ the diagram (4.6) is commutative.

$$
\begin{array}{ccccccccc}
0 & \longrightarrow & \operatorname{coker} M(T_{i',j'}(\rho_r)) & \longrightarrow & H_r(X_{[\theta_{i'},\theta_{j'}]}) & \xrightarrow{\pi'} & \ker M(T_{i',j'}(\rho_{r-1})) & \longrightarrow & 0 \\
 & & \downarrow & & \downarrow{\scriptstyle v} & & \downarrow & & \\
0 & \longrightarrow & \operatorname{coker} M(T_{i,j}(\rho_r)) & \longrightarrow & H_r(X_{[\theta_i,\theta_j]}) & \xrightarrow{\pi''} & \ker M(T_{i,j}(\rho_{r-1})) & \longrightarrow & 0 \\
 & & \downarrow & & \downarrow{\scriptstyle v'} & & \downarrow & & \\
0 & \longrightarrow & \operatorname{coker} M((\rho_r)_u) & \longrightarrow & H_r(X; (\xi_f, u)) & \xrightarrow{\pi} & \ker M((\rho_{r-1})_u) & \longrightarrow & 0
\end{array}
$$

$$(4.5)$$

$$
\begin{array}{ccccccccc}
0 & \longrightarrow & \operatorname{coker} M(T_{i',j'}(\widetilde{\rho}_r)) & \longrightarrow & H_r(\widetilde{X}_{[c_{i'},c_{j'}]}) & \xrightarrow{\pi'} & \ker M(T_{i',j'}(\widetilde{\rho}_{r-1})) & \longrightarrow & 0 \\
 & & \downarrow & & \downarrow{\scriptstyle v} & & \downarrow & & \\
0 & \longrightarrow & \operatorname{coker} M(T_{i,j}(\widetilde{\rho}_r)) & \longrightarrow & H_r(\widetilde{X}_{[c_i,c_j]}) & \xrightarrow{\pi''} & \ker M(T_{i,j}(\widetilde{\rho}_{r-1})) & \longrightarrow & 0 \\
 & & \downarrow & & \downarrow{\scriptstyle v'} & & \downarrow & & \\
0 & \longrightarrow & \operatorname{coker} M(\widetilde{\rho}_r) & \longrightarrow & H_r(\widetilde{X}) & \xrightarrow{\pi} & \ker M(\widetilde{\rho}_{r-1}) & \longrightarrow & 0.
\end{array}
$$

$$(4.6)$$

Proof. Consider $f : X \to \mathbb{S}^1$ and let $\tilde{f} : \tilde{X} \to \mathbb{R}$ be an infinite cyclic cover of f. Denote by:

$$\mathcal{R} := \bigsqcup_{1 \leq i \leq m} f^{-1}(\theta_{i-1}, \theta_i), \quad \tilde{\mathcal{R}} := \bigsqcup_{i \in \mathbb{Z}} \tilde{f}^{-1}(c_{i-1}, c_i),$$

$$\mathcal{X} := \bigsqcup_{1 \leq i \leq m} f^{-1}(\theta_i), \text{ and } \tilde{\mathcal{X}} := \bigsqcup_{i \in \mathbb{Z}} \tilde{f}^{-1}(c_i).$$

Moreover, for $\epsilon < \epsilon(f)/2$ put

$$\mathcal{R}_\epsilon := \bigsqcup_{1 \leq i \leq m} f^{-1}(\theta_{i-1} + \epsilon, \theta_i - \epsilon),$$

$$\tilde{\mathcal{R}}_\epsilon := \bigsqcup_{i \in \mathbb{Z}} \tilde{f}^{-1}(c_{i-1} + \epsilon, c_i - \epsilon),$$

and note that the inclusions $\mathcal{R}_\epsilon \subset \mathcal{R}$ and $\tilde{\mathcal{R}}_\epsilon \subset \tilde{\mathcal{R}}$ are homotopy equivalences.

The short exact sequences (4.2) and (4.3) follow from the long exact sequence

$$\cdots \to H_r(\mathcal{R}) \xrightarrow{M((\rho_r)_u)} H_r(\mathcal{X}) \to H_r(X; (\xi, u)) \to H_{r-1}(\mathcal{R}) \xrightarrow{M((\rho_{r-1})_u)} H_{r-1}(\mathcal{X}) \to \cdots$$
$$(4.7)$$

and the short exact sequence (4.4) from the long exact sequence

$$\cdots \to H_r(\tilde{\mathcal{R}}) \xrightarrow{M(\rho_r)} H_r(\tilde{\mathcal{X}}) \to H_r(\tilde{X}) \to H_{r-1}(\tilde{\mathcal{R}}) \xrightarrow{M(\rho_{r-1})} H_{r-1}(\tilde{\mathcal{X}}) \to \cdots,$$
$$(4.8)$$

where

$$H_r(\mathcal{R}) = \bigoplus_{1 \leq i \leq m} H_r(f^{-1}(\theta_i, \theta_{i+1})), \quad H_r(\tilde{\mathcal{R}}) = \bigoplus_{i \in \mathbb{Z}} H_r(\tilde{f}^{-1}(c_{i-1}, c_i)),$$

$$H_r(\mathcal{X}) = \bigoplus_{1 \leq i \leq m} H_r(f^{-1}(\theta_i)), \quad H_r(\tilde{\mathcal{X}}) = \bigoplus_{i \in \mathbb{Z}} H_r(\tilde{f}^{-1}(c_i)).$$

These two exact sequences remain to be established.

Since both long exact sequences (4.7) and (4.8) are derived in the same way, we will treat only (4.7) for f an angle-valued map, and for simplicity only the case $u = 1$. We define

$$\mathcal{P}' := \bigsqcup_{1 \leq i \leq m} f^{-1}([\theta_i, \theta_{i+1} - \epsilon)), \quad \mathcal{P}'' := \bigsqcup_{1 \leq i \leq m} f^{-1}((\theta_i + \epsilon, \theta_{i+1}])$$

and observe that in view of the choice of ϵ and of the tameness of f, the inclusions $\mathcal{X} \subset \mathcal{P}'$ and $\mathcal{X} \subset \mathcal{P}''$ are homotopy equivalences and $\mathcal{X} \sqcup \mathcal{R}_\epsilon = \mathcal{P}' \cap \mathcal{P}''$.

The Mayer-Vietoris long exact sequence for $X = \mathcal{P}' \cup \mathcal{P}''$ gives the commutative diagram

$$
\begin{array}{ccc}
H_r(\mathcal{R}) & \xrightarrow{\ M(\rho_r(f))\ } & H_r(X) \\
\end{array}
$$

$$
\rightarrow H_{r+1}(\mathcal{T}) \xrightarrow{\partial_{r+1}} H_r(\mathcal{R}_\epsilon) \oplus H_r(\mathcal{X}) \xrightarrow{N} H_r(\mathcal{X}) \oplus H_r(\mathcal{X}) \xrightarrow{(i_r,-i_r)} H_r(X) \rightarrow
$$

with vertical maps pr_1, $(Id,-Id)$, in_2, Δ and bottom row

$$
H_r(\mathcal{X}) \xrightarrow{\ Id\ } H_r(\mathcal{X}).
$$

$$(4.9)$$

In the diagram above Δ denotes the diagonal, in_2 the inclusion of the second component, pr_1 the projection on $H_r(\mathcal{R}_\epsilon)$ followed by the isomorphism $H_r(\mathcal{R}_\epsilon) \to H_r(\mathcal{R})$, and i_r the linear map induced in homology by the inclusion $\mathcal{X} \subset X$. Recall that the matrix $M(\rho_r(f))$ is defined by

$$
\begin{pmatrix}
\alpha_1^r & -\beta_1^r & 0 & \cdots & & 0 \\
0 & \alpha_2^r & -\beta_2^r & \ddots & & \vdots \\
\vdots & \ddots & \ddots & \ddots & & 0 \\
0 & \cdots & 0 & \alpha_{m-1}^r & -\beta_{m-1}^r \\
-\beta_m^r & 0 & \cdots & & 0 & \alpha_m^r
\end{pmatrix},
$$

with $\alpha_i^r \colon H_r(f^{-1}(\theta_{i-1},\theta_i)) \to H_r(f^{-1}(\theta_i))$ and $\beta_i^r \colon H_r(f^{-1}(\theta_i,\theta_{i+1}) \to H_r(f^{-1}(\theta_i))$ induced by the maps a_i and b_i defined in Section 4.1. The block matrix N is defined by

$$
N := \begin{pmatrix} \alpha^r & Id \\ -\beta^r & Id \end{pmatrix},
$$

where α^r and β^r are the matrices

$$
\begin{pmatrix}
\alpha_1^r & 0 & \cdots & 0 \\
0 & \alpha_2^r & \ddots & \vdots \\
\vdots & \ddots & \ddots & 0 \\
0 & \cdots & 0 & \alpha_{m-1}^r
\end{pmatrix}
\quad \text{and} \quad
\begin{pmatrix}
0 & \beta_1^r & 0 & \cdots & 0 \\
0 & 0 & \beta_2^r & \ddots & \vdots \\
\vdots & \vdots & \ddots & \ddots & 0 \\
0 & 0 & \cdots & 0 & \beta_{m-1}^r \\
\beta_m^r & 0 & \cdots & 0 & 0
\end{pmatrix}.
$$

The exact sequence (4.7) is the top sequence in the diagram (4.9).

To derive the exact sequence (4.8) one uses $\tilde{\mathcal{R}}, \tilde{\mathcal{R}}_\epsilon, \tilde{\mathcal{X}}, \tilde{\mathcal{P}}', \tilde{\mathcal{P}}''$ instead of $\mathcal{R}, \mathcal{R}, \mathcal{X}, \mathcal{P}', \mathcal{P}''$ with

$$
\tilde{\mathcal{P}}' := \bigsqcup_{1 \leq i \leq m} \tilde{f}^{-1}\big([\theta_i, \theta_{i+1} - \epsilon)\big), \qquad
\tilde{\mathcal{P}}'' := \bigsqcup_{1 \leq i \leq m} \tilde{f}^{-1}\big((\theta_i + \epsilon, \theta_{i+1}]\big)
$$

and $\tilde{X} = \tilde{\mathcal{P}}' \cup \tilde{\mathcal{P}}''$. $\qquad\qquad\qquad\qquad\qquad\qquad\qquad\qquad$ $\square$

Compatible splittings

A splitting for a surjective linear map $\pi : A \to B$ is a linear map $s : B \to A$ such that $\pi \cdot s = id$.

Definition 4.1.

1. For a tame map $f : X \to \mathbb{R}$, the collection of splittings $\{s_{i,j}, s\}$,

$$s_{i,j} : \ker M(T_{[i,j]}(\rho_{r-1}(f)) \to H_r(X_{[c_i,c_j]}),$$
$$s : \ker M(\rho_{r-1}(f)) \to H_r(X)$$

for the projections

$$\pi_{i,j} : H_r(X_{[c_i,c_j]}) \to \ker M(T_{[i,j]}(\rho_{r-1}(f))),$$
$$\pi : H_r(X) \to \ker M(\rho_{r-1}(f))$$

is said to be collection of *compatible splittings* if for any $i' \leq i \leq j \leq j'$ the following diagram is commutative:

$$
\begin{array}{ccc}
\ker M(T_{i,j}(\rho_{r-1}(f)) & \xrightarrow{s_{i,j}} & H_r(X_{[c_i,c_j]}) \\
\downarrow & & \downarrow \\
\ker M(T_{i',j'}(\rho_{r-1}(f)) & \xrightarrow{s_{i',j'}} & H_r(X_{[c_i,c_j]}) \\
\downarrow & & \downarrow \\
\ker(M\rho_{r-1}(f)_u) & \xrightarrow{s} & H_r(X).
\end{array}
$$

2. For a tame map $f : X \to \mathbb{S}^1$ a collection of splittings $\{s_{i,j}, s\}$

$$s_{i,j} : \ker M(T_{[i,j]}(\rho_{r-1}(f)_u) \to H_r(X_{[c_i,c_j]}),$$
$$s : \ker M(\rho_{r-1}(f)_u) \to H_r(X; (\xi_f, u))$$

for the projections

$$\pi_{i,j} : H_r(X_{[c_i,c_j]}) \to \ker M(T_{[i,j]}(\rho_{r-1}(f)_u)),$$
$$\pi : H_r(X; (\xi_f, u)) \to \ker M(\rho_{r-1}(f)_u)$$

is said to be of *compatible splittings* if for any $i' \leq i \leq j \leq j'$ the following diagram is commutative:

$$\ker M(T_{i,j}(\rho_{r-1}(f)_u)) \xrightarrow{\ s_{i,j}\ } H_r(X_{[c_i,c_j]})$$

$$\downarrow \qquad\qquad\qquad\qquad \downarrow$$

$$\ker M(T_{i',j'}(\rho_{r-1}(f)_u)) \xrightarrow{\ s_{i',j'}\ } H_r(X_{[c_i,c_j]})$$

$$\downarrow \qquad\qquad\qquad\qquad \downarrow$$

$$\ker(M(\rho_{r-1}(f)_u) \xrightarrow{\ s\ } H_r(X,(\xi_f,u)).$$

Note that for $u = 1$ one has $H_r(X,(\xi_f,u)) = H_r(X)$.

In view of Proposition 3.3 in Chapter 3, for any three critical values $s_i < s_j < s_k$ the spaces $\ker(M(T_{i,j}(\rho)))$ and $\ker(M(T_{j,k}(\rho)))$ are contained in $\ker(M(\rho))$ and have intersection zero. Recall that $M(\rho)$ is defined on $\bigoplus_{r \text{ odd}} V_r$ and $M(T_{i,j}(\rho))$ on a subspace of this vector space.

In particular, for $\widetilde{\rho}_{r-1}(f)$ one has the diagram

$$0 \to \ker M(T_{i,j}(\widetilde{\rho}_{r-1})) \xrightarrow{\ s_{i,j}\ } H_r(\widetilde{X}_{[c_i,c_j]}) \xrightarrow{\ \pi_{i,j}\ } \ker M(T_{i,j}(\widetilde{\rho}_{r-1})) \to 0$$

$$\downarrow \qquad\qquad\qquad\qquad\quad \downarrow v \qquad\qquad\qquad\qquad \downarrow \subseteq$$

$$0 \to \ker M(T_{i,k}(\widetilde{\rho}_{r-1})) \xrightarrow{\ s_{i,k}\ } H_r(\widetilde{X}_{[c_i,c_k]}) \xrightarrow{\ \pi_{i,j}\ } \ker M(T_{i,k}(\widetilde{\rho}_{r-1})) \to 0$$

$$\uparrow \qquad\qquad\qquad\qquad\quad \uparrow v \qquad\qquad\qquad\qquad \uparrow \subseteq$$

$$0 \to \ker M(T_{j,k}(\widetilde{\rho}_{r-1})) \xrightarrow{\ s_{j,k}\ } H_r(\widetilde{X}_{[c_j,c_k]}) \xrightarrow{\ \pi_{j,k}\ } \ker M(T_{j,k}(\widetilde{\rho}_{r-1})) \to 0$$

$$(4.10)$$

Therefore one has

Observation 4.4. A splitting $s_{i,j} : \ker M(T_{i,j}(\widetilde{\rho}_{r-1})) \to H_r(\widetilde{X}_{[c_i,c_j]})$ of the projection $\pi_{i,j} : H_r(\widetilde{X}_{[c_i,c_j]}) \to \ker M(T_{i,j}(\widetilde{\rho}_{r-1}))$ and a splitting $s_{j,k} : \ker M(T_{i,j}(\widetilde{\rho})) \to H_r(\widetilde{X}_{[c_j,c_k]})$ of the projection $\pi_{j,k} : H_r(\widetilde{X}_{[c_j,c_k]}) \to \ker M(T_{i,j}(\widetilde{\rho}_{r-1}))$ define a splitting $s_{i,k} : \ker M(Ti,k(\widetilde{\rho}_{r-1})) \to H_r(\widetilde{X}_{[c_i,c_k]})$ of the projection $\pi_{i,k} : H_r(\widetilde{X}_{[c_i,c_k]}) \to \ker M(T_{i,k}(\widetilde{\rho}_{r-1}))$ which makes the diagram (4.10) commutative.

The same remains true for the representation $\rho_r(f)$, f angle-valued, when c_i, c_j, c_k are replaced by $\theta_i, \theta_j, \theta_k$.

As a consequence we can state the following proposition

Proposition 4.2. *There exist collections $\{s_{i,j}, s\}$ of compatible splittings.*

The construction of compatible splittings is done inductively on the difference $j - i$, as follows. If one gives the splittings $s_{i,i+1}$ for any two consecutive critical values $s_i < s_{i+1}$, one constructs inductively $s_{i,j}$ from $s_{i,i+1}, s_{i+1,i+2}, \ldots, s_{j-1,j}$ based on Observation 4.4. One can do this backwards too, namely construct inductively $s_{i-k,i}$ from $s_{i-1,i}, s_{i-2,i-1}, \ldots, s_{i-k,i-k+1}$. If the collection of critical values is infinite one construct inductively $s_{i-n,i+n}$ and passing to limit on $n \to \infty$ one defines s.

In case of the infinite cyclic cover $\widetilde{f}$ one uses the construction only for $1 \leq i \leq m$ and uses the translation by m units for others, in order to achieve periodicity. For more details one can consult [Burghelea, D., Haller, S. (2015)] section 4.

4.3 Barcodes, Jordan cells, and homology

For $f : X \to \mathbb{R}$ and $f : X \to \mathbb{S}^1$ tame maps with X compact, as well as for $\widetilde{f} : \widetilde{X} \to \mathbb{R}$ an infinite cyclic cover of $f : X \to \mathbb{S}^1$ and for any closed interval K in $\mathbb{S}^1$ or $\mathbb{R}$, one can describe

(1) the κ-vector spaces $H_r(X, (\xi_f, u))$, $H_r(f^{-1}(K))$, $H_r(\widetilde{f}^{-1}(K))$,
(2) the $\kappa[t^{-1}, t]$-module $H_r(\widetilde{X})$, and
(3) for $K' \subseteq K$, the inclusion-induced linear maps $H_r(f^{-1}(K)) \to H_r(f^{-1}(K'))$ and $H_r(\widetilde{f}^{-1}(K)) \to H_r(\widetilde{f}^{-1}(K'))$

in terms of barcode and Jordan cells. The description is given in Propositions 4.3, 4.4, 4.5 and ultimately in Theorem 4.1 below. Theorem 4.1 is a consequence of Propositions 4.3, 4.4, 4.5 as explained below and, in turn, these propositions refine the statements of Theorem 4.1.

Recall that for the pair $(X, \xi \in H^1(X; \mathbb{Z}))$ with X a compact ANR and $\pi : \widetilde{X} \to X$ an infinite cyclic cover associated to ξ, the κ-vector space $H_r(\widetilde{X})$ is a f.g $\kappa[t^{-1}, t]$-module. Its torsion part $V_r(X; \xi) := \operatorname{Tor} H_r(\widetilde{X})$, is a finite-dimensional κ−vector space equipped with an isomorphism $T_r : V_r(X, \xi) \to V_r(X, \xi)$ given by multiplication by t, referred to as the homological r-monodromy. The module $H_r^N(X; \xi) := H_r(\widetilde{X})/\operatorname{Tor} H_r(\widetilde{X})$ is a free $\kappa[t^{-1}, t]$-module of rank $\beta_r^N(X; \xi)$, the r-th Novikov-Betti number. This number is the same as the dimension of the $\kappa[t^{-1}, t]]$-vector space $H_r(\widetilde{X}) \otimes_{\kappa[t^{-1}, t]} \kappa[t^{-1}, t]]$, where $\kappa[t^{-1}, t]]$ is the field of Laurent power series.

Theorem 4.1. (cf. [Burghelea, D., Haller, S. (2015)]) *If $f \colon X \to \mathbb{S}^1$ is a tame map and $\xi_f \in H^1(X; \mathbb{Z})$ is the integral cohomology class represented by f, then:*

(1) $\quad \beta_r^N(X, \xi_f) = \operatorname{rank} H_r^N(X; \xi_f) = \sharp \mathcal{B}_r^c(f) + \sharp \mathcal{B}_{r-1}^o(f).$

(2) *The collection $\mathcal{J}_r(f)$ is a homotopy invariant of the pair (X, ξ_f). More precisely, $\bigoplus_{J \in \mathcal{J}_r(f)} J$ is isomorphic to the homological r-monodromy of (X, ξ_f) (as defined in Chapter 2, Subsection 2.3.6, and reviewed below).*

(3) *If $u \in \kappa \setminus 0$, then*
$\dim H_r(X, (\xi_f, u)) = \sharp \mathcal{B}_r^c(f) + \sharp \mathcal{B}_{r-1}^o(f) + \sharp \mathcal{J}_{r,u}(f) + \sharp \mathcal{J}_{r-1,u}(f);$
in particular,
$\dim H_r(X) = \sharp \mathcal{B}_r^c(f) + \sharp \mathcal{B}_{r-1}^o(f) + \sharp \mathcal{J}_{r,1}(f) + \sharp \mathcal{J}_{r-1,1}(f).$

Theorem 4.1 contains also the case of f a real-valued tame map with X compact by regarding $\mathbb{R}$ embedded in $\mathbb{S}^1$. In this case $\mathcal{J}_r(f) = \emptyset$.

For the proof of Theorem 4.1 some preliminaries are necessary.

For $f : X \to \mathbb{R}$ tame map, X compact, and K a closed interval in $\mathbb{R}$, we introduce the sets

(1) $\mathcal{S}_r(f) := \mathcal{B}_r^c(f) \sqcup \mathcal{B}_{r-1}^o(f)$ and

(2) $\mathcal{S}_{r,K}(f) := \begin{cases} I \in \; \mathcal{B}_r(f) \mid I \cap K \text{closed} \neq \emptyset, \\ I \in \; \mathcal{B}_{r-1}^o(f) \mid I \subset K. \end{cases}$

The same notation is used when $f : X \to \mathbb{S}^1$ is a tame map and K is a closed interval contained in $\mathbb{S}^1$. These sets are finite.

For $f : X \to \mathbb{S}^1$ a tame map, $\tilde{f} : \tilde{X} \to \mathbb{R}$ an infinite cyclic cover of f, and K a closed interval in $\mathbb{R}$, denote

(1) $\mathcal{S}_r(\tilde{f}) := \mathcal{B}_r^c(\tilde{f}) \sqcup \mathcal{B}_{r-1}^o(\tilde{f}),$

(2) $\overline{\mathcal{S}}_r(\tilde{f}) := \mathcal{B}_r^c(\tilde{f}) \sqcup \mathcal{B}_{r-1}^o(\tilde{f}) \sqcup \mathcal{J}_r(\tilde{f})$

(3) $\mathcal{S}_{r,K}(\tilde{f}) := \begin{cases} I \in \; \mathcal{B}_r(\tilde{f}) \mid I \cap K \text{ closed} \neq \emptyset \\ I \in \; \mathcal{B}_{r-1}^o(\tilde{f}) \mid I \subset K \end{cases},$

(4) $\overline{\mathcal{S}}_{r,K}(\tilde{f}) := \mathcal{S}_{r,K}(\tilde{f}) \sqcup \mathcal{J}_r(\tilde{f}).$

Recall that for a set S one denotes by $\kappa[S]$ the vector space generated by S, and that for two subsets $S_1, S_2 \subset S$ the *canonical linear map* $\kappa[S_1] \to \kappa[S_2]$ is the linear extension of the map which sends $S_1 \setminus S_2$ into zero and the elements of $S_1 \cap S_2$ into themselves.

The translation of barcodes by 2π defines a free $\mathbb{Z}$-action on $\mathcal{B}_r^c(\widetilde{f})$ and $\mathcal{B}_{r-1}^o(\widetilde{f})$ which provides a structure of a free $\kappa[t^{-1}, t]$-module on $\kappa[\mathcal{S}_r(\widetilde{f})]$. The isomorphism $T_r : \kappa[\mathcal{J}_r(\widetilde{f})] \to \kappa[\mathcal{J}_r(\widetilde{f})]$ described in Chapter 3, in view of the identification of the set $\mathcal{J}_r(\widetilde{f})$ with $\mathcal{J}(\rho_r(\widetilde{f}))$, defines a structure of $\kappa[t^{-1}, t]$-module on $\kappa[\mathcal{J}_r(\widetilde{f})]$.

Therefore, $\kappa[\overline{\mathcal{S}}_r(\widetilde{f})] = \kappa[\mathcal{S}_r(\widetilde{f})] \oplus \kappa[\mathcal{J}_r(\widetilde{f})]$ is a f.g. $\kappa[t^{-1}, t]$-module with torsion submodule $\kappa[\mathcal{J}_r(\widetilde{f})]$ and free part $\kappa[\mathcal{S}_r(\widetilde{f})]$.

Proposition 4.3. *For a tame map $f : X \to \mathbb{R}$, X compact, any $r \geq 0$ and any closed interval $K \subset \mathbb{R}$, the choice of a decomposition into barcodes of each $\rho_r(f)$ and of compatible splittings $\{s_{i,j}, s\}$ induces the isomorphisms*

$$\omega_r : \kappa[\mathcal{S}_r(f)] \to H_r(X)$$

and

$$\omega_{r,K} : \kappa[\mathcal{S}_{r,K}(f)] \to H_r(f^{-1}(K))$$

such that for any $K' \subset K$ closed intervals in $\mathbb{R}$ the diagram (4.11) is commutative

$$
\begin{array}{ccccc}
H_r(f^{-1}(K')) & \longrightarrow & H_r(f^{-1}(K)) & \longrightarrow & H_r(X) \\
\uparrow{\scriptstyle\omega_{r,K'}} & & \uparrow{\scriptstyle\omega_{r;K}} & & \uparrow{\scriptstyle\omega_r} \\
\kappa[\mathcal{S}_{r,K'}(f)] & \longrightarrow & \kappa[\mathcal{S}_{r,K}(f)] & \longrightarrow & \kappa[\mathcal{S}_r]
\end{array}
\qquad (4.11)
$$

In the above diagram the top arrows are inclusion induced linear maps and the bottom arrows are canonical linear maps.

Proposition 4.4. *For a tame map $f : X \to \mathbb{S}^1$ with $\widetilde{f} : \widetilde{X} \to \mathbb{R}$ an infinite cyclic cover of f, any $r \geq 0$, and any closed interval $K \subset \mathbb{R}$, the choice of a decomposition into barcodes and Jordan blocks of each $\rho_r(f)$ with a base for $V_r(f)$ and of compatible splittings $\{s_{i,j}\}$ induces the isomorphisms*

$$\widetilde{\omega}_r : \kappa[\overline{\mathcal{S}}_r(\widetilde{f})] \to H_r(\widetilde{X})$$

and

$$\widetilde{\omega}_{r,K} : \kappa[\overline{\mathcal{S}}_{r,K}(\widetilde{f})] \to H_r(\widetilde{f}^{-1}(K))$$

such that:

(1) $\widetilde{\omega}_r$ is an isomorphism of $\kappa[t^{-1}, t]$-modules,

(2) for any $K' \subset K$ closed intervals in $\mathbb{S}^1$, the diagram

$$H_r(\widetilde{f}^{-1}(K')) \longrightarrow H_r(\widetilde{f}^{-1}(K)) \longrightarrow H_r(\widetilde{X}) \qquad (4.12)$$

$$\widetilde{\omega}_{r,K'} \uparrow \qquad\qquad \widetilde{\omega}_{r;K} \uparrow \qquad\qquad \widetilde{\omega}_r \uparrow$$

$$\kappa[\overline{S}_{r,K'}(\widetilde{f})] \longrightarrow \kappa[\overline{S}_{r,K}(\widetilde{f})] \longrightarrow \kappa[\overline{S}_r]$$

is commutative. Here the top arrows are inclusion-induced linear maps and the bottom arrows are the canonical linear maps.

Proposition 4.3 determines the Betti numbers of a compact ANR in terms of the closed and open barcodes of a tame real-valued map $f : X \to \mathbb{R}$, precisely

$$\beta_r(X) = \sharp \mathcal{B}_r^c(f) + \sharp \mathcal{B}_{r-1}^o(f).$$

It also calculates the dimension of $H_r(f^{-1}(K))$ and the dimensions of the kernel and the cokernel of the inclusion-induced linear map

$$i_{K'}^K(r) : H_r(f^{-1}(K')) \to H_r(f^{-1}(K)), \ K' \subseteq K$$

providing the equalities

$$\dim H_r(f^{-1}(K)) = \sharp S_{r,K}(f),$$
$$\dim \ker i_{K'}^K = \sharp(S_{r,K'}(f) \setminus S_{r,K}(f)),$$
$$\dim \operatorname{coker} i_{K'}^K = \sharp(S_{r,K}(f) \setminus S_{r,K'}(f)).$$

Proposition 4.4 verifies items (1) and (2) in Theorem 4.1. Precisely, it calculates the Novikov-Betti number of (X, ξ_f), where $f : X \to \mathbb{S}^1$ is a tame map, since $\beta_r^N(X, \xi)$ is the rank of $H_r(\widetilde{X})$, and establishes that this rank is equal to $\sharp(\mathcal{B}_r^c(f)) + \sharp(\mathcal{B}_{r-1}^o(f))$. It also identifies the homological r-monodromy, the $\kappa[t^{-1}, t]$-torsion of $H_r(\widetilde{X})$, to the direct sum of the Jordan blocks in $\mathcal{J}_r(f)$.

Proposition 4.5. *For a tame map $f : X \to \mathbb{S}^1$ and $u \in \kappa \setminus 0$, the choice of a decomposition into barcodes and Jordan blocks of each $\rho_r(f)$ with a base for $V_r(f)$ and of a collection of compatible splittings $\{s_{i,j}, s\}$ of the surjective maps*

$$\pi_{i,j} : H_r(X_{[c_i, c_j]}) \to \ker M(T_{[i,j]}(\rho_{r-1}(f)_u)),$$
$$\pi : H_r(X) \to \ker M(\rho_{r-1}(f)_u)$$

provide an isomorphism

$$\widetilde{\omega}_{r,u} : \kappa[S_r(f) \sqcup \mathcal{J}_{r,u}(f) \sqcup \mathcal{J}_{r-1,u}(f)] \to H_r(X, (\xi_f, u)).$$

This proposition establishes Item (3) in Theorem 4.1.

Proof. (Proof of Propositions 4.3, 4.4 and 4.5) In view of Observation 4.3, it suffices to treat the case when the ends of the intervals K and K' are critical values.

The choices of compatible splittings convert the diagrams (1.5) and (1.6) into the diagrams

$$0 \twoheadrightarrow \operatorname{coker} M(T_{i',j'}(\rho_r(f))) \oplus \ker M(T_{i',j'}(\rho_{r-1}(f))) \to H_r(X_{[\theta_{i'},\theta_{j'}]}) \twoheadrightarrow 0$$

$$0 \to \operatorname{coker} M(T_{i,j}(\rho_r(f))) \oplus \ker M(T_{i,j}(\rho_{r-1}(f))) \longrightarrow H_r(X_{[\theta_i,\theta_j]}) \twoheadrightarrow 0$$

$$0 \longrightarrow \operatorname{coker} M((\rho_r(f))_u) \oplus \ker M((\rho_{r-1}(f))_u) \longrightarrow H_r(X; (\xi_f, u))] \twoheadrightarrow 0$$
(4.13)

and

$$0 \twoheadrightarrow \operatorname{coker} M(T_{i',j'}(\widetilde{\rho}_r(f))) \oplus \ker M(T_{i',j'}(\widetilde{\rho}_{r-1}(f))) \twoheadrightarrow H_r(\widetilde{X}_{[c_{i'},c_{j'}]}) \overset{\pi'}{\twoheadrightarrow} 0$$

$$0 \to \operatorname{coker} M(T_{i,j}(\widetilde{\rho}_r)(f)) \oplus \ker M(T_{i,j}(\widetilde{\rho}_{r-1}(f))) \longrightarrow H_r(\widetilde{X}_{[c_i,c_j]}) \twoheadrightarrow 0$$

$$0 \longrightarrow \operatorname{coker} M(\widetilde{\rho}_r(f)) \oplus \ker M(\widetilde{\rho}_{r-1}(f)) \longrightarrow H_r(\widetilde{X}) \longrightarrow 0.$$
(4.14)

The choice of a decomposition of the representations $\rho_r(f)$ into indecomposable representations identifies, by Proposition 3.3 in Chapter 3, $\ker M(\rho_{...}(\cdots))$ and $\operatorname{coker} M(\rho_{...}(\cdots))$ with the vector spaces generated by barcodes and Jordan blocks as specified by Propositions 4.3, 4.4, 4.5. The base for $V_r(f)$ provides the components $\rho_\infty^{\mathcal{Z}}$ in the decomposition of $\widetilde{\rho}_r(f)$.

For the proof of Proposition 4.3 one takes $K' = [c'_i, c'_j]$ and $K = [c_i, c_j]$, while for Proposition 4.4 $K' = [\theta'_i, \theta'_j]$ and $K = [\theta_i, \theta_j]$, and one uses the truncations $T_{i',j'}$ and $T_{i,j}$ and the diagrams (4.13) and (4.14). As long as Proposition 4.4 is concerned, one uses only the bottom line isomorphism in the diagram (4.14).

In all cases the isomorphisms ω's are the isomorphisms provided by the horizontal arrows in the diagrams (4.13) and (4.14) composed with the isomorphism between the vector spaces generated by the barcodes and the Jordan blocks given by Proposition 3.3 in Chapter 3.

For Item (1) in Proposition 4.4 one simply observes that the deck transformation τ on $\widetilde{X}$ induces exactly the translation by 2π of the barcodes of $\rho(\widetilde{f})$ and exactly the linear map T_r on the vector space $\kappa[\mathcal{J}_r(\widetilde{f})]$. $\qquad \square$

Let $f : X \to \mathbb{R}$ be a tame map and for any two closed intervals K_1, K_2, $K_1 \subseteq K_2 \subseteq \mathbb{R}$, denote by

$$\operatorname{coker}_r(K_1, K_2) := \operatorname{coker}(H_r(f^{-1}(K_1) \to H_r(f^{-1}(K_2)))),$$
$$\ker_r(K_1, K_2) := \ker(H_r(f^{-1}(K_1) \to H_r(f^{-1}(K_2))).$$

A straightforward consequence of Propositions 4.3 and 4.4, is Corolloray 4.1 below which calculates in terms of barcodes $\ker_r(K_1, K_2)$ and $\operatorname{coker}_r(K_1, K_2)$ for two closed intervals $K_1 \subseteq K_2$, and the composition of linear maps $\ker_r(K_2, K_3) \to H_r(f^{-1}(K_2)) \to \operatorname{coker}_r(K_1, K_2)$ for three closed intervals $K_1 \subseteq K_2 \subseteq K_3$. As a consequence, one can express in terms of barcodes the homology long exact sequence of the pair $f^{-1}(K_2), f^{-1}(K_1)$ and the homology long exact sequence of the triple $f^{-1}(K_3), f^{-1}(K_2), f^{-1}(K_1)$, see Exercise E. 7 in Section 4.6.

To formulate this corollary it is convenient to use some additional notations.

For a pair of closed intervals $K_1 \subseteq K_2$ in $\mathbb{R}$ denote

$$\mathcal{L}'_{r,K_1,K_2}(f) := \mathcal{S}_{r,K_2}(f) \setminus \mathcal{S}_{r,K_1}(f)$$

and

$$\mathcal{L}''_{r,K_1,K_2}(f) := \mathcal{S}_{r,K_1}(f) \setminus \mathcal{S}_{r,K_2}(f).$$

Corollary 4.1.

(1) *For any two closed intervals $K_1 \subseteq K_2$ the above choices (cf. Propositions 4.3 and 4.4) induce the isomorphisms*

$$\begin{aligned} \omega'_{r,K_1,K_2} &: \kappa[\mathcal{L}'_{r,K_1,K_2}(f)] \to \operatorname{coker}_r(K_1, K_2), \\ \omega''_{r,K_1,K_2} &: \kappa[\mathcal{L}''_{r,K_1,K_2}(f)] \to \ker_r(K_1, K_2). \end{aligned} \tag{4.15}$$

(2) *For any three closed intervals $K_1 \subseteq K_2 \subseteq K_3$, the diagram*

$$\begin{array}{ccc}
\ker_r(K_2,K_3) & & \mathrm{coker}_r(K_1,K_2) \\
\Big\uparrow \omega''_{r,K_2,K_3} & \overset{\subseteq}{\searrow}\quad H_r(f^{-1}(K_2))\quad \overset{\subseteq}{\nearrow} & \Big\uparrow \omega'_{r,K_1,K_2} \\
\kappa[\mathcal{L}''_{r,K_2,K_3}(f)] & \xrightarrow{\quad\partial'_r\quad} & \kappa[\mathcal{L}'_{r,K_1,K_2}(f)]
\end{array}$$

is commutative with the bottom arrow the canonical linear map defined when both $\mathcal{L}''_{r,K_2,K_3}(f)$ and $\mathcal{L}'_{r,K_1,K_2}(f)$ are viewed as subsets of $\mathcal{B}_r(f) \sqcup \mathcal{B}_{r-1}(f) = \mathcal{S}_r(f)$.

The linear map ∂'_r is the relevant part of the boundary homomorphisms $\partial_{r+1} : H_{r+1}(f^{-1}(K_3)), f^{-1}(K_2)) \to H_r(f^{-1}(K_2)), f^{-1}(K_1))$ in the homology long exact sequence of the triple $f^{-1}(K_1) \subseteq f^{-1}(K_2) \subseteq f^{-1}(K_3)$; more precisely, ∂_{r+1} decomposes as

$$\begin{array}{ccc}
H_{r+1}(f^{-1}(K_3)), f^{-1}(K_2) & \xrightarrow{\partial_{r+1}} & H_r(f^{-1}(K_2)), f^{-1}(K_1)) \\
\Big\downarrow & & \Big\uparrow \\
\kappa[\mathcal{L}''_{r,K_2,K_3}(f)] & \xrightarrow{\quad\partial_r{}'\quad} & \kappa[\mathcal{L}'_{r,K_2,K_3}(f)]
\end{array}$$

with the left vertical arrow surjective and the right vertical arrow injective. The arrow ∂'_r is a canonical linear map, nontrivial only on the barcodes in $\mathcal{L}''_{r,K_2,K_3}(f) \cap \mathcal{L}'_{r,K_1,K_2}(f)$, which involves only mixed barcodes. This is because :

$$\mathcal{L}'_{r,K_1,K_2} := \begin{cases}
\mathcal{B}^c_r \ni I = [c',c''] \mid I \cap K_2 \neq \emptyset, I \cap K_1 = \emptyset, \\
\mathcal{B}^o_{r-1} \ni I = (c',c'') \mid I \subset K_2, I \not\subseteq K_1, \\
\mathcal{B}^{c,o}_r \ni I = [c',c'') \mid K_1 < c' \in K_2, c'' > K_2, \\
\mathcal{B}^{o,c}_r \ni I = (c',c''] \mid c' < K_2, K_2 \ni c'' < K_1,
\end{cases}$$

where $c < K$ means $c < x$ for any $x \in K$ and $c > K$ means $c > x$ for any $x \in K$, respectively, and

$$\mathcal{L}''_{r,K_1,K_2} := \begin{cases}
\mathcal{B}^{c,o}_r \ni I = [c',c'') \mid c' \in K_1, c'' \in K_2 \setminus K_1, \\
\mathcal{B}^{o,c}_r \ni I = (c',c''] \mid c'' \in K_1, c' \in K_2 \setminus K_1.
\end{cases}$$

As a general observation the reader should remain aware that $\ker_r(K_1,K_2)$ involves only mixed bar codes.

4.3.1 *Two examples*

Example 1 (real-valued tame map).

Consider the map $f : X \to \mathbb{R}$ in Fig. 4.6 below, with critical values $c_0 < c_1 < c_2 < \cdots < c_5 < c_6$, and choose $t_i \in (c_{i-1}, c_i)$ as indicated.

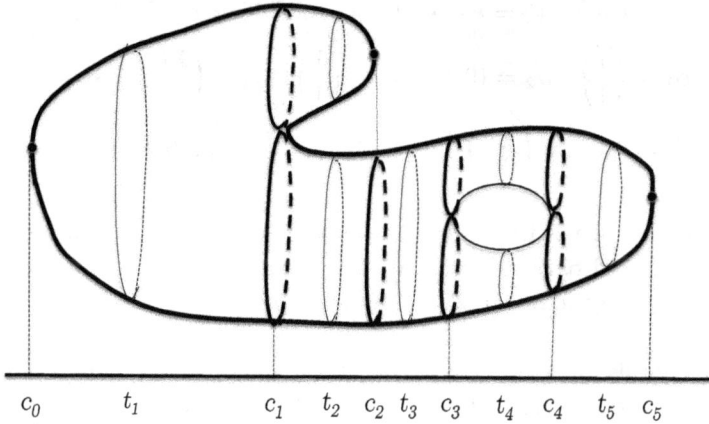

Fig. 4.6   Example of a real-valued map

The representations ρ_r is given by the diagram

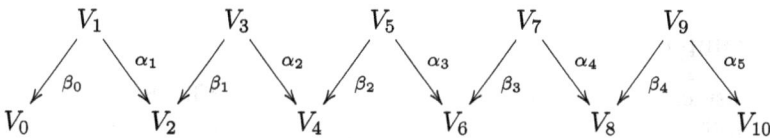

with $V_{2i-1} := H_r(f^{-1}(t_i))$ and $V_{2i} = H_r(f^{-1}(c_i))$. Precisely:

The *representation* ρ_0 is given by:

(i) $V_0 = \kappa,\ V_1 = \kappa,\ V_2 = \kappa,\ V_3 = \kappa \oplus \kappa,\ V_4 = \kappa \oplus \kappa,\ V_5 = \kappa,$
$V_6 = \kappa,\ V_7 = \kappa \oplus \kappa,\ V_8 = \kappa \oplus \kappa,\ V_9 = \kappa,\ V_{10} = \kappa;$

(ii) $\alpha_1 = \begin{pmatrix} 1 \end{pmatrix},\ \alpha_2 = \begin{pmatrix} 1 \end{pmatrix},\ \alpha_3 = \begin{pmatrix} 1 \end{pmatrix},\ \alpha_4 = \begin{pmatrix} 1 \\ 1 \end{pmatrix},\ \alpha_5 = \begin{pmatrix} 1 \end{pmatrix};$

(iii) $\beta_0 = (1)$, $\beta_1 = (1\ 1)$, $\beta_2 = \begin{pmatrix} 0 \\ 1 \end{pmatrix}$, $\beta_3 = \begin{pmatrix} 1 \\ 1 \end{pmatrix}$, $\beta_4 = (1\ 1)$, $\beta_5 = (1)$.

The *representation* ρ_1 is given by:

(i) $V_0 = 0$, $V_1 = \kappa$, $V_2 = \kappa \oplus \kappa$, $V_3 = \kappa \oplus \kappa$, $V_4 = \kappa$, $V_5 = \kappa$,
$V_6 = \kappa \oplus \kappa$, $V_7 = \kappa \oplus \kappa$, $V_8 = \kappa \oplus \kappa$, $V_9 = \kappa$, $V_{10} = 0$;

(ii) $\alpha_1 = \begin{pmatrix} 1 \\ 1 \end{pmatrix}$, $\alpha_2 = (0\ 1)$, $\alpha_3 = \begin{pmatrix} 1 \\ 1 \end{pmatrix}$, $\alpha_4 = \begin{pmatrix} 1 & 0 \\ 0 & 1 \end{pmatrix}$, $\alpha_5 = 0$;

(iii) $\beta_0 = 0$, $\beta_1 = \begin{pmatrix} 1 & 0 \\ 0 & 1 \end{pmatrix}$, $\beta_2 = (1)$ $\beta_3 = \begin{pmatrix} 1 & 0 \\ 0 & 1 \end{pmatrix}$, $\beta_4 = \begin{pmatrix} 1 \\ 1 \end{pmatrix}$.

One can recognize from Fig. 4.6 that $\rho_0(f)$ is a sum of three indecomposables: the closed barcode $\rho^\mathcal{Z}([0,5])$, the open barcode $\rho^\mathcal{Z}((3,4))$, and the open-closed barcode $\rho^\mathcal{Z}((1,2])$. They lead to the intervals $[c_0, c_5]$, (c_3, c_4), $(c_1, c_2]$.

One can also recognize from Fig. 4.6 that $\rho_1(f)$ is a sum of three indecomposables: the closed barcode $\rho^\mathcal{Z}([3,4])$, the open barcode $\rho^\mathcal{Z}((0,5))$, and the closed-open barcode $\rho^\mathcal{Z}([1,2))$. They lead to the intervals $[c_3, c_4]$, (c_0, c_5), $[c_1, c_2)$.

From this, as an application of Theorem 4.1, it follows that $\beta_0(X) = 1$, $\beta_1 = 2$, $\beta_3 = 1$.

By applying Propositions 4.3 one can also calculate the Betti numbers of $f^{-1}((-\infty, (c_1 + c_2)/2])$ and conclude that $\beta_0 = 1$, $\beta_1 = 1$, $\beta_2 = 0$.

Example 2 (angle-valued tame map)

Consider the space X obtained from Y described in the picture below by identifying its right end Y_1 (a union of three circles) to the left end Y_0 (a union of three circles) following the map $\phi \colon Y_1 \to Y_0$ given by the matrix

$$\begin{pmatrix} 1 & 0 & 0 \\ 0 & 2 & 0 \\ 0 & -3 & 2 \end{pmatrix}.$$

The meaning of this matrix as a map ϕ is the following: circle (1) wraps clockwise around circle (1), half of circle (2) covers twice circle (2) and the next half covers counterclockwise three times circle (3), while circle (3) covers clockwise twice circle (3).

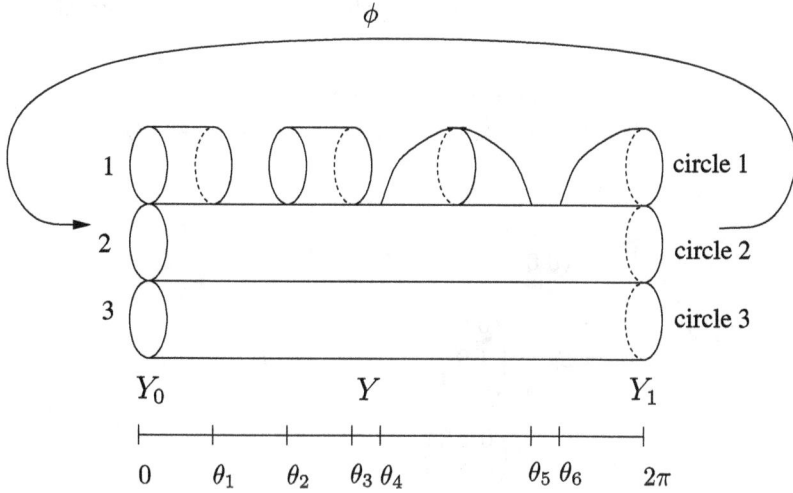

Fig. 4.7 Example of an angle-valued map

The *critical angles* of f are $\{0, \theta_1, \theta_2, \theta_3, \theta_4, \theta_5, \theta_6\}$ and in order to describe the representations $\rho_0(f)$ and $\rho_1(f)$ one chooses the regular values t_i, $1 \leq i \leq 7$, such that

$$0 < t_1 < \theta_1 < t_2 < \theta_2 < t_3 < \theta_3 < t_4 < \theta_4 < t_5 < \theta_5 < t_6 < \theta_6 < t_7 < 2\pi.$$

The *representation* ρ_0 is given by $\rho_0(f) := \{V_r = \kappa, \alpha_i = \beta_i = Id\}$, so in dimension zero there are no barcodes and one Jordan cell $(1; 1)$.

The *representation* ρ_1 is given by $\rho_1 = \{V_r, 1 \leq r \leq 14, \alpha_i, \beta_i, 1 \leq i \leq 7\}$, with

$V_2 = V_4 = V_6 = V_{14} = \kappa^3$, generated by the cycles represented by the circles (1), (2), (3),

$V_8 = V_{10} = V_{12} = \kappa^2$, generated by the cycles represented by the circles (2) and (3),

$V_1 = V_5 = V_9 = V_{13} = \kappa^3$, generated by the cycles represented by the circles (1), (2), (3),

$V_3 = V_7 = V_{11} = \kappa^2$, generated by the cycles represented by the circles (2) and (3).

In terms of this description of V_r, the linear maps α_i, β_i are given by:

$$\alpha_1 = \alpha_3 = Id_3, \quad \alpha_2 = \begin{pmatrix} 0 & 0 \\ 1 & 0 \\ 0 & 1 \end{pmatrix}, \quad \alpha_4 = \alpha_6 = Id_2,$$

$$\alpha_5 = \begin{pmatrix} 0 & 1 & 0 \\ 0 & 0 & 1 \end{pmatrix}, \quad \alpha_7 = \begin{pmatrix} 1 & 0 & 0 \\ 0 & 2 & 0 \\ 0 & -3 & 2 \end{pmatrix}$$

$$\beta_1 = \beta_3 = \begin{pmatrix} 0 & 0 \\ 1 & 0 \\ 0 & 1 \end{pmatrix}, \quad \beta_2 = \beta_7 = Id_3, \quad \beta_5 = Id_2,$$

$$\beta_4 = \beta_6 = \begin{pmatrix} 0 & 1 & 0 \\ 0 & 0 & 1 \end{pmatrix}.$$

It is not hard to recognize from the picture the decomposition of ρ_1 into three barcodes

$$[\theta_2, \theta_3], \quad (\theta_4, \theta_5), \quad (\theta_6, \theta_1 + 2\pi),$$

and the Jordan block
$$\alpha_1 = \begin{pmatrix} 2 & 0 \\ -3 & 2 \end{pmatrix}, \text{ equivalently the Jordan cell } (2; 2).$$
However, the reader can go back to Chapter 3 Subsection 3.3.2, where this decomposition is derived using the algorithm proposed in [Burghelea, D., Dey, T (2013)] and reproduced in that chapter.

4.4　Barcodes and Borel-Moore homology

Let $f : X \to \mathbb{S}^1$ be a tame map, $\widetilde{f} : \widetilde{X} \to \mathbb{R}$ an infinite cyclic cover of f.

In this subsection one denotes $\widetilde{X}_a = \widetilde{f}^{-1}((-\infty, a])$, $\widetilde{X}^a = \widetilde{f}^{-1}([a, \infty))$. All spaces $\widetilde{X}, \widetilde{X}_a, \widetilde{X}^a$ are locally compact ANRs.

One can express the Borel-Moore homology of $\widetilde{X}, \widetilde{X}_a, \widetilde{X}^a$, the relative Borel-Moore homologies for pairs of such spaces, the inclusion-induced linear maps $H_r^{BM}(\widetilde{X}_a)) \to H_r^{BM}(\widetilde{X}_b)$, $a < b$, $H_r^{BM}(\widetilde{X}^b) \to H_r^{BM}(\widetilde{X}^a)$, $a < b$, as well as their kernels and images, in terms of barcodes.

As pointed out in Chapter 2, Subsection 2.3.6, for the Borel-Moore homology of the spaces $\widetilde{X}$, $\widetilde{X}_a$, and $\widetilde{X}^a$ one has the isomorphisms

$$H_r^{\mathrm{BM}}(\widetilde{X}) = \varprojlim_{0<l\to\infty} H_r(\widetilde{X}, \widetilde{X}_{-l} \sqcup \widetilde{X}^l),$$

$$H_r^{\mathrm{BM}}(\widetilde{X}_a) = \varprojlim_{0<l\to\infty} H_r(\widetilde{X}_a, \widetilde{X}_{a-l}), \qquad (4.16)$$

$$H_r^{\mathrm{BM}}(\widetilde{X}^a) = \varprojlim_{0<l\to\infty} H_r(\widetilde{X}^a, \widetilde{X}^{a+l}).$$

To relate them with barcodes one considers the commutative diagram (4.17) whose rows are the long exact sequences of the pairs $(\widetilde{X}_a, \widetilde{X}_{-l})$, $(\widetilde{X}, \widetilde{X}_{-l} \sqcup \widetilde{X}^l)$, $(\widetilde{X}^b, \widetilde{X}^l)$ for a triple of real numbers (l, a, b) with $-l < a$ and $b < l$; in this diagram the vertical arrows are the linear maps induced by the inclusions of pairs

$$(\widetilde{X}_a, \widetilde{X}_{-l}) \subset (\widetilde{X}, \widetilde{X}_{-l} \sqcup \widetilde{X}^l) \supset (\widetilde{X}^b, \widetilde{X}^l).$$

$$
\begin{array}{ccccccccc}
\cdots \longrightarrow & H_r(\widetilde{X}_{-l}) & \xrightarrow{i_{-l,a}(r)} & H_r(\widetilde{X}_a) & \longrightarrow & H_r(\widetilde{X}_a, \widetilde{X}_{-l}) & \longrightarrow & H_{r-1}(\widetilde{X}_{-l}) & \longrightarrow \cdots \\
& \downarrow & & \downarrow & & \downarrow & & \downarrow & \\
\cdots \dashrightarrow & H_r(\widetilde{X}_{-l} \sqcup \widetilde{X}^l) & \xrightarrow{i^l_{-l}(r)} & H_r(\widetilde{X}) & \twoheadrightarrow & H_r(\widetilde{X}, \widetilde{X}_{-l} \sqcup \widetilde{X}^l) & \twoheadrightarrow & H_{r-1}(\widetilde{X}_{-l} \sqcup \widetilde{X}^l) & \twoheadrightarrow \cdots \\
& \uparrow & & \uparrow & & \uparrow & & \uparrow & \\
\cdots \longrightarrow & H_r(\widetilde{X}^l) & \xrightarrow{i^{b,l}(r)} & H_r(\widetilde{X}^b) & \longrightarrow & H_r(\widetilde{X}^b, \widetilde{X}^l) & \longrightarrow & H_{r-1}(\widetilde{X}^l) & \longrightarrow \cdots
\end{array}
$$
$$(4.17)$$

The diagram (4.17) leads to the following commutative diagram (4.18) whose rows are short exact sequences:

$$
\begin{array}{ccccccccc}
0 \to & \mathrm{coker}(i_{-l,a}(r)) & \longrightarrow & H_r(\widetilde{X}_a, \widetilde{X}_{-l}) & \longrightarrow & \ker(i_{-l,a}(r-1)) & \to 0 & (4.18) \\
& \downarrow & & \downarrow & & \downarrow & \\
0 \longrightarrow & \mathrm{coker}(i^l_{-l}(r)) & \longrightarrow & H_r(\widetilde{X}, \widetilde{X}_{-l} \sqcup \widetilde{X}^l) & \longrightarrow & \ker(i^l_{-l}(r-1)) & \to 0 \\
& \uparrow & & \uparrow & & \uparrow & \\
0 \longrightarrow & \mathrm{coker}(i^{b,l}(r)) & \longrightarrow & H_r(\widetilde{X}^b, \widetilde{X}^l) & \longrightarrow & \ker(i^{b,l}(r-1)) & \to 0
\end{array}
$$

Note that there exist compatible linear maps induced by inclusions when we pass from the diagram corresponding to the system (l, a, b) to the diagram corresponding the system (l', a', b') when $l' \geq l$, $a' \geq a$, $b' \leq b$. Note also that for X compact and f tame, the set of barcodes $\mathcal{B}_r(f)$ is finite and therefore there is a maximal length of all barcodes, say $L(f)$. Propositions 4.3 and 4.4 imply the following calculations:

Proposition 4.6. *Let* a, b *fixed and suppose* l *satisfies* $a > -l$, $b < l$. *Then*

(a) $\operatorname{coker}(i_{-l,a}(r)) = \kappa[\mathcal{M}_{-l,a}(r)]$, *with*

$$\mathcal{M}_{-l,a}(r) = \begin{cases} [\alpha, \beta] \in \mathcal{B}_r^c(\widetilde{f}) \mid -l < \alpha \le a, \\ (\alpha, \beta) \in \mathcal{B}_{r-1}^o(\widetilde{f}) \mid -l < \beta \le a, \\ [\alpha, \beta) \in \mathcal{B}_r^{c,o}(\widetilde{f}) \mid -l < \alpha \le a < \beta. \end{cases}$$

(b) $\ker(i_{-l,a}(r)) = \kappa[\mathcal{N}_{-l,a}(r)]$, *with*

$$\mathcal{N}_{-l,a}(r) = \{[\alpha, \beta) \in \mathcal{B}_r^{c,o} \mid \alpha \le -l < \beta \le a\},$$

(c) $\operatorname{coker}(i^{b,l}(r)) = \kappa[\mathcal{M}^{b,l}(r)]$, *with*

$$\mathcal{M}^{b,l}(r) = \begin{cases} [\alpha, \beta] \in \mathcal{B}_r^c(\widetilde{f}) \mid b \le \beta < l, \\ (\alpha, \beta) \in \mathcal{B}_{r-1}^o(\widetilde{f}) \mid b \le \alpha < l, \\ (\alpha, \beta] \in \mathcal{B}_r^{o,c}(\widetilde{f}) \mid \alpha < b \le \beta < l, \end{cases}$$

(d) $\ker(i^{b,l}(r)) = \kappa[\mathcal{N}^{b,l}(r)]$, *with*

$$\mathcal{N}^{b,l}(r) := \{(\alpha, \beta] \in \mathcal{B}_r^{o,c}(\widetilde{f}) \mid b \le \alpha < l \le \beta\},$$

and if $2l > L(f)$ *then:*

(e) $\operatorname{coker}(i_{-l}^l(r)) = \kappa[\mathcal{M}_{-l}^l(r)]$, *with*

$$\mathcal{M}_{-l}^l(r) = \begin{cases} [\alpha, \beta] \in \mathcal{B}_r^c(\widetilde{f}) \mid [\alpha, \beta] \subset (-l, l), \\ (\alpha, \beta) \in \mathcal{B}_{r-1}^o(\widetilde{f}) \mid \alpha < l, \ \beta > -l, \end{cases}$$

(f) $\ker(i_{-l}^l(r)) = \kappa[\mathcal{N}_{-l}^l(r) \sqcup \widetilde{J}_r(f)]$, *with*[4]

$$\mathcal{N}_{-l}^l(r) = \begin{cases} [\alpha, \beta) \in \mathcal{B}_r^{c,o}(\widetilde{f}) \mid (\alpha, \beta) \ni -l, \\ (\alpha, \beta] \in \mathcal{B}_r^{oc}(\widetilde{f}) \mid (\alpha, \beta) \ni l. \end{cases}$$

Clearly, for $l' > l$ one has:

$$\mathcal{M}_{-l',a}(r) \supseteq \mathcal{M}_{-l,a}(r),$$
$$\mathcal{M}^{b,l'}(r) \supseteq \mathcal{M}^{b,l}(r),$$
$$\mathcal{M}_{-l'}^{l'}(r) \supseteq \mathcal{M}_{-l}^l(r).$$

---

[4]In view of the hypothesis (a, b) canot contain both $-l$ and l.

Also clearly, for $l' - l > L(f)$ one has:

$$\mathcal{N}_{-l',a}(r) \cap \mathcal{N}_{-l,a}(r) = \emptyset,$$

$$\mathcal{N}^{b,l'}(r) \cap \mathcal{N}^{b,l}(r) = \emptyset,$$

$$\mathcal{N}^{l'}_{-l'}(r) \cap \mathcal{N}^{l}_{-l}(r) = \emptyset.$$

Note that the sets $\mathcal{M}^{...}_{...}(r)$, $\mathcal{N}^{...}_{...}(r)$, $\mathcal{J}_r(\widetilde{f})$, and $\mathcal{J}_{r-1}(\widetilde{f})$ are all subsets of $S = \mathcal{B}_r(\widetilde{f}) \sqcup \mathcal{B}_{r-1}(\widetilde{f}) \sqcup \mathcal{J}_r(f) \sqcup \mathcal{J}_{r-1}(\widetilde{f})$, and all linear maps between the homologies involved in the diagram (4.17), when identified to vector spaces generated by barcodes, correspond to the canonical linear maps[5] $\kappa[S_1] \to \kappa[S_2]$ associated to the corresponding subsets of S.

To better formulate the calculations of Borel-Moore homology in terms of subsets of $\mathcal{B}^{...}_r(\widetilde{f})$ and $\mathcal{J}_r(\widetilde{f})$, we use the following abbreviations and notations:

For $\alpha, \beta \in \mathbb{R}$

$$\mathcal{S}_{r,\alpha}(\widetilde{f}) := \mathcal{S}_{r,(-\infty,\alpha]}(\widetilde{f}) = \begin{cases} I \in \mathcal{B}^c_r(\widetilde{f}) \mid I \cap (-\infty, \alpha] \neq \emptyset \\ I \in \mathcal{B}^o_{r-1}(\widetilde{f}) \mid I \subset (-\infty, \alpha], \\ I = [a,b) \in \mathcal{B}^{c,o}_r(\widetilde{f}) \mid a \leq \alpha < b, \end{cases}$$

$$\mathcal{S}^\beta_r(\widetilde{f}) := \mathcal{S}_{r,[\beta,\infty)}(\widetilde{f}) = \begin{cases} I \in \mathcal{B}^c_r(\widetilde{f}) \mid I \cap [\beta, \infty) \neq \emptyset, \\ I \in \mathcal{B}^o_{r-1}(\widetilde{f}) \mid I \subset [\beta, \infty), \\ I = (a,b] \in \mathcal{B}^{o,c}_r(\widetilde{f}) \mid a < \beta \leq b. \end{cases} \quad (4.19)$$

One defines

$$\mathbb{I}^{\mathrm{BM},\widetilde{f}}_\alpha(r) := \mathrm{img}(H^{\mathrm{BM}}_r(\widetilde{M}_\alpha) \to H^{\mathrm{BM}}_r(\widetilde{M}))$$

$$\mathbb{I}^{\mathrm{BM},\beta}_{\widetilde{f}}(r) := \mathrm{img}(H^{\mathrm{BM}}_r(\widetilde{M}^\beta) \to H^{\mathrm{BM}}_r(\widetilde{M})).$$

We denote by $\kappa[[S]]$ the set of all κ-valued maps defined on S as opposed to $\kappa[S]$ the set of κ-valued maps with finite support. Consider the commutative diagram (4.20) below whose linear maps $\widetilde{\omega}_{r,\alpha}$, $\widetilde{\omega}_r$, $\widetilde{\omega}^\beta_r$, $\widetilde{\omega}^{\mathrm{BM}}_{r,\alpha}$, $\widetilde{\omega}^{\mathrm{BM}}_r$, and $\widetilde{\omega}^{\mathrm{BM},\beta}_r$ are isomorphisms. Propositions 4.4 and 4.6 leads to the isomorphisms $\widetilde{\omega}_{r,\alpha}$, $\widetilde{\omega}_r$, $\widetilde{\omega}^\beta_r$, in view of the equalities (4.21) and isomorphisms $\widetilde{\omega}^{\mathrm{BM}}_{r,\alpha}$, $\widetilde{\omega}^{\mathrm{BM}}_r$, $\widetilde{\omega}^{\mathrm{BM},\beta}_r$ in view of the equalities (4.16).

---

[5]Definition 3.1 in Chapter 3.

$$\mathbb{I}_\alpha^{\widetilde{f}}(r) \xleftarrow{\widetilde{\omega}_{r,\alpha}} \kappa[(\mathcal{S}_{r,\alpha}(\widetilde{f}) \setminus \mathcal{B}_r^{\mathrm{c},\mathrm{o}}(\widetilde{f})) \sqcup \mathcal{J}_r(\widetilde{f})] \xrightarrow{\pi_{r,\alpha}} \kappa[[\mathcal{S}_{r,\alpha}(\widetilde{f}) \setminus \mathcal{B}_r^{\mathrm{c},\mathrm{o}}(\widetilde{f})]] \xrightarrow{\omega_{r,\alpha}^{\mathrm{BM}}} \mathbb{I}_\alpha^{\mathrm{BM},\widetilde{f}}(r)$$

$$\downarrow \qquad\qquad \downarrow v_{r,\alpha} \qquad\qquad\qquad \downarrow u_{r,\alpha} \qquad\qquad\qquad \downarrow$$

$$H_r(\widetilde{M}) \xleftarrow{\widetilde{\omega}_r} \kappa[\mathcal{S}_r(\widetilde{f}) \sqcup \mathcal{J}_r(\widetilde{f})] \xrightarrow{\pi_r} \kappa[[\mathcal{S}_r(\widetilde{f}) \sqcup \mathcal{J}_{r-1}(\widetilde{f})]] \xrightarrow{\widetilde{\omega}_r^{\mathrm{BM}}} H_r^{\mathrm{BM}}(\widetilde{M})$$

$$\uparrow \qquad\qquad \uparrow v_r^\beta \qquad\qquad\qquad \uparrow u_r^\beta \qquad\qquad\qquad \uparrow$$

$$\mathbb{I}_{\widetilde{f}}^\beta(r) \xleftarrow{\widetilde{\omega}_r^\beta} \kappa[(\mathcal{S}_r^\beta(\widetilde{f}) \setminus \mathcal{B}_r^{\mathrm{o},\mathrm{c}}(\widetilde{f})) \sqcup \mathcal{J}_r(\widetilde{f})] \xrightarrow{\pi_r^\beta} \kappa[[\mathcal{S}_r^\beta(\widetilde{f}) \setminus \mathcal{B}_r^{\mathrm{o},\mathrm{c}}(\widetilde{f})]] \xrightarrow{\widetilde{\omega}_r^{\mathrm{BM},\beta}} \mathbb{I}_{\widetilde{f}}^{\mathrm{BM},\beta}(r)$$

$$\tag{4.20}$$

$$H_r(\widetilde{M}) = \varinjlim_{0 < l \to \infty} H_r(\widetilde{M}_l \cap \widetilde{M}^{-l}),$$

$$H_r(\widetilde{M}_a) = \varinjlim_{-a < l \to \infty} H_r(\widetilde{M}_a \cap \widetilde{M}^{-l}), \tag{4.21}$$

$$H_r(\widetilde{M}^a) = \varinjlim_{a < l \to \infty} H_r(\widetilde{M}_l \cap \widetilde{M}^a).$$

In diagram (4.20) the linear maps $\pi_{r,\alpha}$, π_r, π_r^β, and $v_{r,\alpha}$, v_r^β, $u_{r,\alpha}$, u_r^β are the canonical linear maps (cf. Definition 3.1 in Chapter 3) defined by the sets in brackets, which are all subsets in $S = \mathcal{B}_r(\widetilde{f}) \sqcup \mathcal{B}_{r-1}(\widetilde{f}) \sqcup \mathcal{J}_r(\widetilde{f}) \sqcup \mathcal{J}_{r-1}(\widetilde{f})$. This diagram implies that $\mathbb{I}_\alpha^{\mathrm{BM},\widetilde{f}}(r) \cap \mathbb{I}_{\widetilde{f}}^{\mathrm{BM},\beta}(r)$ identifies to

$$\kappa[(\mathcal{B}_{r,\alpha}^{\mathrm{c}}(\widetilde{f}) \cap \mathcal{B}_r^{\mathrm{c},\beta}(\widetilde{f})) \sqcup (\mathcal{B}_{r-1,\alpha}^{\mathrm{o}}(\widetilde{f}) \cap \mathcal{B}_{r-1}^{\mathrm{o},\beta}(\widetilde{f})) \sqcup \mathcal{J}_r(\widetilde{f})]$$

and $\mathbb{I}_\alpha^{\mathrm{BM},\widetilde{f}}(r) \cap \mathbb{I}_{\widetilde{f}}^{\mathrm{BM},\beta}(r)$ identifies to

$$\kappa[(\mathcal{B}_{r,\alpha}^{\mathrm{c}} \cap \mathcal{B}_r^{\mathrm{c},\beta}(\widetilde{f})) \sqcup (\mathcal{B}_{r-1,\alpha}^{\mathrm{o}}(\widetilde{f}) \cap \widetilde{\mathcal{B}}_{r-1}^{\mathrm{o},\beta})].$$

Note that the set in brackets is finite, so there is no difference between $\kappa[[\cdots]]$ and $\kappa[\cdots]$. It also implies the exactness of the sequence

$$0 \to \kappa[\mathcal{J}_r(\widetilde{f})] \to \mathbb{I}_\alpha^{\widetilde{f}}(r) \cap \mathbb{I}_{\widetilde{f}}^\beta(r) \to \mathbb{I}_\alpha^{\mathrm{BM},\widetilde{f}}(r) \cap \mathbb{I}_{\widetilde{f}}^{\mathrm{BM},\beta}(r) \to 0.$$

To summarize, we have:

Proposition 4.7. *One has:*

(a) $\ker(\mathbb{I}_\alpha^{\widetilde{f}} \cap \mathbb{I}_{\widetilde{f}}^b \to \mathbb{I}_{\widetilde{f}}^{\mathrm{BM},\widetilde{f}} \cap \mathbb{I}_{\widetilde{f}}^{\mathrm{BM},b}) \simeq \kappa[[\mathcal{J}_r(\widetilde{f})]] = \kappa[\mathcal{J}_r(\widetilde{f})]$,

(b) $H_r^{\mathrm{BM}}(\widetilde{X}) \simeq \kappa[[\mathcal{S}_r(\widetilde{f}) \sqcup \mathcal{J}_{r-1}(\widetilde{f})]]$,

(c) $\mathbb{I}_a^{\mathrm{BM},\widetilde{f}}(r) \simeq \kappa[[\mathcal{S}_{r,a}(\widetilde{f}) \setminus \mathcal{B}_r^{\mathrm{c},\mathrm{o}}(\widetilde{f})]]$,

(d) $H_r^{\mathrm{BM}}(\widetilde{X}_a) \simeq \kappa[[\mathcal{S}_{r,a}]]$,

(e) $\mathbb{I}_{\widetilde{f}}^{\mathrm{BM},b}(r) \simeq \kappa[[\mathcal{S}_r^b(\widetilde{f}) \setminus \mathcal{B}_r^{\mathrm{o,c}}(\widetilde{f})]]$,

(f) $H_r^{\mathrm{BM}}(\widetilde{X}^b) \simeq \kappa[[\mathcal{B}_r^b(\widetilde{f})]]$.

In view of diagram (4.20) the canonical maps from homology to Borel-Moore homology, for any of the spaces considered above, are given by the compositions $\widetilde{\omega}_r^{\mathrm{BM}} \cdot \pi_r \cdot \widetilde{\omega}_r^{-1}$ with π_r the canonical linear maps for the sets in brackets. In particular one has:

Proposition 4.8. *The following commutative diagram computes the canonical map from homology to Borel-Moore homology.*

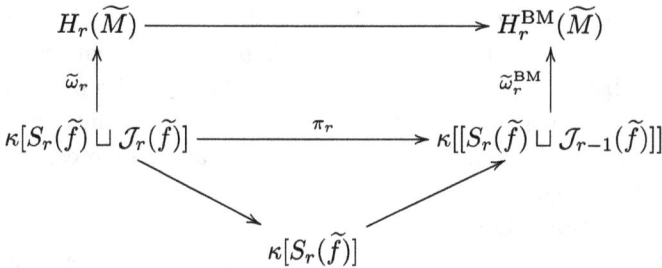

$$
\begin{array}{ccc}
H_r(\widetilde{M}) & \longrightarrow & H_r^{\mathrm{BM}}(\widetilde{M}) \\
\widetilde{\omega}_r \uparrow & & \uparrow \widetilde{\omega}_r^{\mathrm{BM}} \\
\kappa[S_r(\widetilde{f}) \sqcup \mathcal{J}_r(\widetilde{f})] & \xrightarrow{\ \pi_r\ } & \kappa[[S_r(\widetilde{f}) \sqcup \mathcal{J}_{r-1}(\widetilde{f})]] \\
& \searrow \qquad \nearrow & \\
& \kappa[S_r(\widetilde{f})] &
\end{array}
$$

4.5 Calculations of barcodes and Jordan cells

When X is a finite simplicial complex and $f : X \to \mathbb{R}$ or $f : X \to \mathbb{S}^1$ is a simplicial map, the barcodes and the Jordan blocks or Jordan cells are computable by effective computer-implementable algorithms. This is what we mean by "barcodes and Jordan cells are computer friendly". The Novikov-Betti numbers and the homological monodromy, apparently not computer friendly since they involve the infinite cyclic cover of f, are, in view of Theorem 4.1, computer friendly. We will briefly explain below the essential steps in their calculation in the case where X is a finite simplicial complex and f is a simplicial map. The presentation follows closely [Burghelea, D., Dey, T (2013)].

The calculation is realized in three steps.

Step 1. The calculation of the representations $\rho_r(f)$ as a collection of matrices representing the corresponding linear maps α_i and β_i.

Step 2. The calculation of the indecomposable representations of $\rho_r(f)$, i.e. barcodes and Jordan cells representations.

Step 3. The conversion of the barcode representations, indecomposable components of $\rho_r(f)$, into the barcodes of the map f.

Step 2 was treated in Chapter 3, and Step 3 is simply passing from intervals with ends integers i, j to critical values known as being among the values of f on vertices. Step 1 is in principle described in Subsection 2.3.4 but for the reader's convenience we review this material here.

First note that, in view of Observation 4.1, the representations $\rho_r(f)$ can be described using only homology of compact levels rather than of open inter-levels.

Second note that one can assume that f restricted to the set of vertices is injective. If not, a temporary very small modification of the simplicial map into a simplicial map taking different values on different vertices does not change the Jordan cells and preserves the number of barcodes between corresponding modified critical values because of the stability theorems, Theorems 5.5 and 6.1 in Chapter 5 and Chapter 6. Consequently, one can recover the barcodes of the initial map from the barcodes of the modified map. As in [Burghelea, D., Dey, T (2013)] one considers only the case of a simplicial angle-valued map; the case of a real-valued map can be treated similarly or derived from the angle-valued map case.

The calculation relies on the persistence algorithm as treated in [Cohen-Steiner, D., Edelsbrunner, H., Morozov, D. (2006)] and [Zomorodian, A., Carlsson, G. (2005)], properly adjusted. This algorithm is not reviewed in this book and the interested practitioner is directed to consult these references, still, we note the following.

The persistence algorithm takes the *incidence matrix* of a cell complex X, and the submatrix of a subcomplex $Y \subset X$, under the hypothesis that a good total order of the cells of X is provided, and computes first a basis for the homology of Y, then a basis for the homology of X, and then the matrix describing the linear map $H_r(Y) \to H_r(X)$ with resepct to these bases. A *good* total order simply means that the cell σ comes before σ' if σ is a face of σ', and all cells of Y precede those of $X \setminus Y$ in this order. With respect to this order the incidence matrix of Y is located in the upper left corner of the incidence matrix of X. Running the persistence algorithm[6] on the incidence matrix for Y one computes first a basis of the homology group $H_r(Y)$. Then one continues the procedure (by adding the columns and rows of the matrix for X) to obtain a basis of $H_r(X)$. It is straightforward to read off the matrix representation of the linear map $H_r(Y) \to H_r(X)$ with respect to these bases.

In our situation one begins with the incidence matrix of the input simplicial complex X equipped with the map $f : X \to \mathbb{S}^1$ specified by the values

---

[6]Which consists of adding rows and columns until the matrix gets the desired features.

of f on vertices. Let the angles $0 \leq s_1 < s_2 < \cdots < s_m < 2\pi$ be the critical values of f. Choose a collection of regular angles $s_m - 2\pi < t_1 < t_2 < \cdots < t_m < s_m < 2\pi$ with $t_i < s_i < t_{i+1} < s_{i+1}$. Consider a canonical subdivision of X into a cell complex such that $X_i := X_{[t_i, t_{i+1}]} = f^{-1}([t_i, t_{i+1}])$, and $Y_i := f^{-1}(t_i)$ are subcomplexes. This is done as follows:

For any open simplex σ we associate the open cells

(i) $\sigma(i) := \sigma \cap f^{-1}(t_i)$ with $\dim \sigma(i) = \dim \sigma - 1$, if the intersection is nonempty,

(ii) $\sigma\langle i \rangle := \sigma \cap f^{-1}((t_i, t_{i+1}))$ with $\dim \sigma\langle i \rangle = \dim \sigma$, if the intersection is nonempty.

The cells of Y_i are exactly of the form $\sigma(i)$ and their incidences are given as $I(\sigma(i), \tau(i)) = I(\sigma, \tau)$, where $I(\sigma, \tau) = 0, +1,$ or -1 depending on whether σ is a face of τ and whether their orientations match or not.

The cells of X_i consist of cells of Y_i, Y_{i+1}, all cells of the form $\sigma\langle i \rangle$, and all open simplices $\sigma \subset X_i' = X_i \setminus Y_i \sqcup Y_{i+1}$. The incidences are given as $I(\sigma\langle i \rangle, \tau\langle i \rangle) = I(\sigma, \tau)$, $I(\sigma(i), \sigma\langle i \rangle) = 1$, and $I(\sigma(i+1), \sigma\langle i \rangle) = -1$. All other incidences are either the incidences in X_i', or zero. Note that each simplex of X enters as a cell only once if does not intersect any level, and three times if intersects Y_i, as $\sigma(i)$, $\sigma\langle i-1 \rangle$, and $\sigma\langle i \rangle$.

Assume that we are given a good total order for the simplices of X that is in addition compatible with f, this meaning a good total order which makes the cells of Y_i precede the cells of Y_{i+1} precede the cells of $X_i' = X_i \setminus Y_i \sqcup Y_{i+1}$. This induces a total order for the cells in Y_i and Y_{i+1}, and also for the cells in $X_i' = X_i \setminus Y_i \sqcup Y_{i+1}$ for any $1 \leq i \leq m$ with $Y_{m+1} := Y_1$. Clearly, the incidence matrix for X_i can be derived from the incidence matrix of X. Apply the persistence algorithm for $i = 1, 2, \ldots, m$ to $X := X_i$ and $Y = Y_i \sqcup Y_{i+1}$, as described above. This yields the matrices for the linear maps $\alpha_i : H_r(X_{t_i}) \to H_r(X_{s_i})$ and $\beta_i : H_r(X_{t_{i+1}}) \to H_r(X_{s_i})$.

The key observation is that running the algorithm for X_i and for X_{i+1} provides the same basis for $H_r(Y_{i+1})$. The final output is a collection of matrices describing a representation isomorphic to $\rho_r(f)$ to which the algorithm presented in Chapter 3 can be applied.

4.6   Exercises

E.1 Figure 1.1 in (Chapter 1) represents a surface Σ of genus 2. Determine the graph representation for the projection on the horizontal line.

Determine the barcodes and calculate the Betti numbers of this surface in terms of barcodes.

E.2 Suppose that for a tame angle-valued map $f : X \to \mathbb{S}^1$:
- the critical angles are $0 < \theta_1 < \theta_2 < \theta_3 < \theta_4 < \theta_5 \leq 2\pi$,
- the field is $\kappa = \mathbb{R}$, and for some r
- $\mathcal{B}_r^c = \{[\theta_2, \theta_3]\}$, $\mathcal{B}_r^o = \{(\theta_2, \theta_3), (\theta_2, \theta_3 + 2\pi)\}$, $\mathcal{B}_r^{c,o} = \emptyset$, $\mathcal{B}_r^{o,c} = \{(\theta_1, \theta_2 + 4\pi]\}$, and
- $\mathcal{J}_r = \{(1, 1 + i), (1, (1 - i))\}$, $i = \sqrt{-1} \in \mathbb{C}$.

Reconstruct the representation $\rho_r(f) = \{V_r, 1 \leq r \leq 10, \alpha_i, \beta_i, 1 \leq i \leq 5\}$ Precisely, indicate the dimension of the vector spaces V_r and a matrix representation of each of the linear maps α_i, β_i.

E.3 Suppose that X is a 2-dimensional ANR, $f : X \to \mathbb{S}^1$ is a tame map with four critical angles, $\pi/6, \pi/3, \pi, 4/3\pi$, and the underlying field is $\kappa = \mathbb{Z}_3$.

Suppose further that:
$$P_3^f(z) = P_0^f(z) = 0, \ P_2^f(z) = (z - i)(z^2 + 3z + 2), \ P_1^f(z) = z - 2,$$
there are no mixed barcodes, and $\mathcal{J}_0 = \mathcal{J}_2 = \emptyset$, $\mathcal{J}_1 = \{(3, 1)\}$.

Reconstruct the representations $\rho_0(f)$, $\rho_1(f)$, and $\rho_2(f)$.

E.4 Suppose X is a 3-dimensional compact ANR, $f : X \to \mathbb{S}^1$ is a tame angle-valued map with critical values $\pi/6, \pi/3, \pi, 4/3\pi$, and the underlying field is $\kappa = \mathbb{Z}_5$.

Suppose further that:
$$P_0^f(z) = 0, \ P_1^f(z) = z - 2, \ P_2^f(z) = (z - i)(z^2 + 3z + 2), \ P_3^f(z) = 0, \text{ and}$$
$$\mathcal{J}_0 = (1; 1), \mathcal{J}_1 = (-1; 3), \mathcal{J}_2 = (2; 2).$$
1. Determine the Betti numbers of X.
2. Determine the Novikov-Betti numbers of (X, ξ_f).
3. Determine the representations $\rho_0(f)$, $\rho_1(f)$, and $\rho_2(f)$.
4. Repeat the calculation with $\kappa = \mathbb{Z}_3$.

E.5 For a simplicial complex X consider the barycentric subdivision bX and the canonical simplicial map $F : bX \to \mathbb{R}$ which takes the value k on the barycenter of each k-dimensional simplex of X.

Determine the barcodes and the AM complex of F with respect to a field κ when X is:
   a) the 2-dimensional simplex,
   b) the 3-dimensional simplex,
   c) the boundary of the 3-dimensional simplex,

d) the 2-dimensional simplicial complex indicated in the picture below (5 vertices, 6 edges, 1 solid triangle).

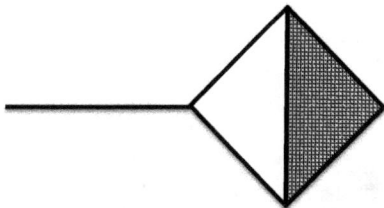

Fig. 4.8   The simplicial complex in item d)

E.6 Consider the radial projection from the center of the ellipse in Fig. 1.2 in Chapter 1. This is an angle-valued map $f : X \to \mathbb{S}^1$ from surface X to a circle of large radius with four critical angles. Describe the associated G_8-representations, calculate the barcodes and Jordan cells, and determine the Novikov-Betti numbers of $(X; \xi_f)$ (ξ_f is the cohomology class defined by f).

E.7 For a tame map $f : X \to \mathbb{R}$ with critical values $\cdots c_i < c_{i+1} < \cdots c_k < c_{k+1} \cdots$ describe in terms of barcodes the long exact sequence in homology of the pair $X_{c_i} \subseteq X_{c_j}$ and of the triple $X_{c_i} \subseteq X_{c_j} \subseteq X_{c_k}$,

E.8 Show that for any smooth tame map $f : M^n \to \mathbb{R}$ or $f : M^n \to \mathbb{S}^1$, M^n a smooth manifold, any given point $x \in M^n$ with $f(x)$ is a regular value, and any $r < n$, one can, by an arbitrarily small perturbation in an arbitrarily small neighborhood of x, introduce exactly one mixed r-barcode.

E.9 Show that for any simplicial tame map $f : X \to \mathbb{R}$ or $f : X \to \mathbb{S}^1$, any value s of f, any integer N, and any $\epsilon > 0$, one can, by an arbitrarily small perturbation of f on $f^{-1}(s - \epsilon, s + \epsilon)$, introduce at least N critical values in the interval $(s - \epsilon, s + \epsilon)$. (Hint: follow the proof of Theorem 5.1 item 4 in the Chapter 5.)

Chapter 5

Configurations δ_r^f and $\widehat{\delta}_r^f$ (Alternative Approach)

Fix a field κ as coefficients for homology.

In this Chapter we provide a new and direct definition of the configuration δ_r^f, without any reference to graph representations. This definition works for any continuous real- or angle-valued map defined on a compact ANR. It permits to recover the closed barcodes $\mathcal{B}_r^c(f)$ and the open barcodes $\mathcal{B}_{r-1}^o(f)$ as points of the support of δ_r^f above or on the diagonal and below diagonal, respectively, in the case of real-valued maps, and outside or on the unit circle and inside the unit circle, respectively, in the case of angle-valued maps.

For a real-valued map f the configuration δ_r^f can be regarded as a refinement of the Betti number $\beta_r(X)$, and for an angle-valued map as a refinement of the Novikov-Betti number $\beta_r^N(X, \xi_f)$. More important, when f is real valued the configuration δ_r^f, which is a configuration of points with multiplicity in $\mathbb{C} = \mathbb{R}^2$, has a refinement to a configuration of vector spaces, $\widehat{\delta}_r^f$, with the same support as δ_r^f, configuration which can be thought as a refinement of the homology $H_r(X)$ because $\bigoplus \widehat{\delta}_r^f(z) \simeq H_r(X)$.

When f is angle-valued the configuration δ_r^f, which is a configuration of points with multiplicity in $\mathbb{C} \setminus 0$, has a refinement to a configuration of free $\kappa[t^{-1}, t]$-modules, $\widehat{\delta}_r^f$, with the same support as δ_r^f which can be regarded as a refinement of the Novikov homology $H_r^N(X; \xi_f)$ because $\bigoplus \widehat{\delta}_r^f(z) \simeq H_r^N(X; \xi_f)$.

When f is a real-valued map and $\kappa = \mathbb{R}$ or $\mathbb{C}$ and $H_r(X)$ is equipped with a Hilbert space structure, the configuration $\widehat{\delta}_r^f$ can be canonically realized as a configuration of mutually orthogonal subspaces of $H_r(X)$ whose sum is $H_r(X)$. The same thing happens in case of angle-valued maps after a canonical completion of the Novikov homology to an $L^\infty(S^1)$-module, known as the L_2-homology. This however will not be discussed in detail in this book; the interested reader can consult [Burghelea, D. (2016a)]. The material of this chapter follows closely the presentation in [Burghelea, D., Haller, S. (2015)] Chapters 5,6,7, [Burghelea, D. (2015b)] and [Burghelea, D. (2016a)].

5.1   General considerations

Let $f : X \to \mathbb{R}$ be a proper continuous map with X an ANR. In consistence with some of the previous notations put

(i) $X_a^f := f^{-1}(-\infty, a])$, the a-sublevel,
$X_f^b := f^{-1}([b, \infty))$, the b-overlevel,
(ii) $i_a^f : X_a^f \to X$, $i_f^b : X^b \to X$, the inclusions,
(iii) $i_a^f(r) : H_r(X_a) \to H_r(X)$, $i_f^b(r) : H_r(X^b) \to H_r(X)$,
the inclusion-induced linear maps.

This notation is illustrated in Fig. 5.1

Further, define

(1) $\mathbb{I}_a^f(r) := \operatorname{img} i_a^f(r) \subseteq H_r(X)$,
(2) $\mathbb{I}_f^b(r) := \operatorname{img} i_f^b(r) \subseteq H_r(X)$,
(3) $\mathbb{F}_r^f(a, b) := \mathbb{I}_a^f(r) \cap I_f^b(r) \subseteq H_r(X)$.

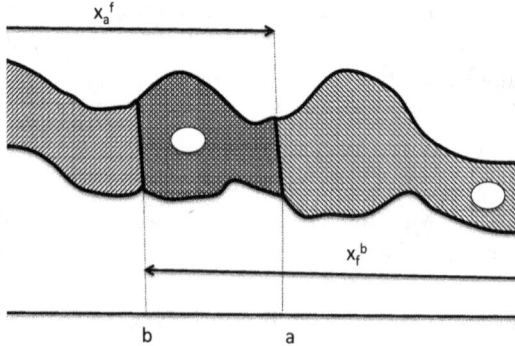

Fig. 5.1   Sublevel and overlevel

Observation 5.1.

   (i) For $a' \leq a$ and $b \leq b'$ one has $\mathbb{F}_r^f(a', b') \subseteq \mathbb{F}_r^f(a, b)$,

   (ii) For $a \leq a$ and $b \leq b'$ one has $\mathbb{F}_r^f(a', b) \cap \mathbb{F}_r^f(a, b') = \mathbb{F}_r^f(a', b')$,

   (iii) $|f(x) - g(x)| < \epsilon$ implies $\mathbb{F}^g(a - \epsilon, b + \epsilon) \subseteq \mathbb{F}_r^f(a, b)$.

Proposition 5.1. *If f is a map as above, then* $\dim \mathbb{F}_r^f(a, b) < \infty$.

Proof. If X is compact the statement is obvious, since $H_r(X)$ has finite dimension. Now suppose X is not compact. In view of Observation 5.1 (i), it suffices to check the statement for $a > b$. In this case if f is weakly tame, then X_a^f, X_f^b, and $X_a^b := X_a^f \cap X_f^b$ are ANR's, with $X_a^f \cap X_f^b$ compact and $X = X_a^f \cup X_f^b$, and we have the Mayer-Vietoris long exact sequence in homology

$$\cdots H_r(X_a^b) \xrightarrow{j} H_r(X_a^f) \oplus H^r(X_f^b) \xrightarrow{i} H_r(X) \xrightarrow{\delta} H_{r-1}(X_a^b). \qquad (5.1)$$

In this sequence $i = i_a^f(r) \pm i_f^b(r)$ and $j = i_{a,a}^b(r) \oplus (\mp i_a^{b,b}(r))$, with $i_{a,a}^b(r)$ and $i_a^{b,b}(r)$ induced by the inclusions $X_a^b \subseteq X_a$ and $X_a^b \subseteq X^b$.

   Observe that $\mathbb{F}_r^f(a, b) := \mathbb{I}_a^f \cap \mathbb{I}_f^b \subseteq i_a^f(r)(\ker(i_a(r) - i^b(r))$. In view of the exactness of the sequence (5.1), $\ker(i_a(r) - i^b(r))$ is isomorphic to a quotient of the vector space of $H_r(X_a^b)$, hence is of finite dimension, which proves the statement.

   If f is not weakly tame one argue as follows. It is known that any locally compact ANR X is properly homotopy dominated with respect to any open cover by some locally finite simplicial complex K, cf. [Ball (1975)][1]. Choose such a cover, for example $f^{-1}(n-1, n+1)_{n \in \mathbb{Z}}$, and a corresponding homotopy domination $X \xrightarrow{i} K \xrightarrow{\pi} X$ for this cover. Choose a proper simplicial (hence tame) map $g : K \to \mathbb{R}$, and $a' > a$ and $b' < b$ such that $\pi(X_a^f) \subset K_{a'}^g$, $\pi(X_f^b) \subset K_g^{b'}$. Then $\mathbb{F}_r^f(a, b)$ is isomorphic to a subspace of $\mathbb{F}_r^g(a', b')$. Since the dimension of $\mathbb{F}_r^g(a', b')$ is finite, so is the dimension of $\mathbb{F}_r^f(a, b)$. $\qquad \square$

   Recall from Chapter 2

Definition 5.1. A real number t is a *w-homologically regular value* if there exists $\epsilon(t) > 0$ such that for any ϵ with $0 < \epsilon < \epsilon(t)$ the inclusions $\mathbb{I}_{t-\epsilon}^f(r) \subseteq \mathbb{I}_t^f(r) \subseteq \mathbb{I}_{t+\epsilon}^f(r)$ and $\mathbb{I}_f^{t-\epsilon}(r) \supseteq \mathbb{I}_f^t(r) \supset \mathbb{I}_f^{t+\epsilon}(r)$ are equalities and a *w-homologically critical value* if it is not a w-homologically regular value.

---

[1]As a replacement for an argument based on an incorrect reference; the above argument and the reference were proposed by the referee.

This concept is relative to the fixed field κ. Denote by $wCr(f)$ the set of all w-homologically critical values. We assume that the field κ is fixed, so for the purpose of a lighter notation we will omit it in this notation. If f is weakly tame, then $wCr(f) \subset Cr(f)$.

Proposition 5.2.

1. If $f : X \to \mathbb{R}$ is a proper map and X is an ANR, then $wCr(f)$ is *discrete*.

2. If $c' < c''$ are two consecutive values in $wCr(f)$, $\alpha \in (c', c'')$, and $a, b \in \mathbb{R}$, then $\mathbb{F}_r^f(c', b) = \mathbb{F}_r^f(\alpha, b)$ and $\mathbb{F}^f(a, \alpha) = \mathbb{F}_r^f(a, c'')$.

Proof. Item 1: The statement is obviously true when X is a simplicial complex and f is a proper simplicial map.

As noticed above, one can find a proper simplicial map $g : K \to \mathbb{R}$ and a proper homotopy domination $X \xrightarrow{i} K \xrightarrow{\alpha} X$ such that $|f \cdot \alpha - g| < M$. If so, for any $a < b$, $a, b \in \mathbb{R}$, one has

$$\dim(\mathbb{I}_b^f(r)/\mathbb{I}_a^f(r)) \le \dim(\mathbb{I}_{b+M}^g(r)/\mathbb{I}_{a-M}^g(r))$$
$$\le \dim(H_r(g^{-1}([a - M, b + M]), g^{-1}(a - M)) < \infty.$$

This implies that there are only finitely many changes of $\mathbb{I}_t^f(r)$ for t between a and b, because there are only finitely many changes of $\mathbb{I}_t^g(r)$ for t between $a - M$ and $b + M$. Similar arguments show that there are only finitely many changes of $\mathbb{I}_f^t(r)$ for t with $a \le t \le b$. This suffices to conclude that $wCr(f) \cap [a, b]$ is a finite set for any $a < b$, which implies that $wCr(f)$ is discrete.

Item 2 follows from Definition 5.1. □

Let $\widetilde{\epsilon}(f) := \inf |c' - c''|$, $c', c'' \in wCr(f), c' \ne c''$.

Definition 5.2. The proper map $f : X \to \mathbb{R}$ is called w-*homologically tame* with respect to κ if $\widetilde{\epsilon}(f) > 0$.

Tame maps are clearly w-homologically tame with respect to any field κ and $\widetilde{\epsilon}(f) > \epsilon(f)$.

Boxes. Consider the sets of the form $B = (a', a] \times [b, b')$ with $a' < a, b < b'$, and refer to B as a *box*, depicted in Fig 5.2.

One assigns to the box B the quotient of subspaces of $H_r(X)$,

$$\mathbb{F}_r^f(B) := \mathbb{F}_r^f(a, b)/\left(\mathbb{F}_r^f(a', b) + \mathbb{F}_r^f(a, b')\right),$$

and the projection

$$\pi_{ab,r}^B : \mathbb{F}_r^f(a, b) \to \mathbb{F}_r^f(B).$$

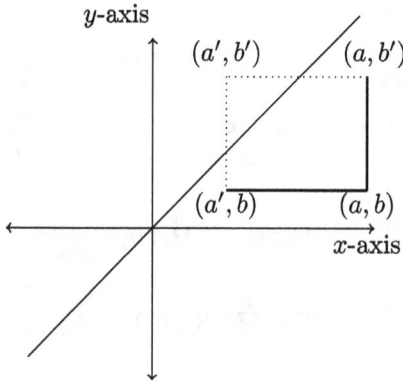

Fig. 5.2 The box $B := (a', a] \times [b, b') \subset \mathbb{R}^2$

It will be convenient to denote

$$\mathbb{F'}_r^f(B) := \mathbb{F}_r^f(a', b) + \mathbb{F}_r^f(a, b') \subseteq \mathbb{F}_r^f(a, b),$$

so that

$$\mathbb{F}_r^f(B) = \mathbb{F}_r^f(a, b) / \mathbb{F'}_r^f(B).$$

Further, denote

$$F_r^f(a, b) := \dim \mathbb{F}_r^f(a, b), \quad F_r^f(B) := \dim \mathbb{F}_r^f(B).$$

In view of Observation 5.1 (ii),

$$F_r^f(B) = F_r^f(a, b) + F^f(a', b') - F_r^f(a', b) - F^f(a, b'). \tag{5.2}$$

If $\kappa = \mathbb{R}$ or $\mathbb{C}$ and $H_r(X)$ is equipped with a non-degenerate positive definite Hermitian inner product, i.e. a Hilbert space structure (when finite dimensional), one denotes by $\mathbf{H}_r(B)$ the orthogonal complement of $(F_r^f)'(B)$ inside $\mathbb{F}_r^f(a, b)$. One has

$$\mathbf{H}_r(B) \subseteq \mathbb{F}_r^f(a, b) \subseteq H_r(X).$$

Proposition 5.3. *Let $a'' < a' < a$, $b < b'$ and consider the boxes $B_1 = (a'', a'] \times [b, b')$, $B_2 = (a', a] \times [b, b')$, and $B = (a'', a] \times [b, b')$ (cf. Fig. 5.3).*
1. The inclusions $B_1 \subset B$ and $B_2 \subset B$ induce the linear maps

$$i_{B_1, r}^B : \mathbb{F}_r^f(B_1) \to \mathbb{F}_r^f(B)$$

and

$$\pi_{B, r}^{B_2} : \mathbb{F}_r^f(B) \to \mathbb{F}_r^f(B_2)$$

such that the sequence

$$0 \longrightarrow \mathbb{F}_r^f(B_1) \xrightarrow{i_{B_1,r}^B} \mathbb{F}_r^f(B) \xrightarrow{\pi_{B,r}^{B_2}} \mathbb{F}_r^f(B_2) \longrightarrow 0$$

is exact, hence $\mathbb{F}_r^f(B)$ *is isomorphic to* $\mathbb{F}_r^f(B_1) \oplus \mathbb{F}_r^f(B_2)$.

2. *If* $\kappa = \mathbb{R}$ *or* $\mathbb{C}$ *and* $H_r(X)$ *is equipped with a Hilbert space structure, then*

$$\mathbf{H}_r(B_1) \perp \mathbf{H}_r(B_2)$$

and

$$\mathbf{H}_r(B) = \mathbf{H}_r(B_1) \oplus \mathbf{H}_r(B_2).$$

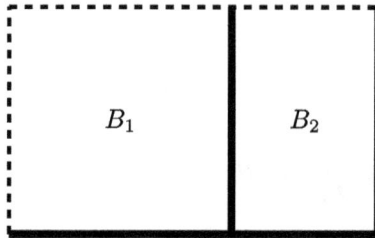

Fig. 5.3   Box divided vertically

Proposition 5.4. *Let* $a' < a$, $b < b' < b''$ *and consider the boxes* $B_1 = (a', a] \times [b', b'')$, $B_2 = (a', a] \times [b, b')$, *and* $B = (a', a] \times [b, b'')$ *(cf. Fig. 5.4).*

1. *The inclusions* $B_1 \subset B$ *and* $B_2 \subset B$ *induce the linear maps*

$$i_{B_1,r}^B : \mathbb{F}_r(B_1) \to \mathbb{F}_r(B)$$

and

$$\pi_{B,r}^{B_2} : \mathbb{F}_r(B) \to \mathbb{F}_r(B_2)$$

such that the sequence

$$0 \longrightarrow \mathbb{F}_r(B_1) \xrightarrow{i_{B_1,r}^B} \mathbb{F}_r(B) \xrightarrow{\pi_{B_2,r}^B} \mathbb{F}_r(B_2) \longrightarrow 0 ,$$

is exact, hence $\mathbb{F}_r^f(B)$ *is isomorphic to* $\mathbb{F}_r^f(B_1) \oplus \mathbb{F}_r^f(B_2)$.

2. *If* $\kappa = \mathbb{R}$ *or* $\mathbb{C}$ *and* $H_r(X)$ *is equipped with a Hilbert space structure, then*

$$\mathbf{H}_r(B_1) \perp \mathbf{H}_r(B_2)$$

and

$$\mathbf{H}_r(B) = \mathbf{H}_r(B_1) \oplus \mathbf{H}_r(B_2).$$

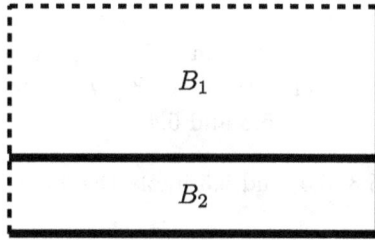

Fig. 5.4   Box divided horizontally

Proposition 5.5. *Let $a'' < a' < a$, $b < b' < b''$, and*
$$B = (a'', a] \times [b, b''),$$
$$B_{1,1} = (a'', a'] \times [b', b''), \quad B_{1,2} = (a'', a'] \times [b, b'),$$
$$B_{2,1} = (a', a] \times [b', b''), \quad B_{2.2} = (a', a] \times [b, b'),$$
$$B_{1,} = B_{1,1} \sqcup B_{1,2}, \quad B_{2,} = B_{2,1} \sqcup B_{2,2},$$
$$B_{,1} = B_{1,1} \sqcup B_{2,1}, \quad B_{,2} = B_{1,2} \sqcup B_{2,2}.$$
(cf. Fig. 5.5)

Then:

1. $i_{B_{1,1}}^B = i_{B_{,1}}^B \cdot i_{B_{1,1}}^{B_{,1}} = i_{B_{1,}}^B \cdot i_{B_{1,1}}^{B_{1,}}$,

   $\pi_B^{B_{2,2}} = \pi_{B_{,2}}^{B_{2,2}} \cdot \pi_B^{B_{,2}} = \pi_{B_{2,}}^{B_{2,2}} \cdot \pi_B^{B_{2,}}$.

2. *If $\kappa = \mathbb{R}$ or $\mathbb{C}$ and $H_r(X)$ has a Hilbert space structure, then*
$$\mathbf{H}_r(B_{i,j}) \perp \mathbf{H}_r(B_{i',j'}), \ (i,j) \neq (i',j'),$$

and
$$\mathbf{H}_r(B) = \mathbf{H}_r(B_{1.1}) \oplus \mathbf{H}_r(B_{1.2}) \oplus \mathbf{H}_r(B_{2.1}) \oplus \mathbf{H}_r(B_{2,2}).$$

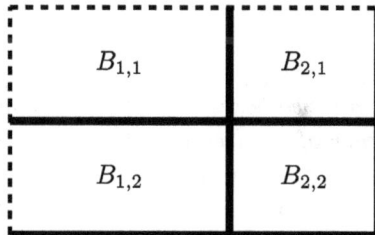

Fig. 5.5   Box divided into four pieces

Proof. Item 1. in Propositions 5.3, 5.4, and 5.5 follows from Observation 5.1 (i) and (ii). To conclude Item 2. in Propositions 5.3 and 5.4, note that $\mathbb{F}_r(a',b) \subset \mathbb{F}'^f_r(B_2)$ and $\mathbb{F}_r(a,b') \subset \mathbb{F}'^f_r(B_2)$. Item 2. in Proposition 5.5 follows from Propositions 5.3 and 5.4. $\qquad\square$

Propositions 5.3, 5.4, and 5.5 imply the following corollary.

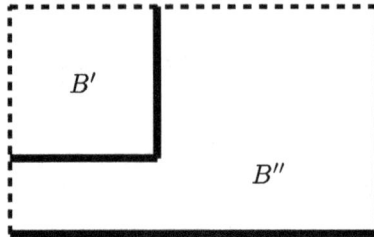

Fig. 5.6 Box in upper left corner

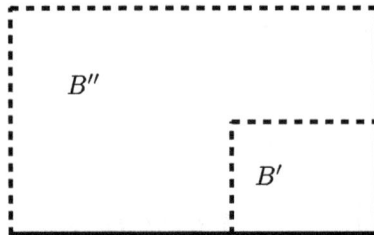

Fig. 5.7 Box in lower right corner

Corollary 5.1.

(i) *If B' and B'' are two boxes with $B' \subseteq B''$ and B' is located in the upper left corner of B'' (see Fig. 5.6), then the inclusion induces the canonical injective linear maps $i^{B''}_{B',r} : \mathbb{F}_r(B') \to \mathbb{F}_r(B'')$.*

(ii) *If B' and B'' are two boxes with $B' \subseteq B''$ and B' is located in the lower right corner of B'' (see Fig. 5.7), then the inclusion induces the canonical surjective linear maps $\pi^{B'}_{B'',r} : \mathbb{F}_r(B'') \to \mathbb{F}_r(B')$.*

(iii) *If B is a finite disjoint union of boxes $B = \bigsqcup B_i$, then $\mathbb{F}_r(B)$ is isomorphic to $\bigoplus_i \mathbb{F}_r(B_i)$.*

(iv) *If, in addition, $\kappa = \mathbb{R}$ or $\mathbb{C}$ and $H_r(X)$ is equipped with a Hilbert space structure, then $\mathbf{H}_r(B) = \bigoplus_i \mathbf{H}_r(B_i)$ and one has a canonical decomposition with $\mathbf{H}_r(B_i)$ mutually orthogonal.*

In view of the above observation, denote $B(a, b; \epsilon) := (a - \epsilon, a] \times [b, b + \epsilon)$ and define

$$\boxed{\widehat{\delta}_r^f(a, b) := \varinjlim_{\epsilon \to 0} \mathbb{F}_r(B(a, b; \epsilon)) = \mathbb{F}_r^f(a, b) / \varinjlim_{\epsilon \to 0} \mathbb{F'}_r^f(B(a, b; \epsilon))} .$$

The limit refers to the direct system of surjective linear maps $\mathbb{F}_r(B(a, b; \epsilon')) \to \mathbb{F}_r(B(a, b; \epsilon''))$ defined by the inclusions $\mathbb{F'}_r^f(B(a, b; \epsilon')) \subseteq \mathbb{F'}_r^f)'(B(a, b; \epsilon''))$ for $\epsilon' > \epsilon''$.

Note that in view of Proposition 5.2 $\widehat{\delta}_r^f(a, b)$ is the quotient of a pair of subspaces of $H_r(X)$, precisely $\mathbb{F}_r^f(a, b) \supseteq \mathbb{F'}_r^f(a, b)$, with

$$\mathbb{F'}_r^f(a, b) := \mathbb{F}_r^f(a_-, b) + \mathbb{F}_r^f(a, b^+),$$

where $a_- = $ the largest w-homologically critical value smaller than a and b^+ the smallest w-homologically critical value larger than b.

If at least one of a, b are w-homologically regular, then the pair is weakly trivial in the sense of Chapter 2 Subsection 2.3.5. Define also

$$\boxed{\delta_r^f(a, b) := \lim_{\epsilon \to 0} F_r(B(a, b; \epsilon))} .$$

Clearly, $\dim \widehat{\delta}_r^f(a, b) = \delta_r^f(a, b)$. Let supp δ_r^f be the set

$$\boxed{\text{supp } \delta_r^f := \{(a, b) \in \mathbb{R}^2 \mid \delta_r^f(a, b) \neq 0\}} .$$

Note that neither $\widehat{\delta}_r$, nor δ_r^f are in general configurations, since their support is not necessary finite (unless X is compact); they are, however, maps from X to the set of finite-dimensional vector spaces or to $\mathbb{Z}$.

Observation 5.2. For any (a, b), $a, b \in \mathbb{R}$, the direct system $\{\mathbb{F}_r^f(B(a, b, \epsilon), \epsilon > 0\}$ stabilizes and $\widehat{\delta}_r^f(a, b) = \mathbb{F}^f(B(a, b; \epsilon))$ for some small enough ϵ. Moreover, $\delta_r^f(a, b) \neq 0$ implies that $a, b \in \text{w}\mathcal{C}r(f)$. In particular, supp δ_r^f is a discrete subset of $\mathbb{R}^2$. If f is w-homologically tame, then for any $(a, b), a, b \in \text{w}\mathcal{C}r(f)$,

$$\widehat{\delta}_r^f(a, b) = \mathbb{F}_r^f(B(a, b; \epsilon))$$

for ϵ with $0 < \epsilon < \widetilde{\epsilon}(f)$.

Recall that for a box $B = (a', a] \times [b, b')$ we have the canonical projections $\pi^B_{ab,r} : \mathbb{F}^f_r(a, b) \to \mathbb{F}^f_r(B)$, and for $B' = (a'', a] \times [b, b'')$, $a'' \leq a' < a$, $b'' \geq b' > b$, we have $\pi^B_{B',r} : \mathbb{F}_r(B') \to \mathbb{F}_r(B)$, the canonical surjective linear map between quotients spaces induced by the inclusion $B \subseteq B'$. Clearly

$$\pi^B_{ab,r} = \pi^B_{B',r} \cdot \pi^{B'}_{ab,r}.$$

Define the surjective linear map

$$\pi_r(a, b) : \mathbb{F}(a, b) \to \varinjlim_{\epsilon \to 0} \mathbb{F}(B(a, b; \epsilon)) = \widehat{\delta}^f_r(a, b)$$

to be

$$\boxed{\pi_r(a, b) := \varinjlim_{\epsilon \to 0} \pi^{B(a,b;\epsilon)}_{ab,r}}.$$

Definition 5.3. A *splitting* is a linear map $s_r(a, b) : \widehat{\delta}^f_r(a, b) \to \mathbb{F}_r(a, b)$ which satisfies $\pi_r(a, b) \cdot s_r(a, b) = id$, and hence is injective. A *special injection* is an injective linear map $i_r(a, b) : \widehat{\delta}^f_r(a, b) \to H_r(X)$ which is a composition of a splitting $s_r(a, b)$ with the inclusion $\mathbb{F}_r(a, b) \subset H_r(X)$.

Then for $B' = (\alpha', a'] \times [b', \beta')$ with $a \in (\alpha' a']$, $b \in [b', \beta')$ and $i_r(a, b) : \widehat{\delta}^f_r(a, b) \to H_r(X)$ a special injection, one has $\mathrm{img}(i_r(a, b)) \subset \mathbb{F}_r(a', b')$ and $\mathrm{img}(s_r(a, b)) \cap \mathbb{F}_r(\alpha', \beta') = 0$, which makes injective the composition

$$i^{B'}_r(a, b) := \pi^B_{ab,r} \cdot i_r(a, b).$$

The following diagram reviews the linear maps considered so far. The reader should have in mind the box $B = B_1 \sqcup B_2$ as in Figures 5.3 and 5.4.

$$(5.3)$$

Propositions 5.3 and 5.4 coupled with the above definitions imply

Corollary 5.2. *For a special injection $i_r(a, b)$, it holds that*

(i) *if $(a, b) \in B_2$, then $\pi^{B_2}_{B,r} \cdot i^B_r(a, b)$ is injective.*

(ii) *if $(a, b) \in B_1$, then $\pi^{B_2}_{B,r} \cdot i^B_r(a, b)$ is zero.*

One should also consider infinite boxes. These are sets of the form $(-\infty, a] \times [b, \infty)$, $(-\infty, a] \times [b, b')$, and $(a', a] \times [b, \infty)$ with the first denoted by $B(a, b; \infty) := (-\infty, a] \times [b, \infty)$. For consistency with the previous notations, we denote

(i) $\mathbb{I}_{-\infty}^f(r) = \bigcap_{a \in \mathbb{R}} \mathbb{I}_a^f(r)$ and $\mathbb{I}_f^\infty(r) = \bigcap_{b \in \mathbb{R}} \mathbb{I}_f^b(r)$,

(ii) $\mathbb{F}_r^f(-\infty, b) := \mathbb{I}_{-\infty}^f(r) \cap \mathbb{I}_f^b(r)$ and $\mathbb{F}_r^f(a, \infty) := \mathbb{I}_a^f(r) \cap \mathbb{I}_f^\infty(r)$,

(iii) $\mathbb{F'}_r^f(B(a, b; \infty)) := \mathbb{F}_r^f(-\infty, b) + \mathbb{F}_r^f(a, \infty)$,

(iv) $\mathbb{F}_r^f(B(a, b; \infty)) := \mathbb{F}_r^f(a, b) / \mathbb{F'}_r^f(B(a, b; \infty))$.

Because of the finite-dimensionality of $\mathbb{F}_r^f(a, b)$, and for m, M large enough it holds that

$$\text{supp}(\delta_r^f) \cap ((-\infty, a] \times [b, \infty)) = \text{supp}(\delta_r^f) \cap ((-m, a] \times [b, M)),$$
$$\text{supp}(\delta_r^f) \cap ((-\infty, a] \times [b, b')) = \text{supp}(\delta_r^f) \cap ((-m, a] \times [b, b')), \quad (5.4)$$
$$\text{supp}(\delta_r^f) \cap ((a', a] \times [b, \infty)) = \text{supp}(\delta_r^f)(\cap((a', a] \times [b, M)),$$

and then

$$\mathbb{F}_r((-\infty, a] \times [b, \infty)) = \mathbb{F}_r((-m, a] \times [b, M)),$$
$$\mathbb{F}_r((-\infty, a] \times [b, b')) = \mathbb{F}_r((-m, a] \times [b, b')), \quad (5.5)$$
$$\mathbb{F}_r((a', a] \times [b, \infty)) = \mathbb{F}_r((a', a] \times [b, M)).$$

Choose a collection of special injections $\{i_r(a, b) \mid (a, b) \in \text{supp}(\widetilde{\delta_r^f})\}$ given by the collection of splittings $S = \{s_r(a, b) \mid (a, b) \in \text{supp}(\widetilde{\delta_r^f})\}$, consider the projection

$$\pi(r) : H_r(\widetilde{X}) \to H_r(\widetilde{X}) / (\mathbb{I}_{-\infty}^f(r) + \mathbb{I}_f^\infty(r)),$$

and denote

$$
\begin{array}{ll}
{}^S I_r : & \bigoplus_{(a,b) \in \text{supp}(\widetilde{\delta_r^f})} \widehat{\delta}_r^f(a, b) \to H_r(\widetilde{X}) \\[2em]
{}^S \widehat{I}_r : & \bigoplus_{(a,b) \in \text{supp}(\widetilde{\delta_r^f})} \widehat{\delta}_r^f(a, b) \to H_r(\widetilde{X}) / \mathbb{I}_{-\infty}^f(r) + \mathbb{I}_f^\infty(r) \\[2em]
{}^S I_r^B : & \bigoplus_{(a,b) \in \text{supp}(\widetilde{\delta_r^f}) \cap B} \widehat{\delta}_r^f(a, b) \to \mathbb{F}_r^f(B)
\end{array}
$$

with

$$^S I_r = \sum_{(a,b)\in\mathrm{supp}(\delta_r^{\tilde{f}})} i_r(a,b)$$

$$^S \widehat{I}_r = \pi(r) \cdot {}^S I_r, \quad \pi(r) : H_r(\widetilde{X}) \to H_r(\widetilde{X})/\mathbb{I}^f_{-\infty}(r) + \mathbb{I}^\infty_f(r)$$

$$^S I_r^B = \sum_{(a,b)\in\mathrm{supp}(\delta_r^{\tilde{f}})\cap B} i_r^B(a,b).$$

For a finite box or, in view of (5.9), for an infinite box B and for $\Sigma \subseteq \mathrm{supp}(\delta_r^f)$ one denotes by

$$\begin{cases} {}^S I_r(\Sigma) \\ {}^S I_r(\Sigma \cap B) \end{cases}, \quad \begin{cases} {}^S \widehat{I}_r(\Sigma) \\ {}^S \widehat{I}_r(\Sigma \cap B) \end{cases}, \quad \text{and } {}^S \widehat{I}_r^B(\Sigma \cap B))$$

the restrictions of $^S I_r$, $^S \widehat{I}_r$, and $^S \widehat{I}_r^B$ to $\bigoplus_{(a,b)\in\mathrm{supp}(\delta_r^{\tilde{f}})\cap P} \widehat{\delta}_r^f(a,b)$, where $P = B$ or $P = B \cap \Sigma$.

One has

Observation 5.3. For $B = B_1 \sqcup B_2$ as in Fig 5.4 or Fig 5.5 and $\Sigma \subseteq \mathrm{supp}\,\delta_r^{\tilde{f}}$ with $\Sigma = \Sigma_1 \sqcup \Sigma_2$, $\Sigma_1 \subseteq B_1$, $\Sigma_2 \subseteq B_2$ the diagram

$$\begin{array}{ccccc}
\mathbb{F}_r(B_1) & \longrightarrow & \mathbb{F}_r(B) & \longrightarrow & \mathbb{F}_r(B_2) \\
\big\uparrow {\scriptstyle I_r^{B_1}(\Sigma_1)} & & \big\uparrow {\scriptstyle I_r^B(\Sigma)} & & \big\uparrow {\scriptstyle I_r^{B_2}(\Sigma_2)} \\
\bigoplus_{(a,b)\in\Sigma_1}\widehat{\delta}_r^{\tilde{f}}(a,b) & \longrightarrow & \bigoplus_{(a,b)\in\Sigma}\widehat{\delta}_r^{\tilde{f}}(a,b) & \longrightarrow & \bigoplus_{(a,b)\in\Sigma_2}\widehat{\delta}_r^{\tilde{f}}(a,b)
\end{array}$$

is commutative.

In view of Corollary 5.2 and of the exactness of both rows, the injectivity of $I_r^{B_1}(\Sigma_1)$ and $I_r^{B_2}(\Sigma_2)$ implies the injectivity of $I_r^{B_1}(\Sigma_1)$ and the following proposition.

Proposition 5.6.

(i) *For any box $B = (a', a] \times [b, b')$, $-\infty \le a' < a, b < b' \le \infty$, the set $\mathrm{supp}\,\delta_r^f \cap B$ is finite.*

(ii) *For any $\Sigma \subseteq \mathrm{supp}\,\delta_r^f$ or $\Sigma \subseteq \mathrm{supp}\,\delta_r^f \cap B$, the linear maps $I_r(\Sigma)$, respectively $^S I_r^B(\Sigma)$, are injective, and for any box B, the linear map $^S I_r^B$ is an isomorphism.*

(iii) *If X is compact, $-m < \inf f$, and $\sup f < M$, then $H_r(X) = \mathbb{F}_r((-m, M] \times [-m, M))$ and $^S I_r$ is an isomorphism.*

(iv) *The linear map $^S\widehat{I}_r$ is an isomorphism.*

Proof. Item (i) is a straightforward consequence of the finite-dimensionality of $\mathbb{F}_r(a,b)$ and of Corollary 5.2. Item (ii) is verified by induction on the cardinality of Σ as follows:

- If the cardinality of Σ is 1, then the statement follows from Observation 5.2.
- If all elements of Σ, (α_i, β_i), $i = 1, \ldots, k$, have the same first component $\alpha_i = a$, then the statement follows by induction on k. Considers the box $B = B_1 \sqcup B_2$ as in Fig. 5.5 such that B_2 contains one element of Σ, say (α_1, β_1), and B_1 contains the remaining $k-1$ elements. The injectivity follows from Observation 5.2 in view of the injectivity of $I_r^{B_2}(\Sigma \cap B_2)$ and of $I_r^{B_1}(\Sigma \cap B_1)$ assumed in the induction step.
- In general, write Σ as the disjoint union $\Sigma = \Sigma_1 \sqcup \Sigma_2 \sqcup \cdots \sqcup \Sigma_k$ such that each Σ_i contains all points of Σ with the same first component a_i, and $a_k > a_{k-1} > \cdots > a_2 > a_1$. Now proceed again by induction on k. Decompose the box B as in Fig. 5.6, $B = B_1 \sqcup B_2$, so that $\Sigma_1 \subset B_2$ and $(\Sigma \setminus \Sigma_1) \subset B_1$. Then the injectivity of $I_r^B(\Sigma)$ follows by using Observation 5.3 from the injectivity of $I_r^{B_2}(\Sigma_1)$ and the injectivity of $I_r^{B_1}(\Sigma \cap B_1)$ assumed in the induction step.

Item (iii) follows from Corollary 5.1 and Item (ii).

Item (iv): Note that for $k < k'$

$$(\mathbb{I}_{-\infty} \cap \mathbb{I}^{-k'} + \mathbb{I}_{k'} \cap \mathbb{I}^\infty) \cap \mathbb{I}^{-k} \cap \mathbb{I}_k = \mathbb{I}_{-\infty} \cap \mathbb{I}^{-k} + \mathbb{I}_k \cap \mathbb{I}^\infty$$

and then

$$H_r(X) = \varinjlim_{k \to \infty} \mathbb{F}_r(k, -k) = \varinjlim_{k \to \infty} \mathbb{I}^{-k} = \varinjlim_{k \to \infty} \mathbb{I}_k \, . \, ^2$$

Then

$$\varinjlim \mathbb{F}(k, -k) / (\mathbb{I}_{-\infty} \cap \mathbb{I}^{-k} + \mathbb{I}_k \cap \mathbb{I}^\infty) = H_r(X) / (\mathbb{I}_{-\infty} + \mathbb{I}^\infty).$$

$\square$

Let $D(a, b; \epsilon) := (a - \epsilon, a + \epsilon] \times [b - \epsilon, b + \epsilon)$. If $x = (a, b)$, we also write $D(x; \epsilon)$ for $D(a, b; \epsilon)$.

---

[2]For simplicity, we omit f and r in the notation.

Proposition 5.7. (cf. [Burghelea, D., Haller, S. (2015)])

    Let $f : X \to \mathbb{R}$ be a tame map and $\epsilon < \epsilon(f)/3$. For any map $g : X \to \mathbb{R}$ which satisfies $\|f - g\|_\infty < \epsilon$ and any critical values $a, b \in \mathcal{C}r(f)$ one has:

$$\sum_{x \in D(a,b;2\epsilon)} \delta_r^g(x) = \delta_r^f(a,b), \tag{5.6}$$

$$\operatorname{supp} \delta_r^g \subset \bigcup_{(a,b) \in \operatorname{supp} \delta_r^f} D(a,b;2\epsilon). \tag{5.7}$$

If, in addition, $H_r(X)$ is equipped with a Hilbert space structure ($\kappa = \mathbb{R}$ or $\mathbb{C}$) then the above statement can be strengthen to

$$x \in D(a,b;2\epsilon) \Rightarrow \widehat{\delta}_r^g(x) \subseteq \widehat{\delta}_r^f(a,b) \text{ and } \bigoplus_{x \in D(a,b;2\epsilon)} \widehat{\delta}_r^g(x) = \widehat{\delta}_r^f(a,b). \tag{5.8}$$

    Proposition 5.7 implies that for any map g in an ϵ-neighborhood of a tame map f (with respect to the $\|\cdots\|_\infty$ norm on the space of continuous functions) the support of δ_r^g lies in a 2ϵ-neighborhood of the support of δ_r^f, and in case X is compact its cardinality (points in the support counted with multiplicities) is equal to $\dim H_r(X)$. Recall that for $h : X \to \mathbb{R}$, $\|h\|_\infty := \sup_{x \in X} |h(x)|$.

Proof of Proposition 5.7

Consider a collection of real numbers $C := \{\cdots c_i < c_{i+1} < c_{i+2} \cdots , i \in \mathbb{Z}\}$ which satisfies the following properties:

    (i) $\mathcal{C}r(f) \subseteq C$,
    (ii) $c_{i+1} - c_i > \epsilon(f)$,
    (iii) $\lim_{i \to \infty} c_i = \infty$,
    (iv) $\lim_{i \to -\infty} c_i = -\infty$.

We establish two intermediate results, Lemmas 5.1 and 5.2.

Lemma 5.1. *For f as in Proposition 5.7 and $c_i, c_j \in C$ one has:*

$$\begin{aligned} \widehat{\delta}_r^f(c_i, c_j) &= \mathbb{F}_r^f((c_{i-1}, c_i] \times [c_j, c_{j+1})) \\ &= \mathbb{F}_r^f(c_i, c_j) / \left(\mathbb{F}_r^f(c_{i-1}, c_j) + \mathbb{F}_r^f(c_i, c_{j+1}) \right), \end{aligned} \tag{5.9}$$

and therefore

$$\begin{aligned} \delta_r^f(c_i, c_j) &= F_r^f((c_{i-1}, c_i] \times [c_j, c_{j+1})) \\ &= F_r^f(c_{i-1}, c_{j+1}) + F_r^f(c_i, c_j) - F_r^f(c_{i-1}, c_j) - F_r^f(c_i, c_{j+1}). \end{aligned} \tag{5.10}$$

This lemma follows from Proposition 5.2 and Observation 5.2.

Lemma 5.2. *Suppose f is tame. Let $a = c_i, b = c_j$, $c_i, c_j \in C$ and $\epsilon < \epsilon(f)/3$. If g is a continuous map with $\|f - g\|_\infty < \epsilon$, then*

$$
\begin{aligned}
\mathbb{F}_r^g(a - 2\epsilon, b + 2\epsilon) &= \mathbb{F}_r^f(c_{i-1}, c_{j+1}), \\
\mathbb{F}_r^g(a + 2\epsilon, b - 2\epsilon) &= \mathbb{F}_r^f(c_i, c_j), \\
\mathbb{F}_r^g(a + 2\epsilon, b + 2\epsilon) &= \mathbb{F}_r^f(c_i, c_{j+1}), \\
\mathbb{F}_r^g(a - 2\epsilon, b - 2\epsilon) &= \mathbb{F}_r^f(c_{i-1}, c_j).
\end{aligned}
\tag{5.11}
$$

Proof. Since $\|f - g\|_\infty < \epsilon$, Observation 5.1 Item 3 shows that

$$
\begin{aligned}
\mathbb{F}_r^f(a - 3\epsilon, b + 3\epsilon) &\subseteq \mathbb{F}_r^g(a - 2\epsilon, b + 2\epsilon) \subseteq \mathbb{F}_r^f(a - \epsilon, b + \epsilon), \\
\mathbb{F}_r^f(a + \epsilon, b - \epsilon) &\subseteq \mathbb{F}_r^g(a + 2\epsilon, b - 2\epsilon) \subseteq \mathbb{F}_r^f(a + 3\epsilon, b - 3\epsilon), \\
\mathbb{F}_r^f(a + \epsilon, b + 3\epsilon) &\subseteq \mathbb{F}_r^g(a + 2\epsilon, b + 2\epsilon) \subseteq \mathbb{F}_r^f(a + 3\epsilon, b + \epsilon), \\
\mathbb{F}_r^f(a - 3\epsilon, b - \epsilon) &\subseteq \mathbb{F}_r^g(a - 2\epsilon, b - 2\epsilon) \subseteq \mathbb{F}_r^f(a - \epsilon, b - 3\epsilon).
\end{aligned}
\tag{5.12}
$$

Further, since $3\epsilon < \epsilon(f)$ it holds that

$$
\begin{aligned}
\mathbb{F}_r^f(a - 3\epsilon, b + 3\epsilon) &= \mathbb{F}_r^f(a - \epsilon, b + \epsilon), \\
\mathbb{F}_r^f(a + \epsilon, b - \epsilon) &= \mathbb{F}_r^f(a + 3\epsilon, b - 3\epsilon), \\
\mathbb{F}_r^f(a + \epsilon, b + 3\epsilon) &= \mathbb{F}_r^f(a + 3\epsilon, b + \epsilon), \\
\mathbb{F}_r^f(a - 3\epsilon, b - \epsilon) &= \mathbb{F}_r^f(a - \epsilon, b - 3\epsilon),
\end{aligned}
\tag{5.13}
$$

which implies that in the relations (5.12) the inclusion "$\subseteq$" is actually an equality.

Note that in view Proposition 5.2, for $\epsilon', \epsilon'' < \epsilon(f)$ one has

$$
\begin{aligned}
\mathbb{F}_r^f(c_{i-1}, c_{j+1}) &= \mathbb{F}_r^f(a - \epsilon', b + \epsilon''), \\
\mathbb{F}_r^f(c_i, c_j) &= \mathbb{F}_r^f(a + \epsilon', b - \epsilon''), \\
\mathbb{F}_r^f(c_i, c_{j+1}) &= \mathbb{F}_r^f(a + \epsilon', b + \epsilon''), \\
\mathbb{F}_r^f(c_{i-1}, c_j) &= \mathbb{F}_r^f(a - \epsilon', b - \epsilon'').
\end{aligned}
\tag{5.14}
$$

Then (5.12) and (5.14) imply the equalities (5.11), hence the statement of Lemma 5.2. $\qquad\square$

To prove Proposition 5.7, observe that Lemma 5.2 gives (for $a = c_i$, $b = c_j$ with $c_i, c_j \in C$) the equality

$\mathbb{F}^g((a - 2\epsilon, a + 2\epsilon] \times [b - 2\epsilon, b + 2\epsilon)) = \mathbb{F}^f((c_{i-1}, c_i] \times [c_j, c_{j+1}))$.

This combined with Lemma 5.1 implies $\mathbb{F}^g((a - 2\epsilon, a + 2\epsilon] \times [b - 2\epsilon, b + 2\epsilon)) = \widehat{\delta}^f(a, b)$. Applying now Proposition 5.6 we obtain the equality (5.6), and this not only for critical values, but for any $a, b \in C$.

To check inclusion (5.7), observe that:

(i) $\|f - g\|_\infty < \epsilon$ implies $X_a^f \subset X_{a+\epsilon}^g \subset X_{a+2\epsilon}^f$ and $X_f^b \subset X_g^{b-\epsilon} \subset X_f^{b-2\epsilon}$, and when $a, b \in C$

$$\mathbb{F}^f(a, b) \subseteq \mathbb{F}^g(a + \epsilon, b - \epsilon) \subseteq \mathbb{F}^f(a + 2\epsilon, b - 2\epsilon). \qquad (5.15)$$

(ii) When $\epsilon < \epsilon(f)/3$ the inclusions (5.15) imply

$$\mathbb{F}^f(a, b) = \mathbb{F}^g(a + \epsilon, b - \epsilon) = \mathbb{F}^f(a + 2\epsilon, b - 2\epsilon),$$

which in view of Proposition 5.6 gives

$$
\sum_{x \in (-\infty, a] \times (b, \infty)} \delta_r^f(x) = \sum_{y \in (-\infty, a-\epsilon] \times (b-\epsilon, \infty)} \delta_r^g(y)
$$
$$
= \sum_{x \in (-\infty, a+2\epsilon] \times (b-2\epsilon, \infty)} \delta_r^f(x). \qquad (5.16)
$$

Since $\mathbb{R}^2 = \bigcup_{i \in \mathbb{Z}} B(c_i, c_{-i}; \infty)$, the equalities (5.6) and (5.16) rule out the existence of $x \in \operatorname{supp} \delta_r^g$ away from $\bigcup_{x \in \operatorname{supp}(\delta_r^f)} D(x; 2\epsilon)$, which finishes the proof of Proposition 5.7.

Let K be a compact ANR and $f : X \to \mathbb{R}$. Denote by $\overline{f}_K; X \times K \to \mathbb{R}$, the composition $f \cdot \pi_K$ with $\pi_K : X \times K \to X$, the projection on the first factor. If f is weakly tame, then so is $\overline{f}_K$, and the sets of critical values of f and of $\overline{f}_K$ coincide. Moreover, by the Künneth theorem about the homology of the cartesian product of two spaces, one has

Observation 5.4.

(i) $\mathbb{F}_r^{\overline{f}_K}(a, b) = \bigoplus_{0 \le k \le r} \mathbb{F}_k^f(a, b) \otimes H_{r-k}(K)$, and therefore

(ii) $\widehat{\delta}_r^{\overline{f}_K}(a, b) = \bigoplus_{0 \le k \le r} \widehat{\delta}_k^f(a, b) \otimes H_{r-k}(K)$, and when K is acyclic

(iii) $\widehat{\delta}_r^{\overline{f}_K}(a, b) = \widehat{\delta}_k^f(a, b)$.

Note that the embedding $I : C(X; \mathbb{R}) \to C(X \times K; \mathbb{R})$ defined by $I(f) = \overline{f}_K$ is an isometry when both spaces are equipped with the norm $\| \cdot \|_\infty$. Note also that when K is acyclic one has $\delta_r^f = \delta_r^{I(f)}$ and $\widehat{\delta}_r^f = \widehat{\delta}_r^{I(f)}$ since $H_r(X) = H_r(X \times K)$.

5.2 The case of real-valued maps $f : X \to \mathbb{R}$

5.2.1 *The main results*

Theorem 5.1. (Topological results) *Suppose X is compact ANR and $f : X \to \mathbb{R}$ a continuous map. Then*

(1) $\delta_r^f(x) \neq 0$ *with* $x = (a, b)$ *implies that both* $a, b \in \mathrm{w}\mathcal{C}r(f)$.

(2) $\sum_{x \in \mathbb{R}^2} \delta_r^f(x) = \dim H_r(X)$ *and* $\bigoplus_{x \in \mathbb{R}^2} \widehat{\delta}_r^f(x) = H_r(X)$. *In particular, $\delta_r^f \in \mathrm{Conf}_{\dim H_r(X)}(\mathbb{R}^2)$ and $\widehat{\delta}_r^f$ is a configuration of subquotients, as considered in Subsection 2.3.5.*

(3) *If $H_r(X)$ is equipped with a Hilbert space structure, then the vector space $\widehat{\delta}_r^f(a, b)$ is realized as a subspace $\widehat{\widehat{\delta}}_r^f(a, b) \subseteq H_r(X)$ such that $\widehat{\widehat{\delta}}_r^f \in \mathrm{CONF}_{H_r(X)}^{\perp}(\mathbb{R}^2)$.*

(4) *If X is homeomorphic to a finite simplicial complex or to a compact Hilbert cube manifold, then for an open and dense set of maps f in the space of continuous maps equipped with the compact-open topology one has $\delta_r^f(x) = 0$ or 1.*

The statements (1) and (2) formulated in terms of barcodes were verified first in [Burghelea, D., Dey, T (2013)] and [Burghelea, D., Haller, S. (2015)] under the hypothesis that f is a tame map.

Let $\underline{D}$ be the metric on $\mathrm{Conf}_{\beta_r(X)}(\mathbb{R}^2)$, $\beta_r(X) = \dim H_r(X)$, induced by the standard Euclidean metric on $\mathbb{R}^2$ which also defines the collision topology. With this metric $\mathrm{Conf}_{\beta_r(X)}(\mathbb{R}^2)$ is a complete metric space isometric to $\mathbb{C}^{\beta_r(X)}$ equipped with the standard metric.

Theorem 5.2. (Stability) *Suppose X is a compact ANR.*

(1) *The assignment $f \rightsquigarrow \delta_r^f$ provides a continuous map from the space of real-valued maps $C(X; \mathbb{R})$ equipped with the compact-open topology to the space of configurations $\mathrm{Conf}_{\beta_r(X)}(\mathbb{R}^2) = \mathbb{C}^{\beta_r}$ equipped with the collision topology (standard topology on $\mathbb{C}^{\beta_r}$).*

Moreover, with respect to the metric $\underline{D}$ on the space of configurations, one has

$$\underline{D}(\delta^f, \delta^g) \leq 2\|(f - g)\|_\infty.$$

(2) *If $\kappa = \mathbb{R}$ or $\mathbb{C}$, then the assignment $f \rightsquigarrow \widehat{\delta}_r^f$ is continuous with respect to both collision topologies considered in Subsection 2.3.5* [3].

---

[3] The continuity with respect to the first collision topology implies continuity with

Theorem 5.2 (1) was first established in [Burghelea, D., Haller, S. (2015)] under the hypothesis that X is homeomorphic to a finite simplicial complex. In view of this theorem the polynomial $P_r^f(z)$ associated to a tame map, as indicated in Chapter 4, makes sense for any continuous map.

Theorem 5.3. (Poincaré Duality)

(1) *Suppose X is a closed triangulable topological manifold[4] (in particular smooth manifold) of dimension n which is κ-orientable and f is a continuous map. Then $\delta_r^f(a, b) = \delta_{n-r}^f(b, a)$.*

(2) *In addition, any collection of isomorphisms[5] $H_k(X) \to H_k(X)^*$, $k = 0, 1, 2, \ldots, n$ induces the isomorphisms of the configurations $\widehat{\delta}_r^f$ and $\widehat{\delta}_{n-r}^f \cdot \tau$ with $\tau : \mathbb{R}^2 \to \mathbb{R}^2$ given by $\tau(a, b) = (b, a)$.*

Item (1) of this theorem was established in [Burghelea, D., Haller, S. (2015)] for tame maps f.

Proposition 5.8. *Suppose that $f : X \to \mathbb{R}$ is a tame real-valued map and $\mathcal{B}_r^c(f)$ and $\mathcal{B}_{r-1}^o(f)$ are the sets of r-closed and $(r-1)$-open barcodes as defined in Chapter 4. Then*

(1) *If $a \leq b$, then $\delta_r^f(a, b) = \{I \in \mathcal{B}_r^c(f) \mid I = [a, b]\}$.*
(2) *If $a > b$, then $\delta_r^f(a, b) = \{I \in \mathcal{B}_{(r-1)}^o(f) \mid I = (b, a)\}$.*

Proof. For f a tame map, Proposition 4.3 in Chapter 4 gives the following equalities.

If $a \leq b$, then $F_r^f(a, b) = \sharp\{I = [\alpha, \beta] \in \mathcal{B}_r^c(f) \mid [\alpha, \beta] \supseteq [a, b]\}$.

If $a > b$, then $F_r^f(a, b) = \sharp \begin{cases} I = [\alpha, \beta] \in \mathcal{B}_r^c(f), \ [\alpha, \beta] \subseteq [a, b], \\ I = (\alpha, \beta) \in \mathcal{B}_{r-1}^o(f), \ (\alpha, \beta) \subset [b, a]. \end{cases}$

Combining this with the equality $\delta_r^f(a, b) = F_r(B(a, b; \epsilon))$ for ϵ small enough and the equality

$$F_r^f(B(a, b; \epsilon)) = F_r^f(a, b) + F_r^f(a - \epsilon.b + \epsilon), -F_r^f(a - \epsilon, b) - F_r^f(a, b + \epsilon)$$

given by (5.2) the statement follows.

$\square$

---

respect to the second.

[4]The result remains true for topological manifolds based on the same arguments; however, a complete proof requires a number of technical results about topological manifolds too lengthy to be described here.

[5]In particular, the isomorphisms induced by Hilbert space structures on $H_r(X)$ for $\kappa = \mathbb{R}$ or $\mathbb{C}$; when M is a smooth closed manifold a Riemannian metric induces such Hilbert space structures.

5.2.2   *Proof of Theorem 5.1*

Items (1) and (2) follow from Observation 5.2 and Proposition 5.6.

Item (3): Any collection of splittings $S = \{s_r(a,b)\}$ realizes the vector spaces $\widehat{\delta}_r^f$ as $\widehat{\widehat{\delta}}_r^f(a,b) := \mathrm{img}(i_r(a,b)) \subseteq H_r(X)$. A Hilbert space structure on $H_r(X)$ provides canonical splittings which make these subspaces orthogonal, hence $\widehat{\widehat{\delta}}_r^f \in \mathrm{CONF}^\perp_{H_r(X)}(\mathbb{R}^2)$.

Item (4): In view of Theorem 5.2, whose proof does not involve Theorem 5.1, it suffices to establish that the set of tame functions f with δ_r^f taking only the values 0 and 1 is dense in the space of all continuous functions. For this purpose we introduce now Property G, supposed to hold for a "generic" set of tame maps.

Property G: We say that a tame map $f : X \to \mathbb{R}$ satisfies property G if there exists a finite sequence of real numbers $a = a_0 < a_1 < \cdots < a_n < a_{n+1} = b$ such that:

(i) $\mathbb{I}_a^f(r) = 0$, $\mathbb{I}_b^f(r) = H_r(X)$;

(ii) For any $i \geq 1$, $\dim(\mathbb{I}_{a_i}^f / \mathbb{I}_{a_{i-1}}^f) \leq 1$.

The verification of Item (4) is based on the Observations 5.5 and 5.6 below.

Observation 5.5. For any tame map f which satisfies property G the configuration δ_r^f takes only the values 0 and 1.

If f has Property G, then $\dim(\mathbb{I}_{a_i}^f / \mathbb{I}_{a_{i-1}}^f) \leq 1$ for $a_i = c_i$, $i = 1, \ldots, n$, because for $\alpha < \beta$ with no critical value in the open interval (α, β) and β a regular value the inclusion $X_\alpha^f \subset X_\beta^f$ induces an isomorphism in homology and for any $a' \leq a \leq b \leq b'$ $\dim(\mathbb{I}_b^f(r) / \mathbb{I}_a^f(r)) \leq \dim(\mathbb{I}_{b'}^f(r) / \mathbb{I}_{a'}^f(r))$.

If so, then for any two consecutive critical values $c_{i-1} < c_i$ and any other critical value c_j, the inclusion $\mathbb{F}_r(c_{i-1}, c_j) \subseteq \mathbb{F}_r(c_i, c_j)$ has cokernel of dimension at most one, which by (5.9) in Lemma 5.1 implies that δ_r^f takes only the values 0 and 1.

Based on this observation, if X is a compact smooth manifold (possibly with boundary), any Morse function $f : X \to \mathbb{R}$ which takes different values at different critical points has property G. Indeed, if $\{\cdots c_i < c_{i+1} < \cdots\}$ is the collection of all critical values, then $X_{c_{i+1}}^f$ is homotopy equivalent to a space obtained from $X_{c_i}^f$ by adding a closed disk D^k along $\partial D^k = S^{k-1}$ or $\partial D_+^k = D^{k-1}$, which insures that Property G is satisfied. Since the set of such Morse functions is dense in the space of all continuous functions equipped with the compact-open topology, item (4) is verified for X the underlying space of a closed smooth manifold.

If X is a compact Hilbert cube manifold, then X homeomorphic to $M \times Q$, with M a compact smooth manifold (possible with boundary), and any continuous map $f : X \to \mathbb{R}$ is arbitrarily close to $\bar{h}_Q$, for $h : M \to \mathbb{R}$ a Morse function. This observation establishes item (4) for all compact Hilbert cube manifolds.

If X is a finite simplicial complex one needs the following observation.

Observation 5.6. If X is a finite simplicial complex and $a < b$, then one can construct a map $h : X \to \mathbb{R}$ that is simplicial on the barycentric subdivision of X and has the following properties:

    (i) $a < h(x) < b$,

    (ii) h takes different values on the barycenters of different simplices,

    (iii) the value of h at the barycenter of a simplex σ is strictly larger than the value of h at the barycenter of any of its faces.

The construction is straightforward. Such map satisfies Property G, since adding a simplex to a finite simplicial complex might change the dimension of the homology by at most one unit, and for any α, X_α^h retracts by deformation to the simplicial complex generated by the barycenters on which h takes values smaller or equal to α.

Let $f : X \to \mathbb{R}$ a simplicial map, X a finite simplicial complex, with critical values $\{\cdots c_{i-1} < c_i < \cdots\}$, and suppose that for some i one has $\dim(\mathbb{I}_{c_i}^f / \mathbb{I}_{c_{i-1}}^f) \geq 2$. We modify f arbitrarily little to a simplicial (for some subdivision) map g. We choose $\epsilon < \epsilon(f)/2$ and a subdivision of X which makes $f^{-1}(c_i \pm \epsilon/2)$, $f^{-1}(c_i)$), and then $f^{-1}([c_i - \epsilon/2, c_i + \epsilon/2])$, $f^{-1}([c_i, c_i + \epsilon])$), into subcomplexes. Next we take the barycentric subdivision of this subdivision. We replace f by a simplicial map g for the new triangulation. The map g takes the same value as f on the barycenters of simplices not contained in $f^{-1}(c_i)$ and as h the map constructed using Observation 5.6 for $a = c_i - \epsilon/2$, $b = c_i + \epsilon/2$. The critical values of the map g are the critical values of f plus the critical values of $h = g|_{f^{-1}(c_i)}$. We leave the reader to check that g satisfies property G based on the fact that the map h does and $\epsilon < \epsilon(f)$. By construction g, differs from f by less than ϵ.

Since simplicial maps (for some subdivision) are dense in the space of continuous maps and any simplicial map is arbitrarily closed to one which satisfies Property G, item (4) is established.

5.2.3 *Proof of Theorem 5.2* (stability property)

The stability theorem is a consequence of Proposition 5.7. In order to explain this, we begin with a few observations.

(1) Consider the space of continuous real-valued maps $C(X, \mathbb{R})$, X a compact ANR, equipped with the compact-open topology which is induced by the metric $D(f, g) := \sup_{x \in X} |f(x) - g(x)| = \|f - g\|_\infty$. This metric is complete.

(2) Observe that if $f, g \in C(X; \mathbb{R})$, then for any $t \in [0, 1]$ the map $h_t := t f(x) + (1 - t) g(x)$ lies in $C(X; \mathbb{R})$, and for any $0 = t_0 < t_1 < \ldots < t_{N-1} < t_N = 1$ it holds that

$$D(f, g) = \sum_{0 \leq i < N} D(h_{t_{i+1}}, h_{t_i}). \tag{5.17}$$

(3) For X a simplicial complex, let $\mathcal{U} \subset C(X; \mathbb{R})$ denote the subset of p.l. maps. Recall that the map f is piecewise linear (p.l.) if f is simplicial with respect to some subdivision of X; moreover, for any two p.l. maps f and g there exists a common subdivision of X which makes f and g simultaneously simplicial. Hence, the map h_t is a simplicial map for any t. Then:

(i) $\mathcal{U}$ is a dense subset in $C(X, \mathbb{R})$,
(ii) if $f, g \in \mathcal{U}$, then $h_t \in \mathcal{U}$, hence $\epsilon(h_t) > 0$; therefore, for any $t \in [0, 1]$ there exists $\delta(t) > 0$ such that $t' \in (t - \delta(t), t + \delta(t))$ implies $D(h_{t'}, h_t) < \epsilon(h_t)/3$.

These two statements are not hard to check. Indeed item (i) follows from the fact that continuous maps can be approximated with arbitrary accuracy by p.l. maps, while item (ii) follows from the continuity in t of the family h_t and from the compactness of X.

(4) Consider $\operatorname{Conf}_{\beta_r}(\mathbb{R}^2) = \mathbb{C}^{\beta_r}$, $\beta_r = \dim(H_r(X)$, the set of configurations of points with multiplicity of cardinality β_r, cf. Subsection 2.3.5 Chapter 2 equipped with the metric $\underline{D}$, which is complete. Since any map in $\mathcal{U}$ is tame, in view of Proposition 5.7, $f, g \in \mathcal{U}$ with $D(f, g) < \epsilon(f)/3$, implies

$$\underline{D}(\delta_r^f, \delta_r^g) \leq 2 D(f, g). \tag{5.18}$$

To prove Theorem 5.2 one first checks that the inequality (5.18) holds for all $f, g \in \mathcal{U}$. To do that we start with $f, g \in \mathcal{U}$ and consider the homotopy h_t, $t \in [0, 1]$, defined above.

Choose a sequence $0 < t_1 < t_3 < t_5 < \cdots < t_{2N-1} < 1$ such that for $i = 1, \ldots, 2N - 1$ the intervals $(t_{2i-1} - \delta(t_{2i-1}), t_{2i-1} + \delta(t_{2i-1}))$ cover $[0, 1]$ and $(t_{2i-1}, t_{2i-1} + \delta(t_{2i-1})) \cap (t_{2i+1} - \delta(t_{2i+1}), t_{2i+1}) \neq \emptyset$. This is possible in view of the compactess of the segment $[0, 1]$. Take $t_0 = 0$, $t_{2N} = 1$, and $t_{2i} \in (t_{2i-1}, t_{2i-1} + \delta(t_{2i-1})) \cap (t_{2i+1} - \delta(t_{2i+1}))$. To simplify the notation, abbreviate h_{t_i} to h_i.

In view of (3) (ii) and the inequality (5.18) above,

$|t_{2i-1} - t_{2i}| < \delta(t_{2i-1})$ implies $D(\delta^{h_{2i-1}}, \delta^{h_{2i}}) < 2D(h_{2i-1}, h_{2i})$, and

$|t_{2i} - t_{2i+1}| < \delta(t_{2i+1})$ implies $D(\delta^{h_{2i}}, \delta^{h_{2i+1}}) < 2D(h_{2i}, h_{2i+1})$.

Then we have

$$\underline{D}(\delta^f, \delta^g) \leq \sum_{0 \leq i < 2N-1} D(\delta^{h_i}, \delta^{h_{i+1}}) \leq 2 \Big(\sum_{0 \leq i < 2N-1} D(h_i, h_{i+1}) \Big) = 2D(f, g).$$

When X is a simplicial complex, then in view of the denseness of $\mathcal{U}$ and of the completeness of the metrics on $C(X; \mathbb{R})$ and on $\mathrm{Conf}_{\beta_r}(\mathbb{R}^2)$ the inequality (5.18) extends to the entire $C(X; \mathbb{R})$. This because the assignment $\mathcal{U} \ni f \rightsquigarrow \delta_r^f \in \mathrm{Conf}_{\beta_r}(\mathbb{R}^2)$ preserves the Cauchy sequences.

Next we verify the inequality (5.18) for $X = K \times Q$, with K a simplicial complex and Q the Hilbert cube. For this purpose one writes $Q := I^k \times Q^{\infty-k}$ and on says that the map $f : K \times Q \to \mathbb{R}$ is a $(\infty - k)$-p.l. map if it equals to the composition $K \times Q = K \times I^k \times Q^{\infty-k} \longrightarrow K \times I^k \xrightarrow{g} \mathbb{R}$ of the projection on $K \times I^k$ with the p.l. map $g : K \times I^k \to \mathbb{R}$. Such composition was denoted in Subsection 2.3.2 by $\bar{g}_{Q^{\infty-k}}$.

Denote by $C_{\mathrm{p.l.}}(K \times Q; \mathbb{R})$ the set of maps in $C(K \times Q; \mathbb{R})$ which are $(\infty - k)$-p.l. for some k.

In view of Observation 2.7 in Chapter 2, $C_{\mathrm{p.l}}(K \times Q; \mathbb{R})$ is dense in $C(K \times Q; \mathbb{R})$. To conclude that (5.18) holds for $K \times Q$, it suffices to check the inequality for $f_1 = (\bar{g}_1)_{Q^{\infty-k}}, f_2 = (\bar{g}_2)_{Q^{\infty-k}} \in C_{\mathrm{p.l}}(K \times Q; \mathbb{R})$. The inequality holds since $\delta_r^{f_i} = \delta_r^{g_i}$.

Since by Theorem 2.7 every compact Hilbert cube manifold is homeomorphic to $K \times Q$ for some finite simplicial complex K, the inequality (5.18) holds for X any compact Hilbert cube manifold. Since for any compact ANR X, by Theorem 2.7, $X \times Q$ is a Hilbert cube manifold, the map $I : C(X; \mathbb{R}) \to C(X \times Q; \mathbb{R})$ defined by $I(f) = \bar{f}_Q$ is an isometric embedding and $\delta_r^f = \delta_r^{\bar{f}_Q}$, the inequality (5.18) holds for any compact ANR X. Both parts 1 and 2 of Theorem 5.2 follow from inequality (5.18).

5.2.4   *Proof of Theorem 5.3* (Poincaré Duality property)

Before we proceed to the proof of Theorem 5.3, we make the following useful elementary observations on linear algebra.

For the commutative diagram

$$E := \left\{ \begin{array}{ccc} C & \xrightarrow{\gamma_2} & A_2 \\ \scriptstyle\gamma_1 \downarrow & & \downarrow \scriptstyle\alpha_2 \\ A_1 & \xrightarrow{\alpha_1} & B \end{array} \right.$$

denote by

$$\ker(E) := \ker(C \xrightarrow{\gamma} A_1 \times_B A_2),$$
$$\mathrm{coker}(E) := \mathrm{coker}(A_1 \oplus_C A_2 \xrightarrow{\alpha} B),$$

with

$$A_1 \times_B A_2 = \{(a_1, a_2) \in A_1 \times A_2 \,|\, \alpha_1(a_1) = \alpha_2(a_2)\},$$
$$A_1 \oplus_C A_2 = A_1 \oplus A_2 / \{(a_1, a_2) \in A_1 \times A_2 \,|\, a_1 = \gamma_1(c),$$
$$a_2 = -\gamma_2(c) \text{ for some } c \in C\},$$

and with $\gamma(c) = (\gamma_1(c), \gamma_2(c))$ and $\alpha(a_1, a_2) = \alpha_1(a_1) + \alpha_2(a_2)$.

If one denotes by E^* the dual diagram

$$E^* := \left\{ \begin{array}{ccc} C^* & \xleftarrow{\gamma_2^*} & A_2^* \\ \scriptstyle\gamma_1^* \uparrow & & \uparrow \scriptstyle\alpha_2^* \\ A_1^* & \xleftarrow{\alpha_1^*} & B^* \end{array} \right.$$

then we have a canonical isomorphism

$$\ker(E) = (\mathrm{coker}(E^*))^*. \tag{5.19}$$

A straightforward calculation of dimensions shows that

(i) If in the diagram E all arrows are injective, then α is injective and $\dim(\mathrm{coker}\, E) = \dim C + \dim B - \dim A_1 - \dim A_2$.

(ii) If in the diagram E all arrows are surjective, then γ is surjective and $\dim(\mathrm{coker}\, E) = \dim C + \dim B - \dim A_1 - \dim A_2$.

To prove Theorem 5.3 we first provide an alternative description (up to isomorphism) of $\mathbb{F}_r(B)$ for a box B.

Consider the quotient space

$$\mathbb{G}_r(a, b) := H_r(X) / \mathbb{I}_a(r) + \mathbb{I}^b(r) \tag{5.20}$$

and for a box $B = (a', a] \times [b, b')$ denote by $\mathcal{G}(B)$ and $\mathcal{F}(B)$ the diagrams

$$\mathcal{G}(B) := \begin{array}{ccc} \mathbb{G}_r(a', b') & \longrightarrow & \mathbb{G}_r(a, b') \\ \downarrow & & \downarrow \\ \mathbb{G}_r(a', b) & \longrightarrow & \mathbb{G}_r(a, b) \end{array} \qquad \mathcal{F}(B) := \begin{array}{ccc} \mathbb{F}_r(a', b') & \longrightarrow & \mathbb{F}_r(a, b') \\ \downarrow & & \downarrow \\ \mathbb{F}_r(a', b) & \longrightarrow & \mathbb{F}_r(a, b) \end{array}$$

whose arrows are induced by the inclusions $\mathbb{I}_{a'}(r) \subseteq \mathbb{I}_a(r)$ and $\mathbb{I}^{b'}(r) \subseteq \mathbb{I}^b(r)$.
Define

$$\mathbb{G}_r^f(B) := \ker(\mathcal{G}(B)$$

and observe that

$$\mathbb{F}_r^f(B) = \operatorname{coker} \mathcal{F}(B).$$

Since

$$\mathbb{G}_r(a', b) \times_{\mathbb{G}_r(a,b)} \mathbb{G}_r(a, b') = H_r(X)/((\mathbb{I}_{a'}(r) + \mathbb{I}^b(r)) \cap (\mathbb{I}_a(r) + \mathbb{I}^{b'}(r))),$$

the vector space $\mathbb{G}_r(B)$ is canonically isomorphic to

$$\boxed{\mathbb{G}_r(B) = ((\mathbb{I}_{a'}(r) + \mathbb{I}^b(r)) \cap (\mathbb{I}_a(r) + \mathbb{I}^{b'}(r)))/(\mathbb{I}_{a'}(r) + \mathbb{I}^{b'}(r))}. \qquad (5.21)$$

Further, since $\mathbb{F}_r(a', b) \oplus_{\mathbb{F}_r(a',b')} \mathbb{F}_r(a, b') = \mathbb{I}_{a'}(r) \cap \mathbb{I}^b(r) + \mathbb{I}_a(r) \cap \mathbb{I}^{b'}(r)$,
the vector space $\mathbb{F}_r(B)$ is canonically isomorphic to

$$\boxed{\mathbb{F}_r(B) = (\mathbb{I}_a(r) \cap \mathbb{I}^b(r))/(\mathbb{I}_{a'}(r) \cap \mathbb{I}^b(r) + \mathbb{I}_a(r) \cap \mathbb{I}^{b'}(r))}. \qquad (5.22)$$

The obvious inclusion $\mathbb{I}_a(r) \cap \mathbb{I}^b(r) \subseteq (\mathbb{I}_{a'}(r) + \mathbb{I}^b(r)) \cap (\mathbb{I}_a(r) + \mathbb{I}^{b'}(r))$
induces the linear map $\theta_r^f(B) : \mathbb{F}_r(B) \to \mathbb{G}_r(B)$ and one has the following
result.

Proposition 5.9. *For any map $f : X \to \mathbb{R}$ and any box B the canonical
linear map $\theta_r^f(B) : \mathbb{F}_r^f(B) \to \mathbb{G}_r^f(B)$ defined above is an isomorphism.*

Proof. To check the injectivity proceed as follows. Suppose $\mathbb{I}_a(r) \cap \mathbb{I}^b(r)$
$\ni x = x_1 + x_2$ with $x_1 \in \mathbb{I}_{a'}(r)$ and $x_2 \in \mathbb{I}^{b'}(r)$. Then $x_1 = x - x_2 \in \mathbb{I}^b(r)$,
hence $x_1 \in (\mathbb{I}_{a'}(r) \cap \mathbb{I}^b(r))$ and similarly $x_2 \in (\mathbb{I}_a(r) \cap \mathbb{I}^{b'}(r))$.

To check the surjectivity, take $x = x_1 + y_1 = x_2 + y_2$ such that $x_1 \in$
$\mathbb{I}_{a'}, y_1 \in \mathbb{I}^b, x_2 \in \mathbb{I}_a, y_2 \in \mathbb{I}^{b'}$. Then $x - x_1 - y_2$ is equivalent to x in $\mathbb{G}_r(B)$.
But $x - x_1 - y_2 = y_1 - y_2 = x_2 - x_1$, hence it lies in both $\mathbb{I}^b$ and $\mathbb{I}^a$.
$\qquad \square$

After these linear algebra considerations, we now turn to the proof of Theorem 5.3.

Let $f : M^n \to \mathbb{R}$ be a continuous map, M^n a κ-orientable closed topological manifold, and a, b be topological regular values, which implies that $f^{-1}(a)$ and $f^{-1}(b)$ are closed codimension-one submanifolds. Both the sublevel M_a^f and the overlevel M_f^b are compact topological submanifolds with boundary in M. Let $i_a : M_a \to M$, $i^b : M^b \to M$, $j_a : M \to (M, M_a)$, and $j^b : M \to (M, M^b)$ denote the obvious inclusions and $i_a(k)$, $i^b(k)$, $j_a(k)$, and $j^b(k)$ be the inclusion-induced linear maps in homology in degree k, and $r_a(k)$, $r^b(k)$, $s_a(k)$, and $s^b(k)$ the inclusion-induced linear maps in cohomology (with coefficients in the field κ) in degree k, as indicated in the diagrams (5.23) and (5.24) below provided by Poincaré Duality, diagrams in which all vertical arrows are isomorphisms.

$$
\begin{array}{ccccc}
H_r(M_a) & \xrightarrow{\ i_a(r)\ } & H_r(M) & \xrightarrow{\ j_a(r)\ } & H_r(M, M_a) \\
\downarrow & & \downarrow & & \downarrow \\
H^{n-r}(M, M^a) & \xrightarrow{\ s^a(n-r)\ } & H^{n-r}(M) & \xrightarrow{\ r^a(n-r))\ } & H^{n-r}(M^a) \\
\downarrow & & \downarrow & & \downarrow \\
(H_{n-r}(M, M^a))^* & \xrightarrow{(j^a(n-r))^*} & (H_{n-r}(M))^* & \xrightarrow{(i^a(n-r))^*} & (H_{n-r}(M^a))^*
\end{array}
$$

$$\tag{5.23}$$

and

$$
\begin{array}{ccccc}
H_r(M^b) & \xrightarrow{\ i^b(r)\ } & H_r(M) & \xrightarrow{\ j^b(r)\ } & H_r(M, M^b) \\
\downarrow & & \downarrow & & \downarrow \\
H^{n-r}(M, M_b) & \xrightarrow{\ s_b(n-r)\ } & H^{n-r}(M)) & \xrightarrow{\ r_b(n-r)\ } & H^{n-r}(M_b) \\
\downarrow & & \downarrow & & \downarrow \\
(H_{n-r}(M, M_b))^* & \xrightarrow{(j_b(n-r))^*} & (H_{n-r}(M))^* & \xrightarrow{(i_b(n-r))^*} & (H_{n-r}(M_b))^*
\end{array}
$$

$$\tag{5.24}$$

Note that as a consequence of these two diagrams the Poincaré Duality provides a canonical isomorphism

$$\mathbb{F}_r^f(a, b) \simeq (\mathbb{G}_{n-r}^f(b, a))^*. \tag{5.25}$$

Indeed, note the following facts.

(i) $\mathbb{F}_r(a, b) = \ker(j_a(r), j^b(r))$ by the exactness of the first rows in the diagrams (5.23) and (5.24). Precisely, $\ker(j_a(r), j^b(r)) = \mathbb{I}_a(r) \cap \mathbb{I}^b(r)$.

(ii) $\ker(j_a(r), j^b(r)) \simeq \ker(r^a(n-r), r_b(n-r))$ by the isomorphism of the upper vertical arrows in these diagrams.

(iii) $\ker(r^a(n-r), r_b(n-r)) \simeq \ker((i^a(n-r))^*, (i_b(n-r))^*))$ by the isomorphism of the lower vertical arrow in these diagrams. The isomorphisms above are induced by the Poincaré Duality and the identification of cohomology with the dual of homology. The composition is still referred to as Poincaré Duality.

(iv) $\ker((i^a(n-r))^*, (i_b(n-r))^*)) = (\operatorname{coker}(i^a(n-r) + i_b(n-r))^* = (\mathbb{G}_{n-r}^f(b, a))^*$, by standard finite-dimensional linear algebra duality.

Putting together these equalities one obtains (5.25).

Suppose f is a topologically tame map (in particular, M is a smooth manifold and f a locally polynomial map[6] which is topologically tame). For $(a, b) \in \mathbb{R}^2$, choose ϵ small enough so that the intervals $(a - \epsilon, a)$, $(a, a + \epsilon)$ as well as $(a-\epsilon, a)$, $(a, a+\epsilon)$ are contained in the set of topologically regular values. Such a choice is possible thanks to the topologically tameness of f. To establish the result we proceed as follows. Observe that:

(i) In view of the tameness of f,
$$\hat{\delta}_r^f(a, b) = \mathbb{F}_r^f((a - \epsilon, a + \epsilon] \times [b - \epsilon, b + \epsilon)). \qquad (5.26)$$

(ii) By definition (the linear algebra consideration above),
$$\mathbb{F}_r^f((a-\epsilon, a+\epsilon] \times [b-\epsilon, b+\epsilon)) = \operatorname{coker} \mathcal{F}_r((a-\epsilon, a+\epsilon] \times [b-\epsilon, b+\epsilon)). \qquad (5.27)$$

(iii) By Proposition 5.13,
$$\operatorname{coker} \mathcal{F}_r((a-\epsilon, a+\epsilon] \times [b-\epsilon, b+\epsilon)) = \ker(\mathcal{G}_r((a-\epsilon, a+\epsilon] \times [b-\epsilon, b+\epsilon)). \qquad (5.28)$$

(iv) By equality (5.19),
$$\ker(\mathcal{G}_r((a - \epsilon, a + \epsilon] \times [b - \epsilon, b + \epsilon))) = (\operatorname{coker}(\mathcal{G}_r((a - \epsilon, a + \epsilon] \times [b - \epsilon, b + \epsilon))^*))^*. \qquad (5.29)$$

(v) By equality (5.25),
$$(\operatorname{coker}(\mathcal{G}_r((a - \epsilon, a + \epsilon] \times [b - \epsilon, b + \epsilon))^*))^* = (\operatorname{coker}(\mathcal{F}_{n-r}((b - \epsilon, b + \epsilon] \times [a - \epsilon, a + \epsilon)))^*. \qquad (5.30)$$

---

[6]This means that in the neighborhood of any point one can find local coordinates in which f is polynomial.

(vi) In view of definition,

$$(\operatorname{coker}((\mathcal{F}_{n-r}((b-\epsilon,b+\epsilon] \times [a-\epsilon,a+\epsilon))))^*$$
$$= (\mathbb{F}_{n-r}((b-\epsilon,b+\epsilon] \times [a-\epsilon,a+\epsilon)))^*. \qquad (5.31)$$

(vii) Due to the tameness of f,

$$(\mathbb{F}_{n-r}^f((b-\epsilon,b+\epsilon] \times [a-\epsilon,a+\epsilon)))^* = (\widehat{\delta}_{n-r}^f(b,a))^*. \qquad (5.32)$$

Putting together the canonical isomorphisms above one derives the canonical isomorphism between $\widehat{\delta}_r^f(a,b)$ and $(\widehat{\delta}_{n-r}^f(b,a))^*$ for f as above, which implies items (1) and (2) in Theorem 5.3 for a topological manifold and f topologically tame. In view of Theorem 5.2 and the fact p.l. maps are dense in the space of all continuous maps, the result holds for all continuous maps on a closed triangulable topological manifold.

5.3 The case of angle-valued map, $f : X \to \mathbb{S}^1$

For $f : X \to S^1$ a continuous map with X a compact ANR one denotes by $\widetilde{f} : \widetilde{X} \to \mathbb{R}$ an infinite cyclic cover of f. The space $\widetilde{X}$ is an ANR and the map $\widetilde{f}$ is proper. Let $\tau : \widetilde{X} \to \widetilde{X}$ be the deck transformation that realizes the action of $1 \in \mathbb{Z}$ on $\widetilde{X}$. Note that for $y \in \widetilde{X}$, $\widetilde{f}(\tau y) = \widetilde{f}(y) + 2\pi$. Denote by $\xi_f \in H^1(X : \mathbb{Z})$ the cohomology class represented by f.

We apply the considerations from the previous section to $\widetilde{f} : \widetilde{X} \to \mathbb{R}$ and note that the deck transformation $\tau : \widetilde{X} \to \widetilde{X}$ induces the isomorphism $t_r : H_r(\widetilde{X}) \to H_r(\widetilde{X})$ and therefore a structure of $\kappa[t^{-1},t]$-module on this κ-vector space. The diagram (5.33) with the vertical arrows induced by t_r implies Observation 5.7 below.

$$(5.33)$$

Observation 5.7.

(i) The isomorphism t_r satisfies $t_r^{\pm 1}(\mathbb{F}_r^{\tilde{f}}(a, b)) = \mathbb{F}_r^{\tilde{f}}(a \pm 2\pi, b \pm 2\pi)$.

(ii) For any box $B = (a', a] \times [b, b')$, consider the shifted box $B + 2\pi = (a' + 2\pi, a + 2\pi] \times [b + 2\pi, b' + 2\pi)$. The isomorphism t_r induces the isomorphisms $t_r(B) : \mathbb{F}_r^{\tilde{f}}(B) \to \mathbb{F}_r^{\tilde{f}}(B + 2\pi)$, and then

$$\widehat{t}_r(a, b) : \widehat{\delta}_r^{\tilde{f}}(a, b) \to \widehat{\delta}_r^{\tilde{f}}(a + 2\pi, b + 2\pi).$$

(iii) $\mathbb{I}_{-\infty}^{\tilde{f}}(r)$ and $\mathbb{I}_{\tilde{f}}^{\infty}(r)$ are invariant, hence $\kappa[t^{-1}, t]$-submodules; therefore, $H_r(\widetilde{M})/(\mathbb{I}_{-\infty}^{\tilde{f}}(r) + \mathbb{I}_{\tilde{f}}^{\infty}(r))$ is a $\kappa[t^{-1}, t]$-module.

Denote by $\text{Tor} \, H_k(\widetilde{X})$ the torsion submodule of $H_r(\widetilde{X})$.

Proposition 5.10. $\mathbb{I}_{-\infty}^{\tilde{f}}(r) = \mathbb{I}_{\tilde{f}}^{\infty}(r) = \text{Tor}(H_r(\widetilde{X}))$.

Proof. If $x \in \text{Tor}(H_r(\widetilde{X}))$, then there exist an integer $k \in \mathbb{Z}$ and a polynomial $P(t) = \alpha_n t^n + \alpha_{n-1} t^{n-1} + \cdots \alpha_1 t + \alpha_0$, $\alpha_i \in \kappa$, $\alpha_0 \neq 0$, such that $P(t)t^k x = 0$. Let $y = t^k x$. Since $H_r(\widetilde{X}) = \bigcup_b \mathbb{I}_{\tilde{f}}^b(r)$, one has $y \in \mathbb{I}_{\tilde{f}}^b(r)$ for some $b \in \mathbb{R}$. Since $P(t)y = 0$, one concludes that $y = -(\alpha_n/\alpha_0)t^{n-1} - \cdots - (\alpha_1/\alpha_0)ty$, and therefore $y \in \mathbb{I}_{\tilde{f}}^{b+2\pi}(r)$. Repeating the argument, one concludes that $y \in \mathbb{I}_{\tilde{f}}^{b+2\pi k}$ for any k, hence $y \in \mathbb{I}_{\tilde{f}}^{\infty}(r)$. Since $x = t^{-k} y$, one has $x \in \mathbb{I}_{\tilde{f}}^{\infty}(r)$. Hence $\text{Tor}(H_r(\widetilde{X})) \subseteq \mathbb{I}_{\tilde{f}}^{\infty}(r)$.

Now let $x \in \mathbb{I}_{\tilde{f}}^{\infty}(r)$. Since $H_r(\widetilde{X}) = \bigcup_a \mathbb{I}_a^{\tilde{f}}(r)$, we have $x \in \mathbb{I}_a^{\tilde{f}}(r)$ for some $a \in \mathbb{R}$, and if, in addition, $x \in \mathbb{I}_{\tilde{f}}^{\infty}(r)$, then by Observation 5.7 (iii) all $x, t^{-1}x, t^{-2}x, \ldots, t^{-k}x, \ldots \in \mathbb{I}_a(r) \cap \mathbb{I}^{\infty}(r)$. Since by Proposition 5.1 the dimension of $\mathbb{I}_a(r) \cap \mathbb{I}^{\infty}(r)$ is finite, there exists $\alpha_{i_1}, \ldots, \alpha_{i_k}$ such that $(\alpha_{i_1} t^{-i_1} + \cdots + \alpha_{i_k} t^{-i_k})x = 0$. This shows that $x \in \text{Tor}(H_r(\widetilde{X}))$. Hence, $\mathbb{I}^{\infty}(r) \subseteq \text{Tor}(H_r(\widetilde{X}))$. Therefore $\mathbb{I}_{\tilde{f}}^{\infty}(r) = Tor(H_r(\widetilde{X}))$. By a similar argument, one concludes that $\text{Tor}(H_r(\widetilde{X})) = \mathbb{I}_{-\infty}^{\tilde{f}}(r)$. $\square$

Recall that $\langle \cdot \rangle : \mathbb{R}^2 \to \mathbb{T} = \mathbb{R}^2/\mathbb{Z}$ denotes the map which assigns to $(a, b) \in \mathbb{R}^2$ its equivalence class $\langle a, b \rangle \in \mathbb{T}$.

For a box $B = (a - \alpha, a] \times [b, b + \beta)$ denote by $(B + c)$ the box $(B + c) = (a + c - \alpha, a + c] \times [b + c, b + c + \beta)$ and by $\langle B \rangle \subseteq \mathbb{T}$ the image of B by the map $\langle \cdot \rangle$.

One calls the box $B = (a - \alpha, a] \times [b, b + \beta)$ *small box* if $0 < \alpha, \beta < 2\pi$, in which case the restriction of $\langle \cdot \rangle$ to B is one-to-one; clearly, if B is a small box, then so is any $(B + c)$, and $(B + 2\pi k) \cap (B + 2\pi(k+1)) = \emptyset$ for $k \neq k'$.

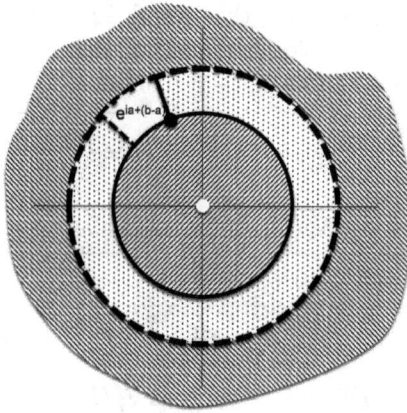

Fig. 5.8   Relevant parts of $\mathbb{T} = \mathbb{C} \setminus \{0\}$

For $(a, b) \in \mathbb{R}^2$ denote by

$$\mathbb{T}\langle a, b \rangle = \langle (-\infty, a] \times [b, \infty) \rangle,$$

the image under $\langle \cdot \rangle$ of the set $\{(x, y) \in \mathbb{R}^2 \mid x \leq a, y \geq b\}$; clearly, for $a' \leq a, b' \geq b$ one has $\mathbb{T}\langle a', b' \rangle \subseteq \mathbb{T}\langle a, b \rangle$.

In Fig. 5.8, after identifying $\mathbb{T}$ with $\mathbb{C} \setminus 0$, the region outside the circle of radius e^{b-a} for $b - a > 1$ corresponds to the set $\mathbb{T}(\langle a, b \rangle)$, the white box corresponds to the small box $B = (a - \alpha, a] \times [b, b + \beta)$ with $\alpha < 2\pi$, and the light dotted plus white area is the image of a box $B = (a - \alpha, a] \times [b, b + \beta)$ with $\alpha > 2\pi$.

Denote by $\pi(r)$ the projection

$$\pi(r) : H_r(\widetilde{X}) \to H_r(\widetilde{X})/\mathrm{Tor}(H_r(\widetilde{X}))$$

and for $\langle a, b \rangle \in \mathbb{T}$ introduce

$$\boxed{\begin{aligned} \mathbb{F}_r^{\widetilde{f}}\langle a, b \rangle &:= \sum_{k \in \mathbb{Z}} \mathbb{F}_r^{\widetilde{f}}(a + 2\pi k, b + 2\pi k) \subseteq H_r(\widetilde{X}) \text{ and} \\ \mathbb{F}_r^{f}\langle a, b \rangle &:= \pi(r)(\mathbb{F}_r^{\widetilde{f}}\langle a, b \rangle) \subseteq H_r(\widetilde{X})/\mathrm{Tor}(H_r(\widetilde{X})) = H_r^N(X, \xi_f) \end{aligned}}. \quad (5.34)$$

The reader should be aware that $\mathrm{Tor}(H_r(\widetilde{X})) \subseteq \mathbb{F}_r^{\widetilde{f}}(a, b) \subseteq \mathbb{F}_r^{\widetilde{f}}\langle a, b \rangle$ for any a, b.

Note the difference between $\mathbb{F}_r^{\widetilde{f}}\langle a, b \rangle$ and $\mathbb{F}_r^{f}\langle a, b \rangle$. The first is a submodule of $H_r(\widetilde{X})$ which, in general, is not a free module since it always contains the torsion of $H_r(\widetilde{X})$; by contrast, the second, $\mathbb{F}_r^f\langle a,b\rangle$, as a submodule of $H_r^N(X;\xi)$ which is always torsion free and f.g., hence free.

For the box $B = (a-\alpha,a] \times [b,b+\beta)$ consider

$$\mathbb{F}'^{\,\widetilde{f}}_r\langle B\rangle := \sum_{k\in\mathbb{Z}} \mathbb{F}'^{\,\widetilde{f}}_r(B+2\pi k) \subseteq \mathbb{F}^{\widetilde{f}}_r\langle a,b\rangle \subseteq H_r(\widetilde{X}) \text{ and}$$

$$\mathbb{F}'^{\,f}_r\langle B\rangle := \pi(r)(\mathbb{F}'^{\,\widetilde{f}}_r\langle B\rangle) \subseteq H_r^N(X;\xi)$$

$$(5.35)$$

and in view of Proposition 5.10,

$$\mathbb{F}^{\widetilde{f}}_r\langle B\rangle = \mathbb{F}^{\widetilde{f}}_r\langle a,b\rangle/\mathbb{F}'^{\,\widetilde{f}}_r\langle B\rangle = \mathbb{F}^f_r\langle a,b\rangle/\mathbb{F}'^{\,f}_r\langle B\rangle = \mathbb{F}^f\langle B\rangle\,,$$

so for a box B there is no difference between $\mathbb{F}^{\widetilde{f}}_r\langle B\rangle$ and $\mathbb{F}^f_r\langle B\rangle$; we will use most of the times the notation $\mathbb{F}^f_r\langle B\rangle$.

Introduce

$$\widehat{\delta}^f_r\langle a,b\rangle = \bigoplus_{k\in\mathbb{Z}} \widehat{\delta}^{\widetilde{f}}_r(a+2\pi k, b+2\pi k)\,,$$

$$(5.36)$$

and for a chosen collection of splittings S consider the diagrams

$$(5.37)$$

and

$$(5.38)$$

with pr^B the projection on the sum of the components in the direct sum $\bigoplus$ which correspond to $(a, b) \in \mathrm{supp}\, \delta_r^f \cap B$.

Definition 5.4. The collection of splittings $S = \{i_r(a, b) : \widehat{\delta}_r^{\widetilde{f}}(a, b) \to \mathbb{F}_r^{\widetilde{f}}(a, b)\}$ of the projections $\mathbb{F}_r^{\widetilde{f}}(a, b) \to \widehat{\delta}^{\widetilde{f}}(a, b)$ (cf. Definition (4.1) Chapter 4) which satisfy $t_r \cdot i_r(a, b) = i_r(a + 2\pi, b + 2\pi) \cdot \widehat{t}_r(a, b)$ is called a collection of $\mathbb{Z}$-*compatible splittings*.

Such collections exist. Indeed, it suffices to chose *splittings* only for $\{(a, b) \in \mathrm{supp}\, \delta_r^{\widetilde{f}}, 0 \le a < 2\pi\}$, observe that any $(a', b') \in \mathrm{supp}\, \delta_r^{\widetilde{f}}$ is of the form $a' = a + 2\pi k$, $b' = b + 2\pi k$ for some integer $k \in \mathbb{Z}$ with $0 \le a < 2\pi$, and take $i_r(a', b') := (\widehat{t}_r)^k \cdot i_r(a, b)(\widehat{t}_r)^{-k}$.

If the splittings are $\mathbb{Z}$-compatible, then all arrows in the diagrams (5.37) and (5.38) are $\kappa[t^{-1}, t]$-linear, and in view of Observation 5.7 one has the following facts:

(i) $\widehat{\delta}_r^f\langle \alpha, \beta \rangle$ is a free $\kappa[t^{-1}, t]$-module with the multiplication by t given by the isomorphism $\bigoplus_{k \in \mathbb{Z}} \widehat{t}_r(\alpha + 2\pi k, \beta + 2\pi k)$;

(ii) the entries in the diagrams (5.37) and (5.38) are all $\kappa[t^{-1}, t]$-modules, with those located on the left and right columns of (5.37) and on the left column of (5.38) free $\kappa[t^{-1}, t]$-modules;

(iii) in view of Proposition 5.6, for B a small box the compositions of horizontal arrows in both diagrams (5.37) and (5.38) are isomorphisms.

To summarize, we have

Proposition 5.11. *The $\kappa[t^{-1}, t]$-modules $\mathbb{F}_r^f\langle a, b \rangle$ and $\mathbb{F'}_r^f\langle B \rangle$ are split free submodules of $H_r^N(X; \xi)$, and $\mathbb{F}_r^f\langle B \rangle$ is a quotient of split free submodules, hence also free. In particular, $\widehat{\delta}_r^f\langle a, b \rangle$, which is canonically isomorphic to $\mathbb{F}_r^f\langle B(a, b; \epsilon) \rangle$ for $\epsilon < \epsilon(f)$, is a quotient of split free submodules.*

Definition of $\delta_r^f, \widehat{\delta}_r^f, \widetilde{\widehat{\delta}}_r^f$, and $P_r^f(z)$.

In view of Proposition 5.11, one considers the configurations δ_r^f defined by the formula

$$\delta_r^f\langle a, b \rangle := \delta_r^{\widetilde{f}}(a, b). \tag{5.39}$$

We use the identification of $\mathbb{T}$ with $\mathbb{C} \setminus 0$ provided by the map $\langle a, b \rangle \rightsquigarrow e^{ia + (b-a)}$ and if $z_1, z_2, \ldots, z_k \in \mathbb{C} \setminus 0$ are the points in the support of δ_r^f, when regarded in $\mathbb{C} \setminus 0$, we define, as for real-valued maps, the monic polynomial with nonzero free coefficient

$$P_r^f(z) := \prod (z - z_i)^{\delta_r^f(z_i)}.$$

It will be shown below, in Theorem 5.4, that this configuration is actually an element of $\mathrm{Conf}_{\beta_r^N(X;\xi_f)}(\mathbb{T})$ hence the degree of this polynomial is $\beta_r^N(X;\xi_f)$.

Consider the assignment $\langle a, b \rangle \rightsquigarrow \widehat{\delta}_r^f \langle a, b \rangle$, with $\widehat{\delta}_r^f \langle a, b \rangle$ a free $\kappa[t^{-1}, t]$-module. Observing that $\widehat{\delta}_r^f \langle a, b \rangle = \mathbb{F}_r^f \langle B \rangle$ for $B = (a_-, a] \times [b, b_+)$, with a_- the largest, strictly smaller than a, w-homologically critical value, and b_+ the smallest, strictly larger than b, w-homologically critical value, we conclude that this assignment comes from a configuration of subquotients, a concept defined in Chapter 2 Subsection 2.3.5.

The choice of a collection of $\mathbb{Z}$−compatible splittings S converts $\widehat{\delta}_r^f$ into a configuration of free split submodules of $H_r^N(X; \xi_f)$ which realizes $\widehat{\delta}_r^f$ as a configuration $^S\widehat{\delta}_r^f$ of free split submodules of $H_r^N(X; \xi_f)$ and refines the configuration δ_r^f. For reasons of notational simplicity "S" will dropped off our notation.

When $\kappa = \mathbb{C}$, a completion procedure not discussed in this book, the von Neumann completion, converts $\mathbb{C}[t^{-1}, t]$ into the von Neumann algebra $L^\infty(\mathbb{S}^1)$, and coupled with a $\mathbb{C}[t^{-1}, t]$-valued inner product on $H_r^N(M; \xi)$ converts $H_r^N(M; \xi)$ into a Hilbert module over the von Neumann algebra $L^\infty(\mathbb{S}^1)$, and converts $\mathbb{F'}_r^f \langle B \rangle$, $\mathbb{F}_r^f \langle B \rangle$ and $\widehat{\delta}_r^f \langle a, b \rangle$ into closed Hilbert submodules. The completed $H_r^N(M; \xi)$ is, up to isomorphism, the L_2-homology $H^{L_2}(\widetilde{X})$. As in the case of a real-valued map f defined on a compact ANR, when a Hilbert space structure on $H_r(M)$ canonically realizes $\widehat{\delta}_r^f$ as a configuration $\widetilde{\widehat{\delta}}_r^f$ of mutually orthogonal subspaces of $H_r(M)$, a $\mathbb{C}[t^{-1}, t]$-valued inner product on $H_r^N(M; \xi)$ makes of $\widehat{\delta}_r^f$ a configuration of mutually orthogonal closed $L^\infty(\mathbb{S}^1)$-Hilbert submodules of $H_r^{L_2}(\widetilde{M})$. If M is the underlying space of a closed Riemannian manifold, or of a finite simplicial complex, then the Riemannian metric or the triangulation, respectively, provides a canonical $\mathbb{C}[t^{-1}, t]$-valued inner product. It turns out that different such metrics on a smooth compact manifold, or different triangulations on a compact ANR, respectively, lead to Hilbert module structures which are isometric by isometries which intertwine these configurations. More details about these constructions can be found in [Burghelea, D. (2016a)], [Burghelea, D. (2016b)].

5.3.1   The main results

Denote by $C(X, \mathbb{S}^1)_\xi$ the connected component of $C(X, \mathbb{S}^1)$ corresponding to the cohomology class $\xi \in H^1(X; \mathbb{Z})$ consisting of continuous angle-valued maps f with $\xi_f = \xi$.

The main results about the configurations $\delta_r^f, \widehat{\delta}_r^f$ are contained in the following theorems.

Theorem 5.4. (Topological results)

Suppose $f : X \to \mathbb{S}^1$ is a continuous map and X a compact ANR.

(1) *If $\delta_r^f(z) \neq 0$, $z = e^{ia + (b-a)}$, then both e^{ia} and e^{ib} $(e^{ia}, e^{ib} \in \mathbb{S}^1)$ are w-homological critical values of f, equivalently a and b are w-homologically critical values of $\widetilde{f}$, hence also critical values when f is weakly tame.*

(2) *$\sum_{z \in \mathbb{C}\backslash 0} \delta_r^f(z) = \beta_r^N(X; \xi_f)$ and $\bigoplus_{z \in \mathbb{C}\backslash 0} \widehat{\delta}_r^f(z) \simeq H_r^N(X; \xi_f)$.*

(3) *If X is homeomorphic to a finite simplicial complex or to a compact Hilbert cube manifold then for an open and dense set of maps $f \in C_\xi(X, \mathbb{S}^1)$ one has $\delta^f(z) = 0$ or 1.*

Items (1) and (2) in Theorem 5.4 were first established in [Burghelea, D., Haller, S. (2015)] for tame maps.

Let $\underline{D}$ be the metric on $\mathrm{Conf}_{\beta_r^N(X;\xi)}(\mathbb{T})$, $\beta_r^N(X; \xi) = \mathrm{rank}\, H_r^N(X; \xi)$, $\mathbb{T} = \mathbb{R}^2/\mathbb{Z}$, induced by the Euclidean metric on $\mathbb{R}^2$ which is translation invariant and induces a complete metric on $\mathbb{T}$. Equipped with this metric $\mathrm{Conf}_{\beta_r^N(X;\xi)}(\mathbb{T})$ is complete metric space.

Theorem 5.5. (Stability property)

Let X is a compact ANR and $\xi \in H^1(X; \mathbb{Z})$. The assignment

$$C(X, \mathbb{S}^1)_\xi \ni f \rightsquigarrow \delta_r^f \in \mathrm{Conf}_{\beta_r^N(X;\xi)}(\mathbb{T})$$

equivalently

$$C(X, \mathbb{S}^1)_\xi \ni f \rightsquigarrow P_r^f(z) \in \mathbb{C}^{\beta_r^N(X;\xi)-1} \times (\mathbb{C} \backslash 0)$$

is a continuous map.

Moreover, with respect to the metric $\underline{D}$ on the space of configurations one has

$$\underline{D}(\delta_r^f, \delta_r^g) < 2D(f, g),$$

where $D(f, g) := \|f - g\|_\infty = \sup_{x \in X} |f(x) - g(x)|$.

In order to formulate the next results we recall the following facts.

A $\kappa[t^{-1},t]$-module is a κ-vector space V equipped with a linear isomorphism $T : V \to V$ which gives the multiplication by t, and therefore the vector space $V^* = \hom_\kappa(V,\kappa)$ is also a $\kappa[t^{-1},t]$-module with multiplication by t given by the isomorphism $T^* : V^* \to V^*$.

For V is a $\kappa[t^{-1},t]$-module denote by $\mathrm{Tor}(V)$ the submodule of torsion elements and by $F(V) := V/\mathrm{Tor}(V)$ the quotient module which is torsion free.

If V is f.g. $\kappa[t^{-1},t]$-module, then $\mathrm{Tor}(V)$ is a finite-dimensional κ-vector space, $F(V)$ a finite-rank free module and V is isomorphic to $F(V) \oplus \mathrm{Tor}(V)$. This because $\kappa[t^{-1},t]$ is a principal ideal domain.

As a consequence, if V is f.g, module, then $\mathrm{Tor}(V^*) \simeq (\mathrm{Tor}(V))^*$ and $V^* \simeq (F(V))^* \oplus (\mathrm{Tor}(V))^*$, although $F(V)^*$ is not f.g. unless isomorphic to 0.

For V_1 and V_2 two $\kappa[t^{-1},t]$-modules, a *nondegenerate bilinear map* compatible with the multiplication by t is a κ-bilinear map $P : V_1 \times V_2 \to \kappa$ which satisfies $P(tv_1,v_2) = P(v_1,tv_2)$ such that the induced linear maps $V_1 \to V_2^*$ and $V_2 \to V_2^*$ are injective $\kappa[t^{-1},t]$-linear maps. If V_1 and V_2 are finite-dimensional κ-vector spaces, then both $V_1 \to V_2^*$ and $V_2 \to V_1^*$ are isomorphisms.

Theorem 5.6. (Poincaré Duality property)

Suppose M is a closed triangulable topological manifold of dimension n which is κ-orientable and $f : M \to \mathbb{S}^1$ is a continuous map with $\xi_f \neq 0$. Then one has

(1) $\delta_r^f \langle a,b \rangle = \delta_{n-r}^f \langle b,a \rangle$, *equivalently* $\delta_r^f(z) = \delta_{n-r}^f(Tz)$, *with* $T(z) = z|z|^{-2}e^{i \ln |z|}$.

(2) *The Poincaré Duality between Borel-Moore homology and the cohomology of $\widetilde{M}$ induces non-degenerate κ-bilinear maps*

$$\mathrm{PD}^N : H_r^N(X;\xi_f) \times H_{n-r}^N(X;\xi_f) \to \kappa,$$
$$\mathrm{PD}^N \langle a,b \rangle : \widehat{\delta}_r^f \langle a,b \rangle \times \widehat{\delta}_{n-r}^f \langle b,a \rangle \to \kappa,$$

compatible with the multiplication by t.

As in the case of Theorem 5.3 the result remains true without triangulability hypothesis but a complete proof will require additional considerations on topological manifolds.

Item (1) in Theorems 5.4 and 5.6 were first established in [Burghelea, D., Haller, S. (2015)] for X homeomorphic to a simplicial complex.

In case $\kappa = \mathbb{R}$ or $\mathbb{C}$ and $H_r^N(X; \xi_f)$ is completed to $H_r^{L^2}(\tilde{X})$ via a geometric structure on X (i.e. Riemannian metric or triangulation) then the above bilinear maps are nondegenerate bilinear maps of Hilbert modules.

Proposition 5.12. *If $f : X \to \mathbb{S}^1$ is a tame map and $\mathcal{B}_r^c(f)$ and $\mathcal{B}_{r-1}^o(f)$ are the sets of r-closed and $(r-1)$-open barcodes defined in Chapter 4, then the following assertions hold:*

(1) *If $a \le b$, then $\delta_r^f \langle a, b \rangle = \#\{I \in \mathcal{B}_r^c(f) \mid I = [a, b]\}$.*
(2) *If $a > b$, then $\delta_r^f \langle a, b \rangle = \#\{I \in \mathcal{B}_{(r-1)}^o(f) \mid I = (b, a)\}$.*

The proof follows from Proposition 5.8 in view of the fact that $\delta_r^f \langle a, b \rangle = \delta_r^{\widetilde{f}}(a, b)$.

5.3.2   Proofs of Theorems 5.4 and 5.5 (topological results and stability)

Theorem 5.4

Proof. Item (1) is verified by Observation 5.2, and item (2) follows from the isomorphism ${}^S \hat{I}_r = \pi(r) \cdot {}^S I_r$ established in Proposition 5.6. Note also that $\widehat{\delta}_r^f$ comes from a *configuration of quotients*, as described in Subsection 2.3.5 with

$$\widehat{\delta}_r^f \langle a, b \rangle = \mathbb{F}_r^f \langle a, b \rangle / \mathbb{F'}_r^f \langle B(a, b; \epsilon) \rangle = \mathbb{F}_r^f \langle a, b \rangle / \mathbb{F'}_r^f \langle B \rangle = \mathbb{F}_r^f \langle B \rangle$$

for $B = (a_-, a] \times [b, b_+)$. Note that the free module $\mathbb{F'}_r^f \langle B(a, b; \epsilon) \rangle = \mathbb{F'}_r^f \langle B \rangle$ for ϵ small enough.

For item (3) one proceeds as in the proof of Theorem 5.1 item (4), given in Subsection 5.2.3. For example, in case X is a smooth manifold, possibly with boundary, any continuous angle-valued map is arbitrarily closed to a Morse angle-valued map f which takes different values at different critical points; and in case of a simplicial complex any such map is arbitrarily closed to one with infinite cyclic cover that satisfies property G defined in the proof of Theorem 5.1 item (4). For such maps δ_r^f takes only the values 0 and 1, and then the same is true for δ_r^f.   □

Theorem 5.5

Proof. The proof is similar to the proof of Theorem 5.2 in Subsection 5.2.3. We replace $f : X \to \mathbb{R}$ by $\widetilde{f} : \widetilde{X} \to \mathbb{R}$, a lift (= infinite cyclic cover) of $f : X \to \mathbb{S}^1$ representing ξ. The basic ingredient is Proposition 5.7 based on Lemmas 5.1 and 5.2 which hold even if X is only locally compact.

The steps of the proof are essentially those described in Subsection 5.2.3 and reviewed below.

(1) For a pair (X, ξ), X a compact ANR, let $C_\xi(X; \mathbb{S}^1)$, be the set of maps in the homotopy class defined by ξ, equipped with the compact-open topology, and $\widetilde{X}$ be an infinite cyclic cover associated to ξ. Observe that:

a) the compact-open topology on $C_\xi(X; \mathbb{S}^1)$ is induced by the complete metric $D(f, g) := \sup_{x \in X} |f(x) - g(x)|$, and $D(f, g) = D(\widetilde{f}, \widetilde{g})$ for appropriate liftings $\widetilde{f}, \widetilde{g}$ of the maps f, g,

b) for $f, g \in C_\xi(X, \mathbb{S}^1)$ with $D(f, g) < \pi$ and any sequence $0 = t_0 < t_1 < \cdots < t_{N-1} < t_N = 1$, the canonical homotopy h_t from f to g satisfies

$$D(f, g) = \sum_{0 \leq i < N} D(h_{t_{i+1}}, h_{t_i}). \tag{5.40}$$

(2) For X a simplicial complex, let $\mathcal{U} \subset C_\xi(X; \mathbb{S}^1)$ be the subset of p.l. maps. One can verify that:

a) $\mathcal{U}$ is a dense subset in $C_\xi(X, \mathbb{S}^1)$,

b) if $f, g \in \mathcal{U}$ satisfy $D(f, g) < \pi$, then for the canonical homotopy each h_t belongs to $\mathcal{U}$, hence $\epsilon(h_t) > 0$, hence for any $t \in [0, 1]$ there exists $\delta(t) > 0$ such that $t', t'' \in (t - \delta(t), t + \delta(t))$ implies $D(h_{t'}, h_t) < \epsilon(h_t)/3$.

Both statements are argued as in Subsection 5.2.3.

(3) Consider the space of configurations $\mathrm{Conf}_{\beta_r}(\mathbb{T})$, $\beta_r = \beta_r^N(X; \xi)$ equipped with the induced metric, $\underline{D}$, which is complete. Since any map in $\mathcal{U}$ is tame, Proposition 5.7 shows that if $f, g \in \mathcal{U}$ satisfy $D(f, g) < \epsilon(f)/3$, then

$$\underline{D}(\delta_r^f, \delta_r^g) \leq 2D(f, g). \tag{5.41}$$

This suffices to conclude the continuity of the assignment $f \rightsquigarrow \delta_r^f$.

To finalize the proof of Theorem 5.5, notice first that the inequality (5.41) extends to all $f, g \in \mathcal{U}$, second that (5.41) extends to all $f, g \in C_\xi(X; \mathbb{S}^1)$ for X a finite simplicial complex, third that (5.41) extends to all $f, g \in C_\xi(X; \mathbb{S}^1)$ for X an arbitrary compact ANR. This is done in the same way as in the case of real-valued maps in the following steps.

- *Step 1*: In view of the continuity of the assignment $f \rightsquigarrow \delta_r^f$, it suffices to verify the inequality (5.41) for $f, g \in \mathcal{U}$ with the additional property $D(f, g) < \pi$; this because for any f, g one has $D(f, g) \leq \pi$.

Start with $f, g \in \mathcal{U}$ and consider the canonical homotopy $\widetilde{h}_t = t\widetilde{f} + (1 - t)\widetilde{g}$, $t \in [0, 1]$, between two lifts $\widetilde{f}$ and $\widetilde{g}$ of f and g which satisfy $D(f, g) = D(\widetilde{f}, \widetilde{g})$. Note that each $\widetilde{h}_t$ satisfies $\widetilde{h}_t(\mu(n, x)) = \widetilde{h}_t(x) + 2\pi n$, hence is a lift, and each h_t is p.l. map.

Choose a sequence $0 < t_1 < t_3 < t_5 < \cdots < t_{2N-1} < 1$ such that for $i = 1, \ldots, 2N - 1$ the intervals $(t_{2i-1} - \delta(t_{2i-1}), t_{2i-1} + \delta(t_{2i-1}))$ cover $[0, 1]$ and $(t_{2i-1}, t_{2i-1} + \delta(t_{2i-1})) \cap (t_{2i+1} - \delta(t_{2i+1}), t_{2i+1}) \neq \emptyset$. This is possible in view of the compactness of $[0, 1]$.

Take $t_0 = 0$, $t_{2N} = 1$, and $t_{2i} \in (t_{2i-1}, t_{2i-1} + \delta(t_{2i-1})) \cap (t_{2i+1} - \delta(t_{2i+1}, t_{2i+1})$. To simplify the notation, abbreviate h_{t_i} to h_i. In view of (2) and (3) above one has

$$|t_{2i-1} - t_{2i}| < \delta(t_{2i-1}) \text{ implies } D(\delta^{h_{2i-1}}, \delta^{h_{2i}}) < 2D(h_{2i-1}, h_{2i})$$

and

$$|t_{2i} - t_{2i+1}| < \delta(t_{2i+1}) \text{ implies } D(\delta^{h_{2i}}, \delta^{h_{2i+1}}) < 2D(h_{2i}, h_{2i+1}).$$

Then we have

$$
\begin{aligned}
\underline{D}(\delta^f, \delta^g) &\leq \sum_{0 \leq i < 2N-1} D(\delta^{h_i}, \delta^{h_{i+1}}) \\
&\leq 2 \sum_{0 \leq i < 2N-1} D(h_i, h_{i+1}) = 2D(f, g).
\end{aligned}
\tag{5.42}
$$

- *Step 2*: Suppose X is a simplicial complex. In view of the denseness of $\mathcal{U}$ and of the completeness of the metrics on $C_\xi(X; \mathbb{S}^1)$ and $\mathrm{Conf}_{\beta_r}(\mathbb{T})$, the inequality (5.42) extends to the entire $C_\xi(X; \mathbb{S}^1)$, since the assignment $\mathcal{U} \ni f \rightsquigarrow \delta_r^f \in \mathrm{Conf}_{\beta_r}(\mathbb{T})$ preserves the Cauchy sequences of elements in $\mathcal{U}$.

- *Step 3*: We verify the inequality (5.41) exactly in the same way as in the proof of Theorem 5.2, first for X a compact Hilbert cube manifold based on the fact that each such space is a product of Q with a simplicial complex, and then for an arbitrary compact ANR X based on the fact that $X \times Q$ is a Hilbert cube manifold. $\qquad\square$

5.3.3   *Proof of Theorem 5.6 (Poincaré Duality property)*

In view of Theorem 5.5, it suffices to establish Theorem 5.6 for a collection of maps which are dense in the space of all continuous maps with the compact-open topology. The collection we consider consists of the maps whose set of values which are not topologically regular is finite. Clearly, this collection satisfies the density requirement when M is a topological manifold which admits a triangulation. Without any additional specification each map in this subsection belongs to this collection.

Let $f : M \to \mathbb{S}^1$ be a continuous map and $\widetilde{f} : \widetilde{M} \to \mathbb{R}$ an infinite cyclic cover, and suppose a, b are topologically regular values. Note that the diagrams (5.23) and (5.24) in Subsection 5.2.4 continue to hold if one replaces M, M_a, M^a by $\widetilde{M}$, $\widetilde{M}_a$, $\widetilde{M}^a$, and $H_r(M)$, $H_r(M_a)$, $H_r(M^a)$, $H_r(M, M_a)$, $H_r(M, M^a)$ by $H_r^{\mathrm{BM}}(\widetilde{M})$, $H_r^{\mathrm{BM}}(\widetilde{M}_a)$, $H_r^{\mathrm{BM}}(\widetilde{M}^a)$, $H_r^{\mathrm{BM}}(\widetilde{M}, \widetilde{M}_a)$, $H_r^{\mathrm{BM}}(\widetilde{M}, \widetilde{M}^a)$. They become the diagrams (5.43) and (5.44) below, with H_r^{BM} denoting Borel-Moore homology.

$$
\begin{array}{ccccc}
H_r^{\mathrm{BM}}(\widetilde{M}_a) & \xrightarrow{i_a(r)} & H_r^{\mathrm{BM}}(\widetilde{M}) & \xrightarrow{j_a(r)} & H_r^{\mathrm{BM}}(\widetilde{M}, \widetilde{M}_a) \\
\downarrow{\scriptstyle PD_a^1} & & \downarrow{\scriptstyle PD} & & \downarrow{\scriptstyle PD_a^2} \\
H^{n-r}(\widetilde{M}, \widetilde{M}^a) & \xrightarrow{s^a(n-r)} & H^{n-r}(\widetilde{M}) & \xrightarrow{r^a(n-r))} & H^{n-r}(\widetilde{M}^a) \\
\downarrow & & \downarrow & & \downarrow \\
(H_{n-r}(\widetilde{M}, \widetilde{M}^a))^* & \xrightarrow{(j^a(n-r))^*} & (H_{n-r}(\widetilde{M}))^* & \xrightarrow{(i^a(n-r))^*} & (H_{n-r}(\widetilde{M}^a))^*
\end{array}
$$

$$(5.43)$$

$$
\begin{array}{ccccc}
H_r^{\mathrm{BM}}(\widetilde{M}^b) & \xrightarrow{i^b(r)} & H_r^{\mathrm{BM}}(\widetilde{M}) & \xrightarrow{j^b(r)} & H_r^{\mathrm{BM}}(\widetilde{M}, \widetilde{M}^b) \\
\downarrow{\scriptstyle PD_1^b} & & \downarrow{\scriptstyle PD} & & \downarrow{\scriptstyle PD_2^b} \\
H^{n-r}(\widetilde{M}, \widetilde{M}_b) & \xrightarrow{s_b(n-r)} & H^{n-r}(\widetilde{M})) & \xrightarrow{r_b(n-r)} & H^{n-r}(\widetilde{M}_b) \\
\downarrow & & \downarrow & & \downarrow \\
(H_{n-r}(\widetilde{M}, \widetilde{M}_b))^* & \xrightarrow{(j_b(n-r))^*} & (H_{n-r}(\widetilde{M}))^* & \xrightarrow{(i_b(n-r))^*} & (H_{n-r}(\widetilde{M}_b))^*
\end{array}
$$

$$(5.44)$$

Recall from Subsection 2.3.5 the following isomorphisms.

$$
\begin{aligned}
H_r^{\mathrm{BM}}(\widetilde{M}) &= \varprojlim_{0 < l \to \infty} H_r(\widetilde{M}, \widetilde{M}_{-l} \sqcup \widetilde{M}^t), \\
H_r^{\mathrm{BM}}(\widetilde{M}_a) &= \varprojlim_{0 < l \to \infty} H_r(\widetilde{M}_a, \widetilde{M}_{a-l}), \\
H_r^{\mathrm{BM}}(\widetilde{M}^a) &= \varprojlim_{0 < l \to \infty} H_r(\widetilde{M}, \widetilde{M}^{a+l}), \\
H_r^{\mathrm{BM}}(\widetilde{M}, \widetilde{M}_a) &= \varprojlim_{0 < l \to \infty} H_r(\widetilde{M}, \widetilde{M}_a \sqcup \widetilde{M}^{a+l}), \\
H_r^{\mathrm{BM}}(\widetilde{M}, \widetilde{M}^a) &= \varprojlim_{0 < l \to \infty} H_r(\widetilde{M}, \widetilde{M}^a \sqcup \widetilde{M}_{a-l}).
\end{aligned}
$$

$$(5.45)$$

The readers not familiar with Borel-Moore homology can regard H_r^{BM} as given by these equalities (5.45).

If one uses $H_r^{\mathrm{BM}}(\cdot)$ instead of $H_r(\cdot)$, one can define $^{\mathrm{BM}}\mathbb{F}_r^{\tilde{f}}(a,b)$ and $^{\mathrm{BM}}\hat{\delta}_r^{\tilde{f}}(a,b)$ instead of $\mathbb{F}_r^{\tilde{f}}(a,b)$ and $\hat{\delta}_r^{\tilde{f}}(a,b)$.

We now focus on the Poincaré Duality isomorphism, the composition of the vertical arrows in the middle of diagram (5.43) or (5.44)

$$\mathrm{PD}_r^{\mathrm{BM}} : H_r^{\mathrm{BM}}(\widetilde{M}) \xrightarrow{\mathrm{PD}_r} H^{n-r}(\widetilde{M}) \xrightarrow{=} (H_{n-r}(\widetilde{M}))^* .$$

Note that all three terms of this sequence are $\kappa[t^{-1},t]$-modules and the two arrows are $\kappa[t^{-1},t]$-linear. The $\kappa[t^{-1},t]$-module structure on $(H_{n-r}(\widetilde{M}))^*$ is given by the κ-linear isomorphism

$$t_{n-r}^* : H_{n-r}(\widetilde{M})^* \to H_{n-r}(\widetilde{M})^*,$$

the dual of $t_{n-r} : H_{n-r}(\widetilde{M}) \to H_{n-r}(\widetilde{M})$.

Recall that:

- for $a, b \in \mathbb{R}$,
  $\mathbb{G}_r^{\tilde{f}}(a,b) := H_r(\widetilde{M})/(\mathbb{I}_a^{\tilde{f}} + \mathbb{I}_{\tilde{f}}^b)$, cf. (5.20) and
  $p^{a,b}(r) : H_r(\widetilde{M}) \to \mathbb{G}_r^{\tilde{f}}(a,b)$ is the projection on the quotient space;

- for a box $B = (a',a] \times [b,b')$,
  $\mathbb{G}_r^{\tilde{f}}(B) := \ker(\mathbb{G}_r^{\tilde{f}}(a',b') \to \mathbb{G}_r^{\tilde{f}}(a',b) \times_{\mathbb{G}_r^{\tilde{f}}(a,b)} \mathbb{G}_r^{\tilde{f}}(a,b'))$,
  $u_r(B) : \mathbb{G}_r^{\tilde{f}}(B) \to \mathbb{G}_r^{\tilde{f}}(a',b')$ denotes the canonical inclusion, and
  $\theta_r^{\tilde{f}}(B) : \mathbb{F}_r^{\tilde{f}}(B) \to \mathbb{G}_r^{\tilde{f}}(B)$ is the canonical isomorphism described in Subsection 5.2.4 of this chapter.

For the box $B = (a',a] \times [b,b')$ denote by B' the box $B' = (b,b'] \times [a',a)$. We have the following proposition.

Proposition 5.13.

(i) *For any a, b regular values, the Poincaré Duality isomorphism restricts to an isomorphism*

$$\mathrm{PD}_r^{\mathrm{BM}}(a, b) : {}^{\mathrm{BM}} \mathbb{F}_r^{\widetilde{f}}(a, b) \to (\mathbb{G}_{n-r}^{\widetilde{f}}(b, a))^*.$$

(ii) *For any box $B = (a', a] \times [b, b')$ with all a, a', b, b' topologically regular values, the Poincaré Duality isomorphism $\mathrm{PD}_r^{\mathrm{BM}}$ induces the isomorphisms $\mathrm{PD}_r^{\mathrm{BM}}(a, b)$, $\mathrm{PD}_r^{\mathrm{BM}}(B)$, making the diagram below commutative.*

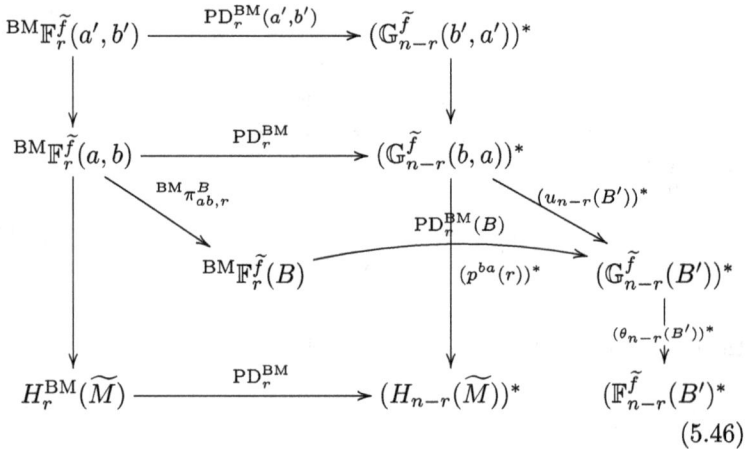

$$(5.46)$$

(iii) *The Poincaré Duality isomorphism $\mathrm{PD}_r^{\mathrm{BM}}(a, b)$ induces the isomorphism*

$$^{\mathrm{BM}} \widehat{\delta}_r^{\widetilde{f}}(a, b) \to (\widehat{\delta}_{n-r}^{\widetilde{f}}(b, a))^*.$$

Proof. Item (i): In view of the diagrams (5.43) and (5.44) one has

$$^{\mathrm{BM}} \mathbb{F}_r^{\widetilde{f}}(a, b) = \mathrm{img}\, i_a(r) \cap \mathrm{img}\, i^b(r) = \ker j_a(r) \cap \ker j^b(r)$$
$$= \ker(i^a(n-r))^* \cap \ker(i_b(n-r))^* = (\mathrm{coker}(i_b(n-r) \oplus i^a(n-r))^*$$
$$= (\mathbb{G}_{n-r}^{\widetilde{f}}(b, a))^*.$$

The second equality holds by exactness of the first rows in the diagrams (5.43) and (5.44), the third by the equality of the top and bottom right horizontal arrows the fourth by linear algebra duality and the fifth by definition.

Item (ii): As in Subsection 5.2.4, the image under $^{\mathrm{BM}}\mathrm{PD}_r$ of the diagram

$$^{\mathrm{BM}}\mathcal{F}(B) := \left\{ \begin{array}{ccc} ^{\mathrm{BM}}\mathbb{F}_r^{\widetilde{f}}(a',b') & \longrightarrow & ^{\mathrm{BM}}\mathbb{F}_r^{\widetilde{f}}(a,b') \\ \downarrow & & \downarrow \\ ^{\mathrm{BM}}\mathbb{F}_r^{\widetilde{f}}(a',b) & \longrightarrow & ^{\mathrm{BM}}\mathbb{F}_r^{\widetilde{f}}(a,b) \end{array} \right.$$

is the diagram

$$\left\{ \begin{array}{ccc} (\mathbb{G}_{n-r}^{\widetilde{f}}(b',a'))^* & \longrightarrow & \mathbb{G}_{n-r}^{\widetilde{f}}(b',a))^* \\ \downarrow & & \downarrow \\ (\mathbb{G}_r^{\widetilde{f}}(b,a'))^* & \longrightarrow & \mathbb{G}_{n-r}^{\widetilde{f}}(b,a))^* \end{array} \right.$$

which is the dual of the diagram

$$\mathcal{G}(B') := \left\{ \begin{array}{ccc} \mathbb{G}_{n-r}^{\widetilde{f}}(b,a) & \longrightarrow & (\mathbb{G}_{n-r}^{\widetilde{f}}(b',a) \\ \downarrow & & \downarrow \\ \mathbb{G}_{n-r}^{\widetilde{f}}(b,a') & \longrightarrow & (\mathbb{G}_{n-r}^{\widetilde{f}}(b',a') \end{array} \right. ,$$

and therefore $^{\mathrm{BM}}\mathrm{PD}_r$ induces a linear isomorphism from $^{\mathrm{BM}}\mathbb{F}_r^{\widetilde{f}}(B) = \operatorname{coker}\mathcal{F}(B)$ to $(\ker(\mathcal{G}(B')))^* = (G_{n-r}^{\widetilde{f}}(B'))^*$.   □

Item (iii): If either a or b are regular values in view of the definitions both $^{\mathrm{BM}}\widehat{\delta}_r^{\widetilde{f}}(a,b) = 0$ and $\widehat{\delta}_r^{\widetilde{f}}(a,b) = 0$ so it remains to check the result for $a = c$ and $b = c'$, for c, c' critical values.

For c and c' critical values of $\widetilde{f}$, choose $a' = c - \epsilon$, $a = c + \epsilon$, $b = c' - \epsilon$, $b' = c + \epsilon$ with $\epsilon < \epsilon(f)$; then a, b, a', b' are topologically regular. In view of Proposition (5.13) item (ii),

$$^{\mathrm{BM}}\widehat{\delta}_r^{\widetilde{f}}(c,c') = ^{\mathrm{BM}}\mathbb{F}_r^{\widetilde{f}}(B),$$

is isomorphic to

$$(\widehat{\delta}_{n-r}^{\widetilde{f}}(c',c))^* = (\mathbb{F}_{n-r}^{\widetilde{f}}(B'))^* = (\mathbb{G}_{n-r}^{\widetilde{f}}(B'))^*.$$

Then from diagram (5.46) one derives the diagram

$$
\begin{array}{ccc}
{}^{\mathrm{BM}}\mathbb{F}_r^{\widetilde{f}}(a,b) & \xrightarrow{\mathrm{PD}_r^{\mathrm{BM}}(a,b)} & (\mathbb{G}_{n-r}^{\widetilde{f}}(b,a))^*
\end{array}
$$

$$
\pi_{ab,r}^B \qquad u_{n-r}^*
$$

$$
\mathrm{PD}_r^{\mathrm{BM}}(c,c')
$$

$$
{}^{\mathrm{BM}}i_a^b(r) \qquad {}^{\mathrm{BM}}\widehat{\delta}_r^{\widetilde{f}}(c,c') \qquad\qquad p_r(a,b)^* \qquad (\widehat{\delta}_{n-r}^{\widetilde{f}}(c',c))^*
$$

$$
\begin{array}{ccc}
H_r^{\mathrm{BM}}(\widetilde{M}) & \xrightarrow{\mathrm{PD}_r^{\mathrm{BM}}} & (H_{n-r}(\widetilde{M}))^*
\end{array}
$$

$$(5.47)$$

with the horizontal arrows isomorphisms, the vertical arrows injective, and the oblique arrows surjective. Here $p_r(a,b)$ is the composition $p_r(a,b) = H_r(\widetilde{M}) \to H_r(\widetilde{M})/(\mathbb{I}_a(r) + \mathbb{I}_b(r))$ which in view of Proposition 5.10 factors through $H^N(M;\xi_f)$.

Let $\lambda_r : H_r(\widetilde{M}) \to H_r^{\mathrm{BM}}(\widetilde{M})$ be the obvious κ-linear map from the homology to the Borel-Moore homology which is $\kappa[t^{-1},t]$-linear. Observe that for $a,b \in \mathbb{R}$ λ_r restricts to the linear map

$$
\lambda_r(a,b) : \mathbb{F}^{\widetilde{f}}(a,b) \to {}^{\mathrm{BM}}\mathbb{F}_r^{\widetilde{f}}(a,b)
$$

and the following facts hold.

Proposition 5.14.

(1) *The linear maps $\lambda_r(a,b)$, $a,b \in \mathbb{R}$ are compatible with the deck transformations, and are surjective.*

(2) *The kernel of the linear map $\lambda_r(a,b)$ is independent of (a,b) and $\ker \lambda_r(a,b) = \ker \lambda_r = \mathrm{Tor}(H_r(\widetilde{M}))$.*

(3) *The $\kappa[t^{-1},t]$-module $\mathrm{Tor}(H_r^{\mathrm{BM}}(\widetilde{M}))$ is a finite-dimensional κ-vector space isomorphic to $(\mathrm{Tor}H_{n-r}(\widetilde{M}))^*$.*

(4) $\lambda_r(H_r(\widetilde{M})) \cap \mathrm{Tor}(H_r^{\mathrm{BM}}(\widetilde{M})) = 0$.

Proof. Observe that in view of Proposition 5.10 one has a short exact sequence

$$
0 \to \mathrm{Tor}(H_r(\widetilde{M})) \to \mathbb{F}_r^{\widetilde{f}}(a,b) \to {}^{\mathrm{BM}}\mathbb{F}_r^{\widetilde{f}}(a,b) \to 0,
$$

which is natural with respect to pairs (a,b) and compatible with the action by the deck transformations, hence making the diagram (5.48) commutative.

$$0 \to \text{Tor}(H_r(\widetilde{M})) \longrightarrow \mathbb{F}_r^{\tilde{f}}(a,b) \longrightarrow {}^{\text{BM}}\mathbb{F}_r^{\tilde{f}}(a,b) \longrightarrow 0$$

$$\downarrow t_r \qquad\qquad \downarrow t_r \qquad\qquad\qquad \downarrow t_r^{\text{BM}}$$

$$0 \to \text{Tor}(H_r(\widetilde{M})) \to \mathbb{F}_r^{\tilde{f}}(a+2\pi, b+2\pi) \to {}^{\text{BM}}\mathbb{F}_r^{\tilde{f}}(a+2\pi, b+2\pi) \to 0$$

$$(5.48)$$

To see this one uses the diagram (5.49) below (where $-l < a' < a$ and $b < b' < t$) whose vertical columns are exact sequences.

$$(5.49)$$

where

$$\text{in}_-(r-1) : H_{r-1}(\widetilde{M}_{-l}) \to H_{r-1}(\widetilde{M}_{-l} \sqcup \widetilde{M}^t) = H_{r-1}(\widetilde{M}_{-l}) \oplus H_r(\widetilde{M}^t)$$

and

$$\text{in}_+(r-1) : H_{r-1}(\widetilde{M}^t) \to H_{r-1}(\widetilde{M}_{-l} \sqcup \widetilde{M}^t) = H_{r-1}(\widetilde{M}_{-l}) \oplus H_r(\widetilde{M}^t)$$

are the obvious inclusions as the left and right components of $H_r(\widetilde{M}_{-l} \sqcup \widetilde{M}^t) = H_r(\widetilde{M}_{-l}) \oplus H_r(\widetilde{M}^t)$. By passing to the limits $l, t \to \infty$, diagram (5.49) induces diagram (5.50) which provides the relation between $\mathbb{F}_r(a,b)$, $\mathbb{F}_r(a',b')$, $H_r(\widetilde{M})$ and their Borel-Moore versions.

$$(5.50)$$

Since $^{\mathrm{BM}}\mathbb{F}_r^{\widetilde{f}}(a,b) = \mathrm{img}(^{\mathrm{BM}}i_a(r)) \cap \mathrm{img}(^{\mathrm{BM}}i^b(r))$ and $\mathrm{img}(in_-(r) \cap \mathrm{img}(in_+(r)) = 0$ for any r, l, t, a careful analysis of the projective limit and of the diagram (5.49) implies that

$$\mathbb{F}_r^{\widetilde{f}}(a,b) \to^{\mathrm{BM}} \mathbb{F}_r^{\widetilde{f}}(a,b)$$

is surjective, with kernel isomorphic to

$$\varprojlim_{0<l,t\to\infty} \mathrm{img}(H_r(\widetilde{M}_{-l} \sqcup \widetilde{M}^t) \to H_r(\widetilde{M})) = \mathbb{I}_{-\infty}^{\widetilde{f}}(r) + \mathbb{I}_{\widetilde{f}}^\infty(r) = \mathrm{Tor}(H_r(\widetilde{M})).$$

This establishes items (1) and (2) in Proposition 5.14. Item (3) follows from the isomorphism $\mathrm{PD}^{\mathrm{BM}} : H_r^{\mathrm{BM}}(\widetilde{M}) \to (H_{n-r}(\widetilde{M}))^*$, which is obviously $\kappa[t^{-1}, t]$-linear and item (4) from Proposition 4.8 in Chapter 4. $\qquad\square$

Proposition 5.14 implies:,

(1) The isomorphism $\mathrm{PD}_r^{\mathrm{BM}} : H_r^{\mathrm{BM}}(\widetilde{M}) \to (H_{n-r}(\widetilde{M}))^*$ induces the isomorphism

$$\mathrm{Tor}(H_r^{\mathrm{BM}}(\widetilde{M})) = (\mathrm{Tor}\,H_{n-r}(\widetilde{M}))^*.$$

(2) The linear map $\lambda_r : H_r(\widetilde{M}) \to H_r^{\mathrm{MB}}(\widetilde{M})$ induces the injective map

$$H_r^N(X; \xi_f) = H_r(\widetilde{M})/\mathrm{Tor}(H_r(\widetilde{M})) \to H_r^{\mathrm{BM}}(\widetilde{M})/\mathrm{Tor}(H_r^{\mathrm{BM}}(\widetilde{M})).$$

(3) The isomorphism $\mathrm{PD}_r^{\mathrm{BM}} : H_r^{\mathrm{BM}}(\widetilde{M}) \to (H_{n-r}(\widetilde{M}))^*$ induces the isomorphisms

$$H_r^{\mathrm{BM}}(\widetilde{M})/\mathrm{Tor}(H_r^{\mathrm{BM}}(\widetilde{M})) \to (H_{n-r}^N(M; \xi_f))^*$$

and then combining with (2) gives

$$\mathrm{PD}_r^N : H_r^N(M; \xi_f) \to H_r^{\mathrm{BM}}(\widetilde{M})/\mathrm{Tor}(H_r^{\mathrm{BM}}(\widetilde{M})) \to (H_{n-r}^N(M; \xi_f))^*$$

which is injective and $\kappa[t^{-1}, t]$-linear.

(4) $^{\mathrm{BM}}\mathbb{F}_r^{\widetilde{f}}(a,b) = \mathbb{F}_r^{\widetilde{f}}(a,b)/\mathrm{Tor}(H_r(\widetilde{M}))$ and $^{\mathrm{BM}}\widehat{\delta}_r^{\widetilde{f}}(a,b) = \widehat{\delta}_r^{\widetilde{f}}(a,b)$.

(5) $^{\mathrm{BM}}i_a^b(r)$ in diagram (5.47) factors through $H_r^N(M; \xi_f) \to H_r^{\mathrm{BM}}(\widetilde{M})$.

(6) The linear map $^{\mathrm{BM}}i_a^b(r)$ in diagram (5.47) factors through the linear map

$$^{\mathrm{BM}}\mathbb{F}_r^{\widetilde{f}}(a,b) = \mathbb{F}_r^{\widetilde{f}}(a,b)/\mathrm{Tor}(H_r(\widetilde{M})) \to H_r^N(M; \xi_f).$$

As a consequence, the diagram (5.47) leads to the diagram

$$
\begin{array}{ccc}
\mathbb{F}_r^{\widetilde{f}}(a,b)/\mathrm{Tor}(H_r(\widetilde{M})) & \xrightarrow{\mathrm{PD}_r(a,b)} & (\mathbb{G}_{n-r}^{\widetilde{f}}(b,a))^* \\
& & \\
\downarrow i_a^b(r) & \widehat{\delta}_r^{\widetilde{f}}(a,b) \quad p_{n-r}(a,b)^* & (\widehat{\delta}_{n-r}^{\widetilde{f}}(b,a))^* \\
& & \\
H_r^N(M;\xi_f) & \xrightarrow{\mathrm{PD}_r^N} & (H_{n-r}^N(M;\xi_f))^*
\end{array}
$$

(5.51)

with the arrows

$$
\mathbb{F}_r^{\widetilde{f}}(a,b)/Tor(H_r(\widetilde{M})) \to (\mathbb{G}_{n-r}^{\widetilde{f}}(b,a))^*
$$

and

$$
\widehat{\delta}_r^{\widetilde{f}}(a,b) \to (\widehat{\delta}_{n-r}^{\widetilde{f}}(b,a))^*
$$

isomorphisms and

$$
H_r^N(M;\xi_f) \to (H_{n-r}^N(M;\xi_f))^*
$$

injective.

Item (1) in Theorem 5.6 follows from the isomorphism $\widehat{\delta}_r^{\widetilde{f}}(a,b) \to (\widehat{\delta}_{n-r}^{\widetilde{f}}(b,a))^*$.

To prove item (2) one considers the diagram 5.52

$$
\begin{array}{ccc}
\mathbb{F}_r^f\langle a,b\rangle & \xrightarrow{\mathrm{PD}_r(\langle a,b\rangle)} & (\mathbb{G}_{n-r}^f)^*\langle b,a\rangle \\
& & \\
\downarrow i_r & \widehat{\delta}_r^f\langle a,b\rangle \quad p_{n-r}^*\langle b,a\rangle & (\widehat{\delta}_{n-r}^{\widetilde{f}})^*\langle a,b\rangle \subset (\widehat{\delta}_{n-r}^f r\langle b,a\rangle)^* \\
& & \\
H_r^N(M;\xi_f) & \xrightarrow{\mathrm{PD}_r^N} & (H_{n-r}^N(M;\xi_f))^*
\end{array}
$$

,

(5.52)

where

$$
\mathbb{F}_r^f\langle a,b\rangle := \bigoplus_{n\in\mathbb{Z}}(\mathbb{F}_r^{\widetilde{f}}(a+2\pi n, b+2\pi n)/Tor(H_r(\widetilde{M})),
$$

$$
\widehat{\delta}_r^f\langle a,b\rangle := \bigoplus_{n\in\mathbb{Z}}\widehat{\delta}_r^{\widetilde{f}}(a+2\pi n, b+2\pi n),
$$

$$(\mathbb{G}_{n-r}^{f})^{*}\langle b, a\rangle = \bigoplus_{n\in\mathbb{Z}}(\mathbb{G}_{n-r}^{\widetilde{f}}(b+2\pi n, a+2\pi n))^{*},$$

$$(\widehat{\delta}_{n-r}^{f})^{*}\langle b, a\rangle = \bigoplus_{n\in\mathbb{Z}}(\widehat{\delta}_{n-r}^{\widetilde{f}}(b+2\pi n, a+2\pi n))^{*},$$

$$i_{r} = \bigoplus_{n\in\mathbb{Z}}{}^{BM}i_{a+2\pi n}^{b+2\pi n}(r),$$

$$p_{n-r}^{*}\langle b, a\rangle = \bigoplus_{n\in\mathbb{Z}}p_{n-r}(b+2\pi n, a+2\pi n)^{*},$$

$$\mathrm{PD}_{r}(\langle a, b\rangle) = \bigoplus_{n\in\mathbb{Z}}\mathrm{PD}_{r}(a+2\pi n, b+2\pi n).$$

In this diagram the multiplication by t, which makes $\mathbb{F}_{r}^{f}\langle a, b\rangle$, $(\mathbb{G}_{n-r}^{f}\langle b, a\rangle)^{*}$, $\widehat{\delta}_{r}^{f}\langle a, b\rangle$, and $(\widehat{\delta}_{n-r}^{f}\langle a, b\rangle))^{*}$ torsion-free $\kappa[t^{-1}, t]$-modules, is induced by the isomorphisms t_{r} and $(t_{n-r})^{*}$. All arrows are $\kappa[t^{-1}, t]$-linear. The first two horizontal maps are isomorphisms, and the third only injective.

Clearly, the second and the third horizontal maps define non-degenerate bilinear maps refining the Poincaré Duality and establish item (2) of the Theorem 5.6, with the diagram (5.52) establishing the claimed compatibility.

Chapter 6

Configurations γ_r^f

In this chapter, in analogy with the configurations δ_r^f whose support consists of r-closed and $(r-1)$-open barcodes, we describe and study the configurations γ_r^f whose support consists of r-mixed barcodes. They are configuration of points with multiplicity in $\mathbb{R}^2 \setminus \Delta$ for real-valued maps, and of points with multiplicity in $(\mathbb{R}^2/\mathbb{Z}) \setminus (\Delta/\mathbb{Z})$ for angle-valued maps.

For f real-valued, the r-closed-open barcode $[a, b)$ corresponds to the point (a, b) above the diagonal Δ, $\Delta = \{(x, y) \in \mathbb{R}^2 \mid x = y\}$, and the r-open-closed barcode $(a, b]$ corresponds to the point (b, a) below Δ.

For f angle-valued, the r-closed-open barcode $[a, b)$ corresponds to the point $\langle a, b \rangle \in \mathbb{T} = \mathbb{R}^2/\mathbb{Z}$ above the diagonal $\Delta/\mathbb{Z}$ and the r-open-closed barcode $(a, b]$ corresponds to the point $\langle b, a \rangle$ in $\mathbb{T}$ below $\Delta/\mathbb{Z}$.

It turns out that when f is real-valued the restriction of γ_r^f to the *above-diagonal* component $(\mathbb{R}^2 \setminus \Delta)^+$ of $\mathbb{R}^2 \setminus \Delta$ is the *persistence diagram of f* proposed in [Edelsbrunner, H., Letscher, D., Zomorodian, A. (2002)], but deprived of the infinite barcodes, while the restriction of γ_r^f to the *below-diagonal* component $(R^2 \setminus \Delta)^-$ of $\mathbb{R}^2 \setminus \Delta$ is the *persistence diagram* (deprived of the infinite barcodes) of $-f$ after composing with the map $\nu : \mathbb{R}^2 \to \mathbb{R}^2$ given by $\nu(x, y) = (-x, -y)$.

The location of the mixed barcodes for a real-valued map is the space $\mathbb{R}^2 \setminus \Delta$, and for angle-valued map is $(\mathbb{R}^2/\mathbb{Z}) \setminus (\Delta/\mathbb{Z}) = \mathbb{C} \setminus \{0 \sqcup \mathbb{S}^1\}$.

In case of angle-valued map the points $z = e^{ia+(b-a)} \in \mathbb{C} \setminus 0$, $|z| > 1$, in the support of γ_r^f correspond to the equivalence class of closed-open barcodes $[a + 2\pi n, b + 2\pi n)$ of $\widetilde{f}$, and the points $z = e^{ia+(b-a)}$, $|z| < 1$ in the support of γ_r^f correspond to the equivalence class of open-closed barcodes $(b + 2\pi n, a + 2\pi n]$ of $\widetilde{f}$.

The definitions of γ_r^f in this section is based on the concept of Fredholm cross-ratio and its properties are derived from the general properties of this concept. The reader is invited to revisit the short Subsection 2.1.2.

Recall from Chapter 2 Subsection 2.1.2 that for $\alpha : A \to B$, $\beta : B \to C$, $\gamma : C \to D$ Fredholm linear maps between κ-vector spaces, the vector space $\widehat{\omega}(\alpha, \beta, \gamma)$ is derived from the diagram

$$
\begin{array}{ccc}
\ker(\gamma\beta\alpha) & \xrightarrow{\ j_1\ } & \ker(\gamma\beta) \\
{\scriptstyle i_1}\big\uparrow & & {\scriptstyle i_2}\big\uparrow \\
\ker(\beta\alpha) & \xrightarrow{\ j_2\ } & \ker(\beta)
\end{array}
\tag{6.1}
$$

and is given by

$$
\begin{aligned}
\widehat{\omega}(\alpha, \beta, \gamma) :&= \ker(\gamma\beta)/\big(\mathrm{img}(j_1) + \mathrm{img}(i_2)\big) \\
&= \mathrm{coker}\,\big(\ker(\gamma\beta\alpha)/\ker(\beta\alpha) \to \ker(\gamma\beta)/\ker(\beta)\big).
\end{aligned}
\tag{6.2}
$$

6.1　General considerations

Let $f : X \to \mathbb{R}$ be a tame map as defined in Subsection 2.3.1, Chapter 2 which implies that $f^{-1}[a, b]$ is a compact ANR for any $a \le b$. In this chapter $X_a^f = f^{-1}((-\infty, a])$ and $X_f^a = f^{-1}(b, \infty))$. For any $(a, b) \in \mathbb{R}^2 \setminus \Delta := \{(a, b) \in \mathbb{R}^2 \mid a \ne b\}$, define $i_{a,b}^f(r)$,

$$
\begin{cases}
i_{a,b}^f(r) : H_r(X_a^f) \to H_r(X_b^f), & \text{if } a < b, \\
i_{a,b}^f(r) : H_r(X_f^a) \to H_r(X_f^b), & \text{if } a > b,
\end{cases}
$$

and then the vector space

$$
\mathbb{T}_r^f(a, b) := \ker i_{a,b}^f(r).
$$

For $a \le b \le c$, one has

$$
\begin{aligned}
i_{a,c}^f(r) &= i_{b,c}^f(r) \cdot i_{a,b}^f(r), \\
i_{c,a}^f(r) &= i_{b,a}^f(r) \cdot i_{c,b}^f(r).
\end{aligned}
\tag{6.3}
$$

Observation 6.1. The linear maps $i_{a,b}^f(r)$ are Fredholm and therefore $\mathbb{T}_r^f(a, b)$ has finite dimension.

Indeed, the exact homology sequence of the pair (X_b^f, X_a^f),

$$
\cdots \longrightarrow H_{r+1}(X_b^f, X_a^f) \longrightarrow H_r(X_a^f) \xrightarrow{\ i_{a,b}(r)\ } H_r(X_b^f) \longrightarrow H_r(X_b^f, X_a^f) \longrightarrow \cdots
$$

coupled with the excision property for homology, $H_*(X_b^f, X_a^f) = H_*(f^{-1}[a,b], f^{-1}(a))$, imply that $i_{a,b}^f(r)$ is a Fredholm map, hence $\mathbb{T}_r^f(a,b)$ is a finite-dimensional vector space. This is the case because $H_*(f^{-1}[a,b], f^{-1}(a))$ is finite-dimensional for any r thanks to the fact that $f^{-1}[a,b]$ and $f^{-1}(a)$ are compact ANRs.

As in Chapter 5, in this chapter one considers "boxes B", but only boxes away from the diagonal i.e., $B \subset \mathbb{R}^2 \setminus \Delta$. There are two type of such boxes:

(i) **Box above diagonal** with relevant vertex (a,b), $a < b$, given by $B = (a', a] \times (b', b]$, $a' < a \le b' < b$, cf. Fig. 6.1.

(ii) **Box below diagonal** with relevant vertex (c,d), $c > d$, given by $B = [c, c') \times [d, d')$, $c' > c \ge d' > d$, cf. Fig. 6.2.

The transformation $t : \mathbb{R}^2 \to \mathbb{R}^2$ given by $\nu(x,y) = (-x, -y)$ sends a box above diagonal into a box below diagonal, mapping the relevant vertex into the relevant vertex, and vice versa.

For the box above diagonal $B = (a', a] \times (b', b]$, $a' < a \le b' < b$, consider

$$H_r(X_{a'}^f) \xrightarrow{i_{a',a}} H_r(X_a^f) \xrightarrow{i_{a,b'}} H_r(X_{b'}^f) \xrightarrow{i_r(b',b)} H_r(X_b^f) \quad \text{and define}$$

$$\mathbb{T}_r^f(B) := \widehat{\omega}(i_{a',a}(r), i_{a,b'}{}'(r), i_{b',b}(r)),$$

derived from the diagram of type (6.1)

$$
\begin{array}{ccc}
\mathbb{T}_r^f(a', b) & \xrightarrow{i_1} & \mathbb{T}_r^f(a, b) \\
{\scriptstyle j_1}\big\uparrow & & {\scriptstyle j_2}\big\uparrow \\
\mathbb{T}_r^f(a', b') & \xrightarrow{i_2} & \mathbb{T}_r^f(a, b').
\end{array}
\qquad (6.4)
$$

For the box below diagonal $B = [c, c') \times [d, d')$, $c' > c \ge d' > d$, consider

$$H_r(X_{f}^{c'}) \xrightarrow{i_{c',c}} H_r(X_f^c) \xrightarrow{i_{c,d'}} H_r(X_f^{d'}) \xrightarrow{i_r(d',d)} H_r(X_f^d) \quad \text{and define}$$

$$\mathbb{T}_r^f(B) := \widehat{\omega}(i_{c,c'}, i_{c',d}, i_{d,d'}),$$

derived from the diagram of type (6.2)

$$
\begin{array}{ccc}
\mathbb{T}_r^f(c', d) & \xrightarrow{i_1} & \mathbb{T}_r^f(c, d) \\
{\scriptstyle j_1}\big\uparrow & & {\scriptstyle j_2}\big\uparrow \\
\mathbb{T}_r^f(c', d') & \xrightarrow{i_2} & \mathbb{T}_r^f(c, d').
\end{array}
\qquad (6.5)
$$

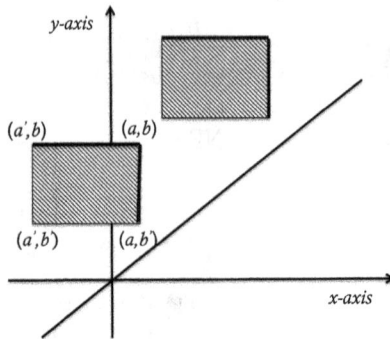

Fig. 6.1   Boxes above diagonal

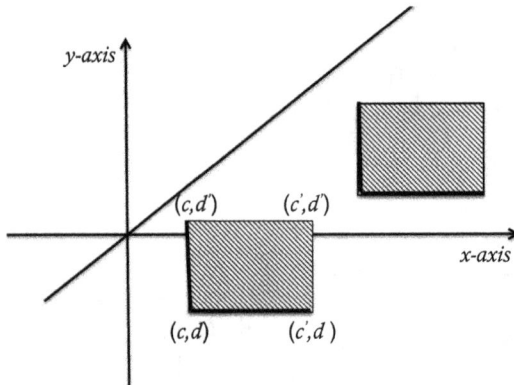

Fig. 6.2   Boxes below diagonal

When implicit from the context, the map f will be discarded from the notation. As in Chapter 5, where we have considered $\mathbb{F}_r(B)$, we have the following propositions about $\mathbb{T}_r(B)$.

Proposition 6.1. *For three boxes B, B_1, B_2 above or below diagonal such that $B = B_1 \sqcup B_2$ with B_1 and B having the same relevant vertex (see Figures 6.3 and 6.4 below), the inclusions $B_1, B_2 \subset B$ induce the surjective linear map*

$$\pi_B^{B_1}(r) : \mathbb{T}_r^f(B) \to \mathbb{T}_r^f(B_1)$$

and the injective linear map

$$i_{B_2}^B(r) : \mathbb{T}_r^f(B_2) \to \mathbb{T}_r^f(B)$$

such that the following sequence is exact:

$$0 \longrightarrow \mathbb{T}_r^f(B_2) \xrightarrow{i_{B_2}^B(r)} \mathbb{T}_r^f(B) \xrightarrow{\pi_B^{B_1}(r)} \mathbb{T}_r^f(B_1) \longrightarrow 0.$$

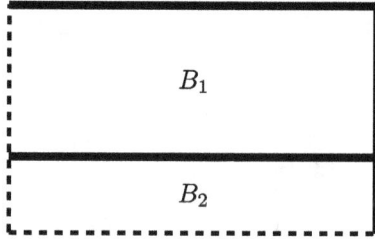

Fig. 6.3 Box above diagonal divided horizontally

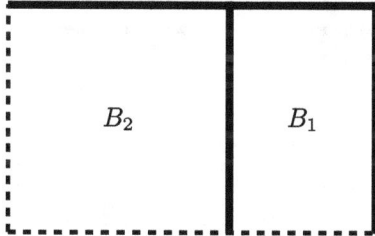

Fig. 6.4 Box above diagonal divided vertically

The Figures 6.3 and 6.4 depict only boxes above diagonal.

Proposition 6.2.

1. Let $a'' < a' < a \leq b'' < b' < b$ and
$B = (a'', a] \times (b'', b]$
$B_{1,1} = (a', a] \times (b', b]$, $B_{1,2} = (a'', a'] \times (b', b]$,
$B_{2,1} = (a', a] \times (b'', b']$, $B_{2,2} = (a'', a'] \times (b'', b']$,
$B_{,1} = B_{1,1} \sqcup B_{2,1}$, $B_{,2} = B_{1,2} \sqcup B_{2,2}$,
$B_{1,} = B_{1,1} \sqcup B_{2,1}$, $B_{2,} = B_{2,1} \sqcup B_{2,2}$.
(*See* Fig. 6.5 below.)

Then one has

$$i^B_{B_{2,2}}(r) = i^B_{B_{,2}}(r) \cdot i^{B_{,2}}_{B_{2,2}}(r) = i^B_{B_{2,}}(r) \cdot i^{B_{2,}}_{B_{2,2}}(r)$$
$$\pi^{B_{1,1}}_B(r) = \pi^{B_{1,1}}_{B_{,1}}(r) \cdot \pi^{B_{,1}}_B(r) = \pi^{B_{1,1}}_{B_{1,}}(r) \cdot \pi^{B_{1,}}_B(r).$$

2. Let $a'' > a' > a \geq b'' > b' > b$ and
$B = [a, a'') \times [b, b'')$,
$B_{1,1} = [a, a') \times [b, b')$, $\quad B_{1,2} = [a, a') \times [b', b'')$,
$B_{2,1} = [a', a'') \times [b, b')$, $\quad B_{2.2} = [a'.a'') \times [b', b'')$,
$B_{1,} = B_{1,1} \sqcup B_{1,2}$, $\quad B_{2,} = B_{2,1} \sqcup B_{2,2}$,
$B_{,1} = B_{1,1} \sqcup B_{2,1}$, $\quad B_{,2} = B_{1,2} \sqcup B_{2,2}$.
(*See* Fig. 6.6 below.)
Then

$$i^B_{B_{2,2}}(r) = i^B_{B_{,2}}(r) \cdot i^{B_{,2}}_{B_{2,2}}(r) = i^B_{B_{2,}}(r) \cdot i^{B_{2,}}_{B_{2,2}}(r)$$
$$\pi^{B_{1,1}}_B(r) = \pi^{B_{1,1}}_{B_{,2}}(r) \cdot \pi^{B_{,2}}_B(r) = \pi^{B_{1,1}}_{B_{2,}}(r) \cdot \pi^{B_{2,}}_B(r).$$

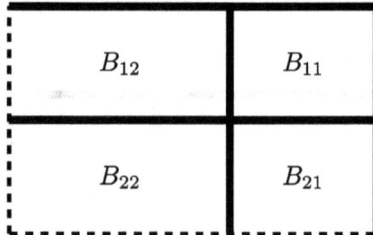

Fig. 6.5   Box above diagonal divided in four pieces

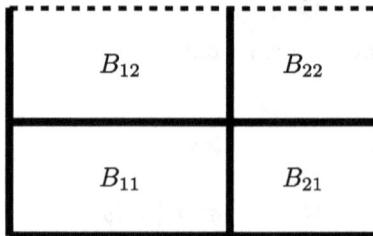

Fig. 6.6   Box below diagonal divided in four pieces

Proof. The verification is a straightforward consequence of the definitions and of Theorem 2.4 (3) in Chapter 2. For example, in the case above diagonal the boxes B, B_1, B_2 are given by either $a'' < a' < a \le b'' < b$ or $a'' < a \le b'' < b' < b$, with $B = (a'', a] \times (b'', b]$ in both cases, $B_1 = (a', a] \times (b'', b]$ and $B_2 = (a'', a'] \times (b'', b]$ in the first case, and $B_1 = (a'', a] \times (b', b]$ and $B_2 = (a'', a] \times (b'', b']$ in the second, as indicated below in 6.3 and 6.4. One applies Theorem 2.4 (3) items (i) and (ii).

$\square$

Corollary 6.1. *If the box B above or below diagonal is a union of finitely many disjoint boxes $B = \bigsqcup_{i=1,\dots,k} B_i$, then*

$$\mathbb{T}_r(B) \simeq \bigoplus_{1 \le i \le k} \mathbb{T}_r(B_i).$$

Proof. The proof follows the same arguments as in Chapter 5, Corollary 5.1. One refines the decomposition of the box into a disjoint union of boxes $B = \bigsqcup_{1 \le i \le k, 1 \le j \le r} B_{i,j}$ with the property that both $B_{i,j} \sqcup B_{i+1,j}$ and $B_{i,j} \sqcup B_{i,j+1}$ are boxes for any i, j, hence with the hypothesis of Proposition 6.1 satisfied. One checks by induction, applying Proposition 6.1, that the result holds, first for $k = 1$ and arbitrary r, and then for arbitrary k and arbitrary r.

$\square$

For $a < b$ denote by $B(a, b; \epsilon)$, $0 < \epsilon < (b - a)$, the box above diagonal

$$B(a, b; \epsilon) := (a - \epsilon, a] \times (b - \epsilon, b],$$

and for $c > d$ denote by $B(c, d; \epsilon)$, $0 < \epsilon < (c - d)$, the box below diagonal

$$B(c, d; \epsilon) := [c, c + \epsilon) \times [d, d + \epsilon).$$

Observe that, in view of Proposition 6.2, if $B'' \subset B' \subset B$ are boxes above or below diagonal with the same relevant corner, then

$$\pi_B^{B''}(r) = \pi_{B'}^{B''}(r) \cdot \pi_B^{B'}(r). \tag{6.6}$$

Then for any $(a, b) \in \mathbb{R}^2 \setminus \Delta$, and for ϵ, ϵ' with $0 < \epsilon' < \epsilon$, one obtains the linear maps $\pi_{B(a,b;\epsilon)}^{B(a,b;\epsilon')}(r) : \mathbb{T}_r(B(a, b; \epsilon)) \to \mathbb{T}_r(B(a, b; \epsilon'))$, which for $\epsilon \to 0$ provide a direct system of vector spaces, and then define

$$\boxed{\widehat{\gamma}_r^f(a, b) := \varinjlim_{\epsilon \to 0} \mathbb{T}_r^f(B(a, b; \epsilon))},$$

$$\boxed{\gamma_r^f(a, b) := \dim(\widehat{\gamma}_r^f(a, b))}.$$

For $a \in \mathbb{R}$ we denote

$$a_- := \sup\{c \in \mathcal{C}r(f) \mid c < a\},$$
$$a_+ := \inf\{c \in \mathcal{C}r(f) \mid c > a\}. \tag{6.7}$$

Since f is tame, $X_{a_-}^f$ is a deformation retract of $X_{a'}^f$ for $a_- \leq a' < a$, and $X_f^{a_+}$ is a deformation retract of $X_f^{a''}$ for $a < a'' \leq a_+$. Then one has:

Observation 6.2.

   If $a < b$ and $a_- \leq a - \epsilon < a$, $b_- \leq b - \epsilon < b$ then $\widehat{\gamma}_r^f(a,b) = \mathbb{T}_r^f(a - \epsilon, a] \times (b - \epsilon, b])$.

   If $c > d$ and $c < c + \epsilon \leq c_+$, $d < d + \epsilon \leq d_+$ then $\widehat{\gamma}_r^f(c,d) = \mathbb{T}_r^f([c, c + \epsilon) \times [d, d + \epsilon))$.

One can summarize Corollary 6.1 and Observation 6.2 as the following result.

Proposition 6.3.

   (1) $\gamma_r^f(x,y) \neq 0$ *implies* $x, y \in \mathcal{C}r(f)$.
   (2) *For any box B above or below diagonal,*

$$\mathbb{T}_r^f(B) = \bigoplus_{(a,b) \in B} \widehat{\gamma}_r^f(a,b).$$

For $a < b$ and $\epsilon > 0$ with $a + \epsilon < b - \epsilon$ denote

$$D(a,b;\epsilon) := (a - \epsilon, a + \epsilon] \times (b - \epsilon, b + \epsilon].$$

In this case $D(a,b;\epsilon)$ is a box above diagonal, cf. Fig. 6.7.
For $c > d$ and $\epsilon > 0$ with $c - \epsilon > d + \epsilon$ denote

$$D(c,d;\epsilon) := [c - \epsilon, c + \epsilon) \times [d - \epsilon, d + \epsilon).$$

In this case $D(c,d;\epsilon)$ is a box below diagonal, cf. Fig. 6.8.

   Consider the distance $d(x,y) := \sup\{|x_1 - y_1|, |x_2 - y_2|\}$, $x = (x_1, x_2)$, $y = (y_1, y_2)$, and for $\epsilon > 0$, let $\Delta_\epsilon := \{x = (x_1, x_2) \in \mathbb{R}^2 \mid |x_1 - x_2| < \sqrt{2}\epsilon\}$ and $\mathbb{R}_\epsilon^2 = \mathbb{R}^2 \setminus \Delta_\epsilon$, cf. Fig 6.9 below.
Let $\epsilon > 0$ be small enough to have $d(x,y) > 2\epsilon$ and $d(x, \Delta) > 2\epsilon$ for any $x, y \in \operatorname{supp}\gamma_r^f$. Denote

$$(\operatorname{supp} \gamma^f)_\epsilon := \bigcup_{(a,b) \in \operatorname{supp} \gamma_r^f} D(a,b;\epsilon) \subset \mathbb{R}_\epsilon^2 = \mathbb{R}^2 \setminus \Delta_\epsilon.$$

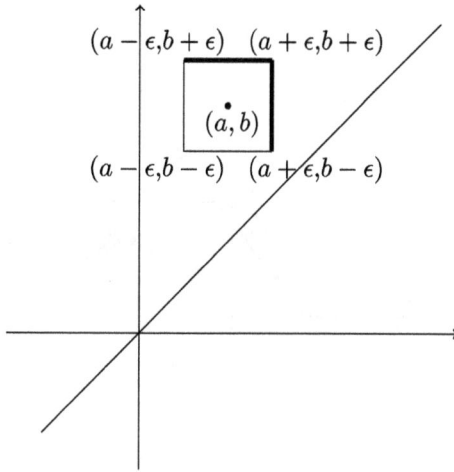

$(a - \epsilon, b + \epsilon)$ $(a + \epsilon, b + \epsilon)$

$(\overset{\bullet}{a}, b)$

$(a - \epsilon, b - \epsilon)$ $(a + \epsilon, b - \epsilon)$

Fig. 6.7   Box $D(a, b; \epsilon)$ above diagonal

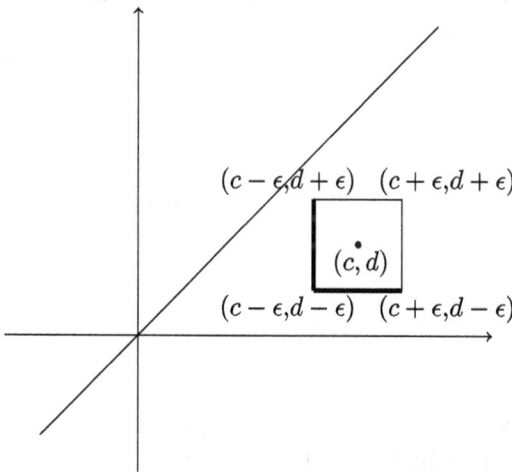

$(c - \epsilon, d + \epsilon)$ $(c + \epsilon, d + \epsilon)$

$(\overset{\bullet}{c}, d)$

$(c - \epsilon, d - \epsilon)$ $(c + \epsilon, d - \epsilon)$

Fig. 6.8   Box $D(a, b; \epsilon)$ below diagonal

Suppose $f : X \to \mathbb{R}$ is a tame map whose set of critical values is

$$\cdots < c_{i-1} < c_i < c_{i+1} < \cdots$$

and it holds that $|c_{i+1} - c_i| > 2\epsilon$, $d(x, \Delta) > 2\epsilon$ $x \in$ supp γ^f.
Similarly to Proposition 5.7 in Chapter 5, we have Proposition 6.4.

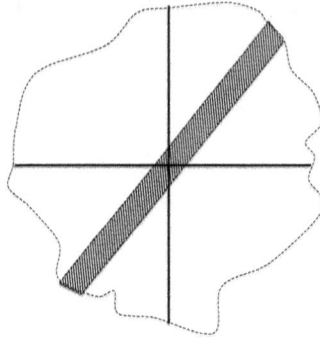

Fig. 6.9   Neighborhood Δ_ϵ

Proposition 6.4. *For any tame map* $g : X \to \mathbb{R}$ *such that* $\sup_{x \in X} |f(x) - g(x)| < \epsilon/3$ *the following holds:*

(1) *for* $x = (a, b) \in \operatorname{supp} \gamma_r^f$ *one has*

$$\sum_{y \in \operatorname{supp} \gamma_r^f \cap D(a, b; \epsilon)} \gamma_r^g(y) = \gamma_r^f(x);$$

(2) $\sharp(\operatorname{supp} \gamma_r^g \cap (\mathbb{R}^2 \setminus \Delta_\epsilon)) = \sharp \operatorname{supp} \gamma_r^f$.

Proof. The proof is a manipulation of the properties of the Fredholm cross-ratio stated in Theorem 2.4 and Proposition 2.3 in Chapter 2.

Item (1). We verify only the case where $x = (a, b)$ is above the diagonal, i.e., $a < b$. The case below the diagonal is dealt with in the same way.

Consider the sequence of inclusions

$$X_{a-\epsilon}^f \subseteq X_{a-(2/3)\epsilon}^g \subseteq X_{a-(1/3)\epsilon}^f \subseteq X_{a+(1/3)\epsilon}^f \cdots \subseteq X_{b+(1/3)\epsilon}^f \subseteq X_{b+(2/3)\epsilon}^g \subseteq X_{b+\epsilon}^f$$

guaranteed by the assumption $|f - g| < \epsilon/3$.

Then we have the inclusion-induced linear maps

$$H_r(X_{a-\epsilon}^f)\xrightarrow{\alpha_1}H_r(X_{a-(2/3)\epsilon}^g)\xrightarrow{\alpha_2}H_r(X_{a-\epsilon/3}^f)\xrightarrow{\alpha_3}H_r(X_{a+(1/3)\epsilon}^f)$$

$$H_r(X_{a+(2/3)\epsilon}^g)\xleftarrow{\beta_2}H_r(X_{a+\epsilon}^f)\xrightarrow{\beta_3}H_r(X_{b-\epsilon}^f)\xrightarrow{\gamma_1}H_r(X_{b-(2/3)\epsilon}^g)$$

$$H_r(X_{b-(1/3)\epsilon}^f)\xleftarrow{\gamma_3}H_r(X_{b+(1/3)\epsilon}^f)\xrightarrow{\delta_1}H_r(X_{b+(2/3)\epsilon}^g)\xrightarrow{\delta_2}H_r(X_{b+\epsilon}^f).$$

The hypothesis on f, precisely that $d(x,y) > 2\epsilon$ for any two different points in the support of γ_r^f, ensures that the maps $\alpha_2 \cdot \alpha_1$, $\beta_2 \cdot \beta_1$, $\gamma_2 \cdot \gamma_1$, and $\delta_2 \cdot \delta_1$ are isomorphisms, so the factorization hypotheses in Proposition 2.3 in Chapter 2 hold. Then $\widehat{\omega}(\alpha,\beta,\gamma)$, with $\alpha = \alpha_1\alpha_2\alpha_3$, $\beta = \beta_1\beta_2\beta_3$, $\gamma = \gamma_1\gamma_2\gamma_3$, is canonically isomorphic to $\widehat{\omega}(\alpha',\beta',\gamma')$ with $\alpha' = \alpha_2\alpha_3\beta_1$, $\beta' = \beta_2\beta_3\gamma_1$, $\gamma' = \gamma_2\gamma_3\delta_1$. This leads to

$$\sum_{y\in D(a,b;2\epsilon/3)\cap\text{supp}\,\gamma_r^g}\gamma_r^g(y) =$$

$$\dim \mathbb{T}_r^g((a - 2\epsilon/3, a + 2\epsilon/3] \times (b - 2\epsilon/3, b + 2\epsilon/3]) =$$

$$\dim \mathbb{T}_r^f((a - \epsilon, a + \epsilon/3] \times (b - \epsilon, b + \epsilon/3]) = \gamma_r^f(x),$$

(6.8)

where the first equality is a consequence of Proposition 6.3, the second of Proposition 2.3 in Chapter 2, and the third a consequence of Observation 6.2. This verifies item (1).

Item (2). One has to show that for any point $x = (a,b) \in \mathbb{R}^2 \setminus (\Delta_\epsilon \sqcup \bigcup_{(a,b)\in\text{supp}\,\delta_r^f} D(a,b;2\epsilon'))$ one has $\gamma_r^g(a,b) = 0$.

For such $x = (a,b)$ at least one component a or b satisfies $|a - c_i| > 2\epsilon'$ or $|b - c_i| > 2\epsilon'$, respectively.

Let $\epsilon' = \epsilon/3$ and consider the inclusions

$$X_{a-2\epsilon'}^f \subseteq X_{a-\epsilon'}^g \subseteq X_{a+\epsilon'}^f \subseteq X_{a+2\epsilon'}^g \subseteq X_{b-2\epsilon'}^f \subseteq X_{b-\epsilon'}^g \subset X_{b+\epsilon'}^f \subseteq X_{b+2\epsilon'}^g$$

and the inclusion-induced linear maps

$$H_r(X_{a-2\epsilon'}^f)\xrightarrow{\alpha_1}H_r(X_{a-\epsilon'}^g)\xrightarrow{\alpha_2}H_r(X_{a+\epsilon'}^f)\xrightarrow{\beta_1}H_r(X_{a+2\epsilon'}^g)$$

$$H_r(X_{a-2/3\epsilon'}^f)\xleftarrow{\gamma_1}H_r(X_{b-\epsilon'}^g)\xrightarrow{\gamma_2}H_r(X_{b+\epsilon'}^f)\xrightarrow{\delta_1}H_r(X_{b+2\epsilon'}^g)$$

$$H_r(X_{b+\epsilon}^f).$$

We want to show that $\widehat{\omega}(\beta_1\alpha_2, \gamma_1\beta_2, \delta_1\gamma_2) = \mathbb{T}^g((a - \epsilon', a + 2\epsilon') \times (b - \epsilon', b + 2\epsilon']) = 0$, which in view of Observation 6.2 implies $\gamma^g((a, b)) = 0$. In view of the definition of Fredholm cross-ratio, cf. (6.1), since δ_1 is injective we have that $\widehat{\omega}(\beta_1\alpha_2, \gamma_1\beta_2, \delta_1\gamma_2) = \widehat{\omega}(\beta_1\alpha_2, \gamma_1\beta_2, \gamma_2)$.

Applying Theorem 2.4 in Chapter 2 we obtain that $\widehat{\omega}(\beta_1\alpha_2, \gamma_1\beta_2, \gamma_2) = 0$ iff:

(i) $\widehat{\omega}(\alpha_2, \gamma_1\beta_2\beta_1, \gamma_2) = 0$, and

(ii) $\widehat{\omega}(\beta_1, \gamma_1\beta_2, \gamma_2) = 0$.

Since $\beta_2 \cdot \beta_1$ is an isomorphism, one has $\widehat{\omega}(\beta_2\beta_1, \gamma_2, \gamma_1) = 0$, which by Theorem 2.4 in Chapter 2 implies $\widehat{\omega}(\beta_1, \gamma_1\beta_2, \gamma_2) = 0$, so (ii) above holds.

To check (i) above note that $0 = \widehat{\gamma}_r^f(a, b) = \mathbb{T}_r^f((a-2\epsilon', a+\epsilon'] \times (b-\epsilon', b+2\epsilon']) = \widehat{\omega}(\alpha_2\alpha_1, \beta_2\beta_1, \gamma_2\gamma_1) = 0$ implies by Theorem 2.4 (3) (i) Chapter 2 that $\widehat{\omega}(\alpha_2, \beta_2\beta_1, \gamma_2\gamma_1) = 0$, which in turn, by Proposition 2.3 Chapter 2, implies that $\widehat{\omega}(\alpha_2, \gamma_1\beta_2\beta_1, \gamma_2) = 0$, hence (i). $\qquad \square$

Now suppose $f : X \to \mathbb{R}$ is a proper tame map. For $m < M$, $m, M \in \mathbb{R}$, denote by g the restriction of f to $f^{-1}[m, M]$.

Proposition 6.5. (Localization property) *For a box B away from the diagonal defined by four real numbers a, b, c, d, all lying in the interval (m, M), it holds that*

$$\mathbb{T}_r^g(B) = \mathbb{T}_r^f(B)$$

and therefore

$$\widehat{\gamma}_r^g(a, b) = \widehat{\gamma}_r^f(a, b).$$

Proof. We verify the result for B above diagonal $B = (a, b] \times (c, d]$ only. The case of B a box below diagonal can be derived from the above case by applying the transformation ν, $\nu(a, b) = (-a, -b)$.

In view of the excision property in homology one has

$$H_*(f^{-1}((-\infty, a]), f^{-1}([m, a])) = H_*(f^{-1}((-\infty, m]), f^{-1}(m)).$$

Then the homology exact sequence of the pair $(f^{-1}((-\infty, a]), f^{-1}([m, a]))$, $a > m$, becomes

$$\cdots H_{r+1}(f^{-1}((-\infty, m]), f^{-1}(m)) \longrightarrow H_r(f^{-1}(m, a])$$

$$H_r(f^{-1}(-\infty, a]) \longrightarrow H_r(f^{-1}((-\infty, m]), f^{-1}(m)) \cdots$$

We apply Theorem 2.4 Chapter 2 Item (2) to A, B, C, D given by $H_r(X_{\ldots}^f)$ and A', B', C', D' given by $H_r(X_{\ldots}^g)$, $\cdots = a, b, c, d$ and conclude the result.

$\square$

6.2   The case of real-valued maps

Suppose $f : X \to \mathbb{R}$ is a tame map with X compact. By Proposition 6.2, the assignment $(R^2 \setminus \Delta) \ni (a, b) \rightsquigarrow \gamma_r^f(a, b) \in \mathbb{Z}_{\geq 0}$ is a configuration of points with multiplicity in $\mathbb{R}^2 \setminus \Delta$.

The support of this assignment consists of the the mixed bar codes introduced in Chapter 4 Section 4.2 as stated the next proposition.

Proposition 6.6. *The points $(a, b) \in \operatorname{supp} \gamma_r^f \cap (\mathbb{R}^2 \setminus \Delta)^+ = \{(a, b) \mid a < b\}$ are in a bijective correspondence with the r-closed-open barcodes $[a, b)$ in $\mathcal{B}_r^{co}(f)$. Similarly, the points $(a, b) \in \operatorname{supp} \gamma_r^f \cap (\mathbb{R}^2 \setminus \Delta)^- = \{(a, b) \mid a > b\}$ are in a bijective correspondence with the r-open-closed barcodes $(b, a]$ in $\mathcal{B}_r^{oc}(f)$. The correspondence preserves the multiplicity.*

Proof. We check this statement for $a < b$ only; the case $b > a$ can be treated similarly. We use Observation 6.2. If either a or b is a regular value, then either i_2 or j_2 in the diagram (6.1) which defines $\mathbb{T}_r^f((a-\epsilon, a] \times (b-\epsilon, b])$ is an isomorphism for ϵ small, and then $\mathbb{T}_r^f((a - \epsilon, a] \times (b - \epsilon, b]) = 0$. If $a, b \in Cr(f)$, then the vector space $\mathbb{T}_r^f((a_-, a] \times (b_-, b])$ defined by (6.1) can be calculated by Corollary 4.1 in Chapter 4, Section 4.3, and turns out to be exactly the vector space generated by the barcodes $[a, b) \in \mathcal{B}_r^{c,o}(f)$. $\square$

Suppose X is compact. Denote by $C(X; \mathbb{R})_{\text{tame}} \subset C(X; \mathbb{R})$ the subspace of tame maps equipped with the compact-open topology, and consider $\operatorname{Conf}_{\text{bn}}(\mathbb{R}^2; \Delta)$, the space of configurations of points in $\mathbb{R}^2 \setminus \Delta$ equipped with the bottleneck topology described in Subsection 2.3.5.

Theorem 6.1. (CEH stability) *The assignment*
$$C(X; \mathbb{R})_{\text{tame}} \rightsquigarrow \operatorname{Conf}_{\text{bn}}(\mathbb{R}^2; \Delta)$$
is a continuous map.

This is a weak form of what we call CEH-stability cf. [Cohen-Steiner, D., Edelsbrunner, H., Harer, J. (2007)] [1].

---

[1] A stronger metric space version of this result is known as the Cohen-Steiner Edelsbrunner, Harer stability theorem; despite of this stronger version, the assignment cannot be extended to all continuous maps (to permit the definition of the configuration γ_r for all continuous maps) by lack of completeness of the bottleneck metric.

Proof. Let $f : X \to \mathbb{R}$ be a tame map with X compact. Consider the configuration γ_r^f and let

$$\underline{\epsilon}(f) := \min\{\inf\{d(x,y) \mid x,y \in \operatorname{supp} \gamma_r^f, x \neq y\}, d(\operatorname{supp} \gamma_r^f, \Delta)\},$$

which is a strictly positive real number. For any $\epsilon < \underline{\epsilon}(f)$ let $g : X \to \mathbb{R}$ be a tame map which satisfies $\sup |f - g| < \epsilon/3$. By Proposition 6.4 (1) for any $x \in \operatorname{supp} \gamma_r^f$ one has

$$\sum_{y \in D(x;2\epsilon/3)} \gamma_r^g(y) = \gamma_r^f(x)$$

and in view of Proposition 6.4 (2) the points in $\operatorname{supp} \gamma_r^g \setminus \Delta_\epsilon$ counted with multiplicity, have the same cardinality as the points in $\operatorname{supp} \gamma_r^f$ counted with multiplicity, and the points of the support of γ_r^g not contained in Δ_ϵ are contained in the boxes $D(x; 2/3\epsilon)$, $x \in \operatorname{supp} \gamma_r^f$. In each such box there are precisely $\gamma_r^f(x)$ such points counted with multiplicity. This is exactly what is needed to conclude the continuity of the assignment $C(X; \mathbb{R})_{\text{tame}} \rightsquigarrow \operatorname{Conf}_{\text{bn}}(\mathbb{R}^2; \Delta)$ at $f \in C(X; \mathbb{R})_{\text{tame}}$ with respect to the bottleneck topology on $\operatorname{Conf}(\mathbb{R}^2 \setminus \Delta)$. $\qquad \square$

The Poincaré Duality for compact κ-orientable manifolds with boundary $(N^n, \partial N^n)$ provides natural canonical isomorphisms $H_r(N) \to (H_{n-r}(N, \partial N))^*$, which induce an identification (Poincaré Duality) between the assignments $\widehat{\gamma}_r^f$ and $\widehat{\gamma}_{n-1-r}^f$, and then between γ_r^f and γ_{n-1-r}^f.

Theorem 6.2. *For a κ-oriented closed topological triangulable manifold M^n and a topologically tame map $f : M \to \mathbb{R}$, the Poincaré Duality induces a natural isomorphism between $\widehat{\gamma}_r^f(a,b)$ and $(\widehat{\gamma}_{n-1-r}^{-f}(-a,-b))^* = (\widehat{\gamma}_{n-r-1}^f(b,a))^*$. In particular, $\gamma_r^f(a,b) = \gamma_{n-r-1}^{-f}(-a,-b) = \gamma_{n-1-r}^f(b,a)$.*

To prove this statement we first assume that $a < b < c < d$ are all *topologically regular values* for $f : M^n \to \mathbb{R}$, which ensures that $f^{-1}(a)$, $f^{-1}(b)$, $f^{-1}(c)$, and $f^{-1}(d)$ are codimension-one submanifolds, and consider the *Poincaré Duality* diagram below.

$$
\begin{array}{ccccccc}
(1) & H_r(M_a^f) & \longrightarrow & H_r(M_b^f) & \longrightarrow & H_r(M_c^f) & \longrightarrow & H_r(M_d^f) \\
& \Big\downarrow \text{PD} & & \Big\downarrow \text{PD} & & \Big\downarrow \text{PD} & & \Big\downarrow \text{PD} \\
(2) & H^{n-r}(M, M_f^a) & \longrightarrow & H^{n-r}(M, M_f^b) & \longrightarrow & H^{n-r}(M, M_f^c) & \longrightarrow & H_r^{n-r}(M, M_f^d) \\
& \Big\uparrow \partial & & \Big\uparrow \partial & & \Big\uparrow \partial & & \Big\uparrow \partial \\
(3) & H^{n-r-1}(M_f^a) & \longrightarrow & H^{n-r-1}(M_f^b) & \longrightarrow & H^{n-r-1}(M_f^c) & \longrightarrow & H^{n-r-1}(M_f^d) \\
& \Big\Vert & & \Big\Vert & & \Big\Vert & & \Big\Vert \\
(4) & (H_{n-r-1}(M_f^a))^* & \longrightarrow & (H_{n-r-1}(M_f^b))^* & \longrightarrow & (H_{n-r-1}(M_f^c))^* & \longrightarrow & (H_{n-r-1}(M_f^d))^* \\
& \Big\uparrow & & \Big\uparrow & & \Big\uparrow & & \Big\uparrow \\
(5) & (H_{n-r-1}(M_{-a}^{-f}))^* & \longrightarrow & (H_{n-r-1}(M_{-b}^{-f}))^* & \longrightarrow & (H_{n-r-1}(M_{-c}^{-f}))^* & \longrightarrow & (H_{n-r-1}(M_{-d}^{-f}))^*
\end{array}
$$

$$(6.9)$$

The key observation is that $\hat{\omega}(1) = \hat{\omega}(2) = \hat{\omega}(3) = \hat{\omega}(4) = \hat{\omega}(5)$, where $\hat{\omega}(i)$ denotes the Fredholm cross-ratio of the row (i). This because the rows (1) and (2) are isomorphic by Poincaré Duality, the rows (3) and (4) by the isomorphism between cohomology and the dual of homology, the rows (4) and (5) by the tautology $M_f^a = M_{-a}^{-f}$, and finally $\hat{\omega}(2)$ is canonically isomorphic to $\hat{\omega}(3)$ in view of Theorem 2.4 in Chapter 2 applied to the diagram whose columns are long exact sequence of the pairs (M, M_f^a), (M, M_f^b), (M, M_f^c), and (M, M_f^d).

Let $\nu, \chi : \mathbb{R}^2 \to \mathbb{R}^2$ be given by $\nu(\alpha, \beta) = (-\alpha, -\beta)$, $\chi(\alpha, \beta) = (\beta, \alpha)$. Clearly, for B a box above diagonal $\nu(B)$ and $\chi(B)$ are boxes below diagonal, and vice versa.

As a consequence, the diagram (6.9) induces for the box $B = (a, b] \times (c, d]$ above diagonal the canonical isomorphism

$$
PD_r : \mathbb{T}_r^f(B) \longrightarrow (\mathbb{T}_{n-1-r}^f(\chi(B)))^* \xrightarrow{\;=\;} (\mathbb{T}_{n-1-r}^{-f}((\nu B))^* \,.
$$

Similarly, starting with a box below diagonal B one obtains a canonical isomorphism

$$
PD_r : \mathbb{T}_r^f(B) \longrightarrow (T_{n-r-1}^f(\chi(B)))^* \xrightarrow{\;=\;} (\mathbb{T}_{n-1-r}^{-f}((\nu B))^* \,.
$$

Since f is topologically tame, for any (a, b) one can choose a sequence $\epsilon_i > 0$ with $\epsilon_i \to 0$ and such that $a \pm \epsilon_i$ and $b \pm \epsilon_i$ are topologically regular values. Clearly, if one takes the box $D(a, b; \epsilon_i)$, since $\nu(D(a, b; \epsilon_i)) =$

$D(-a, -b; \epsilon_i)$ and $\chi(D(a, b; \epsilon_i)) = D(b, a; \epsilon_i)$, Observation 6.2 shows that letting ϵ_i go to zero one obtains the isomorphism

$$PD_r : \widehat{\gamma}_r^f(a, b) \longrightarrow \widehat{\gamma}_{n-r-1}^{-f}(-a, -b) \xrightarrow{=} \widehat{\gamma}_{n-r-1}^f(b, a)$$

and then the equality $\gamma_r^f(a, b) = \gamma_{n-r-1}^{-f}(-a, -b) = \gamma_{n-r-1}^f(b, a)$.

To the extent the reader is interested only in the configurations γ_r^f, Theorem 6.2 can be obtained also from the calculation of $T_r^f(B)$, $T_r^{-f}(\nu(B))$, and $T_r^f(\chi(B))$ in terms of barcodes, cf. Exercise E.4 in Section 8.1.

6.3   The case of angle-valued maps

For $f : X \to \mathbb{S}^1$ a tame map one considers an infinite cyclic cover $\widetilde{f} : \widetilde{X} \to \mathbb{R}$ which is also tame, and the assignments

$$\mathbb{R}^2 \setminus \Delta \ni (a, b) \rightsquigarrow \widehat{\gamma}_r^{\widetilde{f}}(a, b)$$

and

$$\mathbb{R}^2 \setminus \Delta \ni (a, b) \rightsquigarrow \gamma_r^{\widetilde{f}}(a, b) = \dim(\widehat{\gamma}_r^{\widetilde{f}}(a, b)) \in \mathbb{Z}_{\geq 0}.$$

If $\tau : \widetilde{X} \to \widetilde{X}$ is the generator of the group of deck transformations, as in Chapter 5, one obtains $\gamma_r^{\widetilde{f}}(a + 2\pi, b + 2\pi) = \gamma_r^{\widetilde{f}}(a, b)$, which defines a finite configuration of points with multiplicity, $\gamma_r^f : \mathbb{T} \setminus (\Delta/\mathbb{Z}) \to \mathbb{Z}_{\geq 0}$ by the formula

$$\gamma_r^f \langle a, b \rangle := \gamma_r^{\widetilde{f}}(a, b).$$

Indeed, in view of the tameness, which in turn implies the discreteness of the critical values of $\widetilde{f}$, and of the periodicity $\gamma_r^{\widetilde{f}}(a + 2\pi, b + 2\pi) = \gamma_r^{\widetilde{f}}(a, b)$, the support of γ_r^f is finite. As discussed before, $\mathbb{T} \setminus (\Delta/\mathbb{Z})$ identifies to $\mathbb{C} \setminus \{0 \sqcup \mathbb{S}^1\}$, so in the case of an angle-valued map, γ_r^f can be also regarded as a configuration of points in $\mathbb{C} \setminus \{0 \sqcup \mathbb{S}^1\}$. As in the case of the real valued map, one has the following results.

Proposition 6.7. *The points $\langle a, b \rangle \in \operatorname{supp} \gamma_r^f \cap (\mathbb{T}^2 \setminus \Delta/\mathbb{Z})^+$ are in bijective correspondence with the closed-open r-barcodes $\mathcal{B}_r^{co}(f)$ and the points $\langle a, b \rangle \in \operatorname{supp} \gamma_r^f \cap (\mathbb{T}^2 \setminus \Delta/\mathbb{Z})^-$ with the open-closed r-barcodes $\mathcal{B}_r^{oc}(f)$. The correspondence preserves the multiplicity* [2].

---

[2]Here $(\mathbb{T}^2 \setminus \Delta/\mathbb{Z})^+$ and $(\mathbb{T}^2 \setminus \Delta/\mathbb{Z})^-$ denote the components above and below diagonal which correspond to the components outside and inside the unit circle in $\mathbb{C} \setminus \{0 \sqcup \mathbb{S}^1\}$.

The proof is derived form Proposition 6.6 based on Proposition 6.5 (localization property). If one regards γ_r^f as an element of $\text{Conf}_{bn}(\mathbb{T}; (\Delta/\mathbb{Z})) = \text{Conf}_{bn}(\mathbb{C} \setminus 0; \mathbb{S}^1)$ one has the following stability theorem.

Theorem 6.3. (CEH stability theorem) *The assignment*

$$C(X; \mathbb{S}^1)_{\text{tame}} \rightsquigarrow \text{Conf}_{bn}(\mathbb{T}; (\Delta/\mathbb{Z}))$$

is a continuous map.[3]

Theorem 6.4. (cf. [Burghelea, D., Haller, S. (2015)]) *For any κ-orientable closed topological manifold M^n and any topologically tame angle-valued map $f : M \to \mathbb{S}^1 = \{z \in \mathbb{C} \mid |z| = 1\}$ one has $\gamma_r^f(\langle a, b \rangle) = \gamma_{n-r-1}^{f^{-1}}(\langle -a, -b \rangle) = \gamma_{n-r-1}^f(\langle b, a \rangle)$.*

Theorem 6.3 can be recovered from the case of real-valued maps when applied to $\widetilde{f}$, thanks to the localization property stated in Proposition 6.5.

To prove Theorem 6.4 one proceeds like in Chapter 5. First one considers the infinite cyclic cover $\widetilde{f} : \widetilde{M} \to \mathbb{R}$ of $f : M \to \mathbb{S}^1$ and one observes that $\widetilde{f^{-1}} = -\widetilde{f}$. Next one observes that the Poincaré Duality diagram (6.9) remains valid when one replaces the first row with Borel-Moore homologies. Next, in view of the formulae (5.45) in Chapter 5 and of Theorem 2.4 in Chapter 2, the Fredholm cross-ratio of the three Fredholm maps

$$H_r^{BM}(M_a^f) \longrightarrow H_r^{BM}(M_b^f) \longrightarrow H_r^{BM}(M_c^f) \longrightarrow H_r^{BM}(M_d^f)$$

is canonically isomorphic to the Fredholm cross-ratio of the three Fredholm maps

$$H_r(M_a^f) \longrightarrow H_r(M_b^f) \longrightarrow H_r(M_c^f) \longrightarrow H_r(M_d^f) ,$$

which is $\mathbb{T}^{\widetilde{f}}((a, b] \times (c, d])$. Now one uses the same arguments as in the proof of Theorem 6.2 to show that $\gamma_r^{\widetilde{f}}(a, b) = \gamma_{n-r-1}^{-\widetilde{f}}(-a, -b) = \gamma_{n-r-1}^{\widetilde{f}}(b, a)$, which implies the result.

Note that the Poincaré Duality for the configurations γ_r^f's, for both real-valued and angle-valued Morse maps defined on a smooth closed manifold can be derived by comparing the Morse (Novikov) complex of a vector field X which admits the real-valued (respectively, angle-valued) map f as Lyapunov map to the Morse (respectively, Novikov) complex of the vector field $-X$ which admits $-f$ (respectively, f^{-1}) as a Lyapunov map. Indeed,

---

[3]A stronger, metric-space version of this result in the spirit of [Cohen-Steiner, D., Edelsbrunner, H., Harer, J. (2007)] can be obtained.

the rest points of Morse index k of X are the same as the rest points of Morse index $n - k$ of $-X$, and each instanton for X from a rest point x to a rest point y for X is an instanton for $-X$ from y to x. The relation with the mixed barcodes comes via the isomorphism of the complex in question with the AM (respectively, AN) complex established in Chapter 9 and the description of the boundary map in the AM (respectively, AN) complex in terms of mixed barcodes, cf Exercise E.4 Section 8.3.

6.4   Exercises

E.1 Suppose $f : X \to \mathbb{R}$ is a tame map and $g : f^{-1}([a, b]) \to \mathbb{R}$ $a < b$ denotes the restriction of f to $f^{-1}([a, b])$. Supposed that an interval I with endpoints α and β, $\alpha \leq \beta$, is a closed (respectively, open, open-closed, or closed-open) barcode of f. Discuss what kind of barcodes the barcode I induces for the map g. Consider all possible relations between a, b, α, β.

E.2 Suppose $f : M^n \to \mathbb{R}$ is a Morse function on a compact smooth manifold M^n. Show that f is a perfect Morse function with respect to the field κ (i.e., for any r the number c_r of critical points of index r is exactly the Betti number β_r) iff all configurations γ_r^f have empty support. What happens if the equality $c_r = \beta_r$ holds only for one r?

E.3 Show that if $f : X \to \mathbb{R}$ is a tame map and X is compact, then there exists $\epsilon > 0$ depending on f, such that for any tame map g with $\sup_{x \in X} |g(x) - f(x)| < \epsilon$ the AM complex (described in Chapter 4 Subsection 4.2.2) of f is a direct summand in the AM complex of g. (The same question for an angle-valued map $f : X \to \mathbb{S}^1$.)

E.4 Let $f : M^n \to \mathbb{R}$ be a tame map, $n \geq 2$. Show that for any $x \in M$, any $r = 0, 1, 2, \ldots, n-1$, and any open neighborhood $U \ni x$ one can modify f on U into a tame map $g : M^n \to \mathbb{R}$ such that f and g are equal on the complement of U and the collection of r-barcodes of g is the collection of r-barcodes of f with one additional barcode.

Chapter 7

Monodromy and Jordan Cells

The purpose of this chapter is to describe the monodromy of a pair (X, ξ), X a compact ANR and $\xi \in H^1(X; \mathcal{Z})$, from a geometric perspective and in a way that makes its relevant invariants (Jordan blocks or Jordan cells) effectively computable. This monodromy will be referred to as the *geometric monodromy*, as opposed to the *homological monodromy* defined in Subsection 2.3.6 in Chapter 2. As expected, the invariants of these two different descriptions of monodromies are the same. We will use the linear algebra of linear relations, which is a natural approach to the geometric monodromy and has considerable computational merits. Since we want to provide the definition of the geometric monodromy in as large generality as this perspective permits, for the reader's convenience we begin by recalling some definitions from Chapter 2, some with minor relaxations. We suggest the reader to review Section 2.2 in Chapter 2 on linear relations. The discussion in this section follows closely ([Burghelea, D., Haller, S. (2015)]) and ([Burghelea, D. (2015a)]).

7.1 General considerations

All real-valued or angle-valued maps $f : X \to \mathbb{R}$ or $f : X \to \mathbb{S}^1$ are assumed to be proper continuous maps with X an ANR.

    – A number $t \in \mathbb{R}$ or an angle $\theta \in \mathbb{S}^1$ is *weakly regular* if $f^{-1}(t)$ or $f^{-1}(\theta)$ is an ANR.

    – A map f whose set of weakly regular values is not empty is treated as a *good map*, and when all values are weakly regular is called a *weakly tame* map. The concepts are relevant since there are compact ANRs with no good maps but the constant ones; cf. [Daverman, R.J., Walsh, J.J. (1981)], where examples of compact ANR, actually homological n-manifolds, with no codimension-one ANRs are provided. For X a (locally finite) simplicial complex, any real- or angle-valued simplicial map f that is tame is weakly tame.

– For the spaces homeomorphic to simplicial complexes, finite-dimensional topological manifolds, or Hilbert cube manifolds, the set of tame maps, and then that of weakly tame maps, is dense in the space of all continuous maps equipped with the compact open topology. The first because any continuous map can be approximated by simplicial maps with resepct to a convenient subdivision, the latter by reasons explained in Chapter 2 Subsection 2.3.2.

The concept of geometric r-*monodromy*, cf. Definition 7.1 below, will be first considered for good maps, since it will involve an angle θ that is a weakly regular value. It will be shown that different choices of this angle lead to the same geometric r-monodromy, and that this r-monodromy depends only on the cohomology class ξ_f associated with the map f.

Once some elementary properties of this concept will be established, using results on Hilbert cube manifolds, it will be shown that the geometric r-monodromy can be associated to any continuous angle-valued map and is a homotopy invariant for any pair $(X, \xi \in H^1(X; \mathbb{Z}))$, with X any compact ANR. In fact it will suffice to define the geometric monodromy only for simplicial complexes and simplicial maps in order to have it extended to all continuos maps defined on arbitrary compact ANRs. The following proposition will be useful.

Proposition 7.1.

(i) *Two maps $f, g : X \to \mathbb{S}^1$ with $D(f, g) = \sup_{x \in X} d(f(x), g(x)) < \pi$ [1] are homotopic by a canonical homotopy, the "convex combination" homotopy.*

(ii) *Suppose X is a good ANR, $f, g : X \to \mathbb{S}^1$ two homotopic continuous angle-valued maps and $\epsilon > 0$. There exists a finite collection of maps $f_0, f_1, \ldots, f_k, f_{k+1}$, such that:*
a) $f_0 = f$, $f_{k+1} = g$,
b) f_i are weakly tame maps for $i = 1, 2, \ldots, k$,
c) $D(f_i, f_{i+1}) < \epsilon$.

Indeed, if f and g are viewed as maps with values in $\mathbb{C}$, then the map
$$h_t(x) = \frac{tg(x) + (1-t)f(x)}{|tg(x) + (1-t)f(x)|}, \quad 0 \leq t \leq 1,$$

---

[1]For $0 < \theta_1, \theta_2 \leq 2\pi$, $d(\theta_1, \theta_2) := \inf\{|\theta_2 - \theta_1|, 2\pi - |\theta_2 - \theta_1|\}$.

provides the homotopy claimed in item (i). The condition $D(f(x), g(x)) < \pi$ ensures that $|tg(x) + (1 - t)f(x)| \neq 0$. Item (ii) follows from the local contractibility of the metric space of continuous maps when equipped with the distance D equivalently, the compact open topology when X is compact.

Real-valued maps

As in Chapters 5 and 6 for a proper continuous real-valued map $f : X \to \mathbb{R}$ and $a \in \mathbb{R}$ denote:

X_a^f, the sublevel $X_a^f := f^{-1}((-\infty, a])$; if a is a weakly regular value, then $X_a^f := f^{-1}((-\infty, a])$ is an ANR.

X_f^a, the overlevel $X_f^a := f^{-1}([a, \infty))$; if a is a weakly regular value, then $X_a^f := f^{-1}([a, \infty))$ is an ANR.

For two proper continuous real-valued maps $f : X \to \mathbb{R}$ and $g : X \to \mathbb{R}$, and $a < b$ such that $f^{-1}(a) \subset g^{-1}(-\infty, b)$ denote $X_{a,b}^{f,g} := X_b^g \cap X_f^a$; if b is a weakly regular value for g and a is a weakly regular value for f, then $X_{a,b}^{f,g}$ is a compact ANR, cf. Chapter 2 Section 2.3. This ensures that the vector spaces $H_r(g^{-1}(a))$, $H_r(f^{-1}(b))$, and $H_r(X_{a,b}^{f,g})$ have finite dimension.

Denote by $R_{a,b}^{f,g}(r)$ the linear relation defined by the linear maps $i_1(r)$ and $i_2(r)$ induced by the inclusions $f^{-1}(a) \subset X_{a,b}^{f,g}$ and $g^{-1}(b) \subset X_{a,b}^{f,g}$,

$$H_r(f^{-1}(a)) \xrightarrow{i_1(r)} H_r(X_{a,b}^{f,g}) \xleftarrow{i_2(r)} H_r(g^{-1}(b)) .$$

Proposition 7.2. *Let $t_1 < t_2 < t_3$. Suppose that t_1 and t_3 are weakly regular for f, t_2 is weakly regular for g, and $g^{-1}(t_2) \subset f^{-1}((t_1, t_3))$. Then one has*

$$R_{t_2,t_3}^{g,f}(r) \cdot R_{t_1,t_2}^{f,g}(r) = R_{t_1,t_3}^{f,f}(r).$$

Proof. The verification is a consequence of the exactness of the following piece of the Mayer-Vietoris sequence in homology

$$H_r(g^{-1}(t_2)) \xrightarrow{i_1' \oplus i_2'} H_r(X_{t_1,t_2}^{f,g}) \oplus H_r(X_{t_2,t_3}^{g,f}) \xleftarrow{i_1 - i_2} H_r(X_{t_1,t_3}^{f,f}), \quad (7.1)$$

whose linear maps involved in the commutative diagram (7.2) below are induced by the inclusions[2].

$$H_r(X_{t_1,t_3}^{f,f})$$

$$I_1 \quad i_1 \quad i_2 \quad I_2$$

$$H_r(f^{-1}(t_1)) \xrightarrow{j_1} H_r(X_{t_1,t_2}^{f,g}) \xleftarrow{i_1'} H_r(g^{-1}(t_2)) \xrightarrow{i_2'} H_r(X_{t_2,t_3}^{g,f}) \xleftarrow{j_2} H_r(f^{-1}(t_3))$$

$$(7.2)$$

Indeed, the commutativity of the diagram (7.2) implies that $xR_{t_1,t_2}^{f,f}y$ for $x \in H_r(f^{-1}(t_1))$ and $y \in H_r(f^{-1}(t_3))$ iff $i_1(j_1(x)) - i_2(j_2(y)) = 0$.

By the exactness of the sequence (7.1) one has $i_1(j_1(x)) - i_2(j_2(y)) = 0$ iff there exists $u \in H_r(g^{-1}(t_2))$ such that $(i_1' \oplus i_2')(u) = (j_1(x), j_2(y))$. This happens iff $xR_{t_1,t_2}^{f,g}u$ and $uR_{t_2,t_3}^{g,f}y$, which means $xR_{t_1,t_2}^{f,f}y$.   □

Angle-valued maps

Let $f : X \to \mathbb{S}^1$ be an angle-valued map. Let $u \in H^1(\mathbb{S}^1;\mathbb{Z}) \simeq \mathbb{Z}$ be the generator defining the orientation of $\mathbb{S}^1$. Here $\mathbb{S}^1$ is regarded as an oriented one-dimensional manifold. Let $f^* : H^1(\mathbb{S}^1;\mathbb{Z}) \to H^1(X;\mathbb{Z})$ be the homomorphism induced by f in integral cohomology and $\xi_f = f^*(u) \in H^1(X;\mathbb{Z})$. As already pointed out, the assignment $f \rightsquigarrow \xi_f$ establishes a bijective correspondence between the set of homotopy classes of continuous maps from X to $\mathbb{S}^1$ and $H^1(X;\mathbb{Z})$.

The cut at θ (with respect to the map $f : X \to \mathbb{S}^1$)

For $\theta \in \mathbb{S}^1$ a weakly regular value for f, define **the cut at** $\theta = e^{it}$ to be the space $\overline{X}_\theta^f$, the two-sided compactification of $X \setminus f^{-1}(\theta)$ with sides $f^{-1}(\theta)$. Precisely, as a set $\overline{X}_\theta^f$ is a disjoint union of three parts,

$$\overline{X}_\theta^f = f^{-1}(\theta)(1) \sqcup f^{-1}(\mathbb{S}^1 \setminus \theta) \sqcup f^{-1}(\theta)(2),$$

with $f^{-1}(\theta)(1)$ and $f^{-1}(\theta)(2)$ two copies of $f^{-1}(\theta)$.

The topology on $\overline{X}_\theta^f$ is the only topology which makes $\overline{X}_\theta^f$ compact and makes the map from $\overline{X}_\theta^f$ to X defined by the identity on each part continuous, with the restriction to each part being a homeomorphism onto its image. The compact space $\overline{X}_\theta^f$ is a compact ANR.

---

[2]For the sake of simplicity, in both (7.1) and (7.2), we discard "r" in the notations i's and i''s for the inclusion-induced linear maps in r-homology.

The obvious inclusions i_1, i_2, $f^{-1}(\theta) \xrightarrow{i_1} \overline{X}_\theta \xleftarrow{i_2} f^{-1}(\theta)$ induce in homology in dimension r the linear maps $i_1(r)$ and $i_2(r)$,

$$H_r(f^{-1}(\theta)) \xrightarrow{i_1(r)} H_r(\overline{X}_\theta) \xleftarrow{i_2(r)} H_r(f^{-1}\theta)) \ .$$

These linear maps define the linear relation $R_\theta^f(r) := R(i_1(r), i_2(r))$ and then, by Proposition 2.4 in Chapter 2, provide the linear relation $(R_\theta^f(r))_{\mathrm{reg}}$ and the linear isomorphism $T_\theta^f(r) : V_\theta^f(r) \to V_\theta^f(r)$, such that $(R_\theta^f(r))_{\mathrm{reg}} = R(T_\theta^f(r))$. Note that if $\widetilde{f} : \widetilde{X} \to \mathbb{R}$ is the infinite cyclic cover of f, then the relation $R_\theta^f(r)$ is isomorphic to the relation $R_{\theta,\theta+2\pi}^{\widetilde{f},\widetilde{f}}(r)$.

7.2 Geometric r-monodromy via linear relations

Definition 7.1. The *geometric r-monodromy* of $f : X \to \mathbb{S}^1$ at $\theta \in \mathbb{S}^1$, for θ a weakly regular value, is the similarity class of the relation $R_\theta^f(r)_{\mathrm{reg}}$, equivalently, the similarity class of the linear isomorphism $T_\theta^f(r) : V_\theta^f(r) \to V_\theta^f(r)$ associated to $R_\theta^f(r)$ by Proposition 2.4 in Chapter 2. One denotes these similarity classes by $[(R_\theta^f(r))_{\mathrm{reg}}]$ or $[T_\theta^f(r)]$.

For a map $f : X \to \mathbb{S}^1$ and K a compact ANR, denote by

$$\overline{f}_K : X \times K \to \mathbb{S}^1$$

the composition of f with the projection of $X \times K$ on X. Note that if θ is a weakly regular value for f, then it remains a weakly regular value for $\overline{f}_K$, and $(\overline{X \times K})_\theta^{\overline{f}_K} = \overline{X}_\theta^f \times K$. Therefore, in view of the Kunneth formula (expressing the homology of the product of two spaces), one has

$$V_\theta^{\overline{f}_K}(r) = \bigoplus_l V_\theta^f(r-l) \otimes H_l(K),$$

$$T_\theta^{\overline{f}_K}(r) = \bigoplus_l T_\theta^f(r-l) \otimes Id_{H_l(K)}, \tag{7.3}$$

where $Id_{H_l(K)}$ denotes the identity map on $H_l(K)$.

In particular, if K is contractible, then

$$[T_\theta^{\overline{f}_K}(r)] = [T_\theta^f(r)], \tag{7.4}$$

and if $K = \mathbb{S}^1$, then

$$[T_\theta^{\overline{f}_K}(r)] = \begin{cases} [T_\theta^f(0)], & \text{if } r = 0, \\ [T_\theta^f(r) \oplus T_\theta^f(r-1)], & \text{if } r \geq 1. \end{cases} \tag{7.5}$$

Proposition 7.3.

(1) *If θ_1 and θ_2 are two different weakly regular values for f, then $[T_{\theta_1}^f(r)] = [T_{\theta_2}^f(r)]$.*

(2) *If X is a good ANR and $f, g : X \to \mathbb{S}^1$ are two homotopic maps with θ_1 a weakly regular value for f and θ_2 a weakly regular value for g, then $[T_{\theta_1}^f(r)] = [T_{\theta_2}^g(r)]$.*

(3) *If $f : X \to \mathbb{S}^1$ and $g : Y \to \mathbb{S}^1$ are two maps with θ_1 a weakly regular value for f and θ_2 a weakly regular value for g, then $[T_{\theta_1}^f(r)] = [T_{\theta_2}^g(r)]$ iff $[T_{\theta_1}^{\overline{f}_{\mathbb{S}^1}}(r)] = [T_{\theta_2}^{\overline{g}_{\mathbb{S}^1}}(r)]$.*

(4) *If $f : X \to \mathbb{S}^1$ and $g : Y \to \mathbb{S}^1$ are two maps with θ_1 a weakly regular value for f and θ_2 a weakly regular value for g, and if $\omega : X \to Y$ is a homeomorphisms such that $g \cdot \omega$ and f are homotopic, then $[T_{\theta_1}^f(r)] = [T_{\theta_2}^g(r)]$.*

Proof. *Proof of (1):* For X a compact ANR and $\xi \in H^1(X; \mathbb{Z})$, consider $\pi : \widetilde{X} \to X$ an infinite cyclic cover associated to ξ. Recall that, as described in Chapter 2 Subsection 2.3.3, an infinite cyclic cover is a map $\pi : \widetilde{X} \to X$ together with a free action $\mu : \mathbb{Z} \times \widetilde{X} \to \widetilde{X}$ such that $\pi(\mu(n, x)) = \pi(x)$ and the map from $\widetilde{X}/\mathbb{Z}$ to X induced by π is a homeomorphism. The above cover is said to be *associated to* ξ if any $\widetilde{f} : \widetilde{X} \to \mathbb{R}$ which satisfies $\widetilde{f}(\mu(n, x)) = \widetilde{f}(x) + 2\pi n$ induces a map from X to $\mathbb{R}/2\pi\mathbb{Z} = \mathbb{S}^1$ representing the cohomology class $\xi_f = \xi$. Recall also from Chapter 2 the same subsection that given two infinite cyclic covers $\pi_i : \widetilde{X}_i \to X$ representing ξ, there exist homeomorphisms $\omega : \widetilde{X}_1 \to \widetilde{X}_2$ which intertwine the free actions μ_1 and μ_2 and satisfy $\pi_2 \cdot \omega = \pi_1$ associated to ξ.

Any map $f : X \to \mathbb{S}^1$ such that $f^*(u) = \xi$, with u the canonical generator of $H^1(\mathbb{S}^1)$, has lifts $\widetilde{f} : \widetilde{X} \to \mathbb{R}$, which make the diagram below a pull-back diagram

$$
\begin{array}{ccc}
\mathbb{R} & \xrightarrow{\ p\ } & \mathbb{S}^1 \\
\Big\uparrow{\scriptstyle \widetilde{f}} & & \Big\uparrow{\scriptstyle f} \\
\widetilde{X} & \xrightarrow{\ \pi\ } & X
\end{array}
\qquad (7.6)
$$

Here $p(t)$ is given by $p(t) = e^{it} \in \mathbb{S}^1 \subset \mathbb{C}$.

Consider two different weakly regular values $\theta_1 = e^{it_1}$, $\theta_2 = e^{it_2} \in \mathbb{S}^1$ for f (i.e., with $t_2 - t_1 \leq \pi$, hence $t_1 < t_2 < t_1 + 2\pi < t_2 + 2\pi$). We apply the discussion in Section 7.1 to the real-valued map $\widetilde{f} : \widetilde{X} \to \mathbb{R}$ and note that

$$
R_{\theta_1}^f(r) = R_{t_1, t_1 + 2\pi}^{\widetilde{f}, \widetilde{f}}(r) = R_{t_2, t_1 + 2\pi}^{\widetilde{f}, \widetilde{f}}(r) \cdot R_{t_1, t_2}^{\widetilde{f}, \widetilde{f}}(r)
$$

and

$$R_{\theta_2}^f(r) = R_{t_2,t_2+2\pi}^{\tilde{f},\tilde{f}}(r) = R_{t_1+2\pi,t_2+2\pi}^{\tilde{f},\tilde{f}}(r) \cdot R_{t_2,t_1+2\pi}^{\tilde{f},\tilde{f}}(r).$$

Using the linear isomorphisms induced by π in the homology of the levels, the linear relations $R_{t_1,t_2}^{\tilde{f},\tilde{f}}(r)$ and $R_{t_1+2\pi,t_2+2\pi}^{\tilde{f},\tilde{f}}(r)$ can be identified with the linear relation $R' = R_{\theta_1,\theta_2}^f(r)$: $H_r(f^{-1}(\theta_1)) \rightsquigarrow H_r(f^{-1}(\theta_2))$, while $R_{t_2,t_1+2\pi}^{\tilde{f},\tilde{f}}(r)$ can be identified with the linear relation $R'' = R_{\theta_2,\theta_1+2\pi}^f(r)$: $H_r(f^{-1}(\theta_2)) \rightsquigarrow H_r(f^{-1}(\theta_2))$, $\theta_1 < \theta_2 < \theta_1 + 2\pi$.

Therefore, $R_{\theta_1}^f(r) = R'' \cdot R'$ and $R_{\theta_2}^f(r) = R' \cdot R''$, which in view of Theorem 2.4 (iii) in Chapter 2 implies that $(R_{\theta_1}^f(r))_{\text{reg}}$ is similar to $(R_{\theta_2}^f(r))_{\text{reg}}$.

Proof of (2): In view of Proposition 7.1 item (ii) it suffices to prove the statement under the following additional hypotheses:

(i) At least one of the maps f or g is weakly tame.
(ii) $\pi/2 < D(f,g) < \pi$.

Since f and g are homotopic, $\xi_f = \xi_g$. For any infinite cyclic cover $\tilde{X} \to X$ associated with $\xi = \xi_f = \xi_g$ both f and g have lifts $\tilde{f}$ and $\tilde{g}$, as indicated in the diagrams below:

$$
\begin{array}{ccc}
\mathbb{R} \xrightarrow{p} \mathbb{S}^1 & \qquad & \mathbb{R} \xrightarrow{p} \mathbb{S}^1 \qquad\qquad (7.7) \\
\uparrow{\tilde{f}} \qquad \uparrow{f} & & \uparrow{\tilde{g}} \qquad \uparrow{g} \\
\tilde{X} \xrightarrow{\pi} X & & \tilde{X} \xrightarrow{\pi} X.
\end{array}
$$

Choose $t_1, t_2 \in [0, 2\pi)$ such that the angles θ_1 and θ_2 are given by these real numbers. Under the additional hypotheses above one can find lifts $\tilde{f}$ and $\tilde{g}$ such that $\tilde{g}^{-1}(t_2) \subset \tilde{f}^{-1}(t_1, t_1+2\pi)$ and $\tilde{f}^{-1}(t_1+2\pi) \subset \tilde{g}^{-1}(t_2, t_2+2\pi)$. We apply the considerations in Section 7.1 to the real-valued maps $\tilde{f}, \tilde{g} : \tilde{X} \to \mathbb{R}$ and conclude that

$$R_{\theta_1}^f(r) = R_{t_1,t_1+2\pi}^{\tilde{f},\tilde{f}}(r) = R_{t_2,t_1+2\pi}^{\tilde{g},\tilde{f}}(r) \cdot R_{t_1,t_2}^{\tilde{f},\tilde{g}}(r)$$

and

$$R_{\theta_2}^g(r) = R_{t_2,t_2+2\pi}^{\tilde{g},\tilde{g}}(r) = R_{t_1+2\pi,t_2+2\pi}^{\tilde{f},\tilde{g}}(r) \cdot R_{t_2,t_1+2\pi}^{\tilde{g},\tilde{f}}(r).$$

Let $R' := R_{t_2,t_1+2\pi}^{\tilde{g},\tilde{f}}(r)$ and $R'' := R_{t_1,t_2}^{\tilde{f},\tilde{g}}(r) = R_{t_1+2\pi,t_2+2\pi}^{\tilde{f},\tilde{g}}(r)$. Then $R_{\theta_1}^f(r) = R'' \cdot R'$ and $R_{\theta_2}^g(r) = R' \cdot R''$ which, by Theorem 2.4 item (iii) in Chapter 2, implies that $(R_{\theta_1}^f(r))_{\text{reg}}$ and $(R_{\theta_2}^g(r))_{\text{reg}}$ are similar.

Proof of (3): Recall that for a linear isomorphism $T : V \to V$ one denotes by $\mathcal{J}(T)$ the set of Jordan cells, which is a *similarity class invariant* with the property that $\mathcal{J}(T_1) = \mathcal{J}(T_2)$ is equivalent to $[T_1] = [T_2]$.

First observe that if $T_1 : V_1 \to V_1$ and $T_2 : V_2 \to V_2$ are two linear isomorphisms, then $\mathcal{J}(T_1 \oplus T_2) = \mathcal{J}(T_1) \sqcup \mathcal{J}(T_2)$.

If so $[T_1 \oplus T_2] = [T'_1 \oplus T'_2]$, hence $\mathcal{J}([T_1]) \sqcup J([T_2]) = \mathcal{J}([T'_1]) \sqcup \mathcal{J}([T'_2])$, and $[T_1] = [T'_1]$, hence $\mathcal{J}([T_1]) = \mathcal{J}([T'_1])$ imply that $\mathcal{J}([T_2]) = \mathcal{J}([T'_2])$, hence $[T_2] = [T'_2]$.

We apply this observation to $T_1 = T^f_{\theta_1}(r-1)$, $T'_1 = T^g_{\theta_2}(r-1)$, and $T_2 = T^f_{\theta_1}(r)$, $T'_2 = T^g_{\theta_2}(r)$. Then, by induction on r, formula (7.5) implies Item (3).

Proof of (4): In view of item (2), one has $[T^{g\cdot\omega}_{\theta_2}(r)] = [T^f_{\theta_1}(r)]$. Since ω induces a homeomorphism between $\overline{X}^{g\cdot\omega}_{\theta_2}$ and $\overline{Y}^g_{\theta_2}$, it follows that the relations $R^{g\cdot\omega}_{\theta_2}(r)$ and $R^g_{\theta_2}(r)$ are similar, which implies $[T^{g\cdot\omega}_{\theta_2}] = [T^g_{\theta_2}]$, whence $[T^f_{\theta_1}(r)] = [T^g_{\theta_2}(r)]$.

$\square$

In view of Proposition 7.3 (1) $[T^f_\theta(r)]$ is independent on θ, so for a *weakly tame map* f one can write $[T^f(r)]$ instead of $[T^f_\theta(r)]$. In view of Proposition 7.3 (2) if f_1 and f_2 are two good maps with $D(f_1, f_2)) < \pi$ then one has $[T^{f_1}(r)] = [T^{f_2}(r)]$.

For X a *good* ANR and f a continuous map, choose a weakly tame maps f' with $D(f, f') < \pi/2$. By Proposition 7.3 (2), $[T^{f'}(r)]$ provides an unambiguously defined geometric r-monodromy for the map f. Indeed, for two such maps f'_1 and f'_2 one has $D(f'_1, f'_2)) < \pi$, and then Proposition 7.3 (2) guarantees that $[T^{f'_1}(r)] = [T^{f'_2}(r)]$. Define $T^f(r) := T^{f'}(r)$. Moreover, in view of (1), if f and g are homotopic, then $[T^f(r)] = [T^g(r)]$. Then for X a *good* ANR and $\xi \in H^1(X; \mathbb{Z})$ one chooses f with $\xi_f = \xi$, and one defines

$$[T^{(X;\xi)}(r)] := [T^f(r)].$$

In order to show that $[T^{(X,\xi)}(r)]$ can be extended to any compact ANR and is a homotopy invariant of the pair (X, ξ)[3] one uses Proposition 7.3 (3) and (4) and the following results discussed in Chapter 2, Subsection 2.2.2 about compact Hilbert cube manifolds.

Theorem (R. Edwards and T. Chapman). If $Q = I^\infty$ denotes the Hilbert cube, then the following properties hold:

---

[3]This means that for pairs (X_1, ξ_1) and (X_2, ξ_2) with X_i, $i = 1, 2$, compact ANRs and $\xi_i \in H^1(X_i; \mathbb{Z})$, $i = 1, 2$, the existence of a homotopy equivalence $\omega : X_1 \to X_2$ satisfying $\omega^*(\xi_2) = \xi_1$ implies $[T^{(X_1, \xi_1)}] = [T^{(X_2, \xi_2)}]$.

1. For any compact ANR X the product $X \times Q$ is a compact Hilbert manifold, hence a good compact ANR.

2. Given a homotopy equivalence $\omega : X \to Y$ between two compact ANRs, the map

$$\omega \times Id_{Q \times \mathbb{S}^1} : X \times Q \times \mathbb{S}^1 \to Y \times Q \times \mathbb{S}^1$$

is homotopic to a homeomorphism $\omega' : X \times Q \times \mathbb{S}^1 \to Y \times Q \times \mathbb{S}^1$.

Extension of the geometric r-monodromy to all pairs (X, ξ)

For any pair (X, ξ) with X a compact ANR and $\xi \in H^1(X; \mathbb{Z})$, and any $r \in \mathbb{Z}_{\geq 0}$, one defines the r-monodromy by

$$[T^{X,\xi}(r)] := [T^{X \times Q, \bar{\xi}}(r)],$$

where $\bar{\xi}$ is the pull-back of ξ by the projection of $X \times Q \to X$. In view of the equality (7.5), if X was already a good ANR, then $[T^{X,\xi}(r)] = [T^{X \times Q, \bar{\xi}}(r)]$.

To verify the homotopy invariance, consider $f_i : X_i \to \mathbb{S}^1$ representing the cohomology class ξ_i. Since $\omega^*(\xi_2) = \xi_1$, the composition $f_2 \cdot \omega$ and f_1 are homotopic, and then in view of Item 2. of the theorem above one has the homeomorphism ω' homotopic to $\omega \times Id_{Q \times \mathbb{S}^1}$. Hence, $(\overline{f_2}) \cdot \omega'$ is homotopic to $(\overline{f_1})$, where $\overline{f_1}$ and $\overline{f_2}$ denote the composition of f_1 and f_2 with the projections $X_i \times Q \times \mathbb{S}^1 \to X_i$, $i = 1, 2$. Now by Proposition 7.3 (4) one has $[T^{X_1, \xi_{\overline{f_1}}}(r)] = [T^{X_2, \xi_{\overline{f_2}}}(r)]$. In view of Proposition 7.3 (3), $[T^{X_1, \xi_1}(r)] = [T^{X_2, \xi_2}(r)]$.

Summing up one has the following results.

Theorem 7.1. *To any pair (X, ξ), with X a compact ANR and $\xi \in H^1(X; \mathbb{Z})$, and any $r = 0, 1, 2, \ldots$, one can associate the similarity class of linear isomorphisms $[T^{(X,\xi)}(r)]$ which is a homotopy invariant of the pair. When $f : X \to \mathbb{S}^1$ is a good map with $\xi_f = \xi$, this is the geometric r-monodromy defined for a good map f and a weakly regular value.*

The collection $\mathcal{J}_r(X; \xi)$ consisting of the pairs with multiplicity, (λ, k), $\lambda \in \overline{\kappa}$, $k \in \mathbb{Z}_{>0}$, which determine the similarity class $[T^{(X;\xi)}(r)] = [\bigoplus_{(\lambda,k) \in \mathcal{J}_r(\xi)} T(\lambda, k)]$, is referred to as *the Jordan cells of the r-monodromy* of (X, ξ).

Theorem 7.2. *If f is a tame map, then the set $\mathcal{J}_r(f)$ defined in Chapter 4 is the same as the set $\mathcal{J}(T^{X, \xi_f}(r))$.*

Proof. As described in Chapter 4, for $f : X \to \mathbb{S}^1$ a tame map with m critical angles $0 \leq \theta_1 < \theta_2 < \cdots < \theta_m < 2\pi$ and $t_1, t_2, \ldots, t_m$ regular values such that $0 \leq \theta_1 < t_2 < \cdots < \theta_{m-1} < t_m < \theta_m < t_{m+1} = t_1 < 2\pi$ one associates the G_{2m}-representation $\rho_r(f)$ with $V_{2i-1} = H_r(f^{-1}(\theta_{i-1}, \theta_i)) = H_r(f^{-1}(t_i))$, $V_{2i} = H_r(f^{-1}(\theta_i)) = H_r(f^{-1}(t_t, t_{i+1}))$, and α_i^r and β_{i-1}^r the linear maps induced in homology by the homotopy classes $\widehat{a}_i$ and $\widehat{b}_i$, or equivalently the inclusions $f^{-1}(t_i) \subset f^{-1}((\theta_{i-1}, \theta_i])$ and $f^{-1}(t_i) \subset f^{-1}([\theta_{i-1}, \theta_i))$.

Let $\widetilde{f} : \widetilde{X} \to \mathbb{R}$ be the canonical infinite cyclic cover of the tame map $f : X \to \mathbb{S}^1$. Put $t_{m+1} = t_1 + 2\pi$ and observe that $V_{2i} = H_r(\widetilde{f}^{-1}(t_t, t_{i+1}))$ and the relation $R_{t_i, t_{i+1}}^{\widetilde{f}, \widetilde{f}}(r)$ is exactly

$$R_{t_i, t_{i+1}}^{\widetilde{f}, \widetilde{f}}(r) = R(\alpha_i^r, \beta_i^r) = R_i(\rho_r(f))^4$$

and $[T^{X, \xi_f}(r)] = [(R^i(\rho_r(f))_{\text{reg}}]$. The result follows from Proposition 3.1 in Chapter 3. $\square$

Theorems 7.1 and 7.2 are implicit in [Burghelea, D., Haller, S. (2015)]. Theorems 7.2 and 4.1 (b) imply that the Jordan cells of $\mathcal{J}([T^{X, \xi_f}(r)])$ and $\mathcal{J}_r(f)$ are the same.

An example

The picture below is taken from [Burghelea, D., Dey, T (2013)] but considered for a different gluing map.

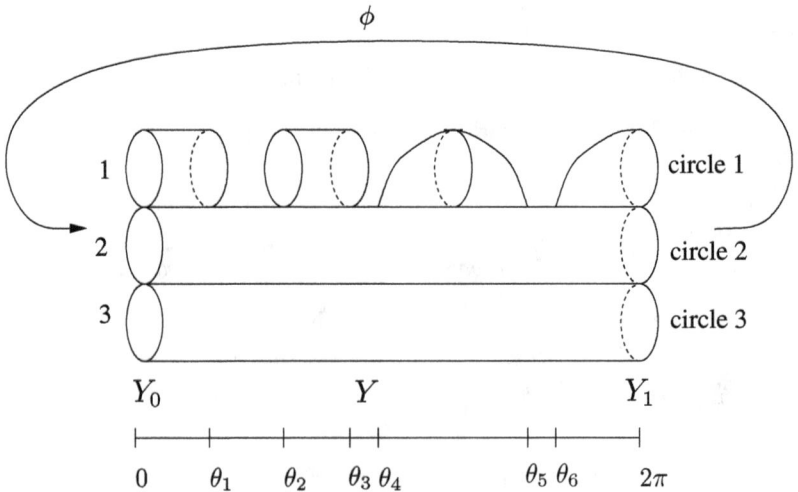

Consider the space X obtained from Y by identifying its right end Y_1 (a union of three circles) to the left end Y_0 (a union of three circles) by using

---

[4]Cf. subsection 3.2.1 for the definition of $R_i(\rho)$

the map $\phi \colon Y_1 \to Y_0$ defined by the matrix

$$\begin{pmatrix} 3 & 3 & 0 \\ 2 & 3 & -1 \\ 1 & 2 & 3 \end{pmatrix}.$$

The meaning of this matrix as a map is the following: circle (1) is divided into 6 parts, circle (2) into 8 parts, and circle (3) into 4 parts; the first 3 parts of circle (1) wrap clockwise around circle (1) to cover it three times, the next 2 wrap clockwise around circle (2) to cover it twice and around circle (3) to cover it 3 times. Similarly, circles (2) and (3) wrap over circles (1), (2) and (3) as indicated by the matrix. The first part of circle (2) wraps counterclockwise on circle (2) and 3 time clockwise around circle (3), ...

The map $f \colon X \to S^1$ is induced by the projection of Y on the interval $[0, 2\pi]$, which becomes S^1 when 0 and 2π are identified. This map has all values weakly regular.

In this example $\mathcal{J}_0(f) = \{(\lambda = 1, k = 1)\}$, $\mathcal{J}_1(f) = \{(\lambda = 2, k = 2)\}$, and $\mathcal{J}_2(f) = \emptyset$. The first and last calculations are obvious. The second is derived by applying the algorithm described in Subsection 3.3.2 of Chapter 3, or in Subsection 2.2.1, for the calculation of $R(A, B)_{\text{reg}}$. Inspecting the picture and using the description of the gluing map the relation $R_0(f) = R_{2\pi}(f) = R(A, B)$ with A and B given by

$$A = \begin{pmatrix} 3 & 3 & 0 \\ 2 & 3 & -1 \\ 1 & 2 & 3 \\ 0 & 0 & 0 \end{pmatrix} \text{ and } B = \begin{pmatrix} 0 & 0 & 0 \\ 0 & 1 & 0 \\ 0 & 0 & 1 \\ 0 & 0 & 0 \end{pmatrix}.$$

Note that the homology $H_1(Y)$ is generated by the fundamental cycles of the circles (1), (2), (3), and of the small cylinder on the top of Y. The example of calculation of $R(A, B)_{\text{reg}}$ worked out in Subsection 2.2.1 involves the matrices A and B described above.

7.3 The calculation of Jordan cells of a simplicial angle-valued map: an algorithm

In this section we describe a way to calculate the geometric monodromy for a simplicial map $f \colon X \to S^1$ for X a finite simplicial complex and θ an angle different from the values of f on vertices. This calculation is carried out in two steps.

STEP 1 inputs (X, f, θ) and outputs for each r, $r \geq 0$, two matrices with the same numbers of rows and columns, A_r and B_r, with the property that $R(A_r, B_r)$ represents the relation $R_\theta^f(r)$.

STEP 2 inputs the two matrices (A_r, B_r) and outputs the invertible matrix T_r with $R(T_r) = R(A_r, B_r)_{\text{reg}}$, whose similarity class is $[T^{X,\xi_f}(r)$. STEP 2 was discussed in Chapter 2 Section 2.2.

We begin with some notations. Let σ be a k-dimensional simplex with vertices $e_0, e_1, \ldots, e_k$, i.e., a convex k-cell generated by $k + 1$ linearly independent points located in some vector space. Let $f : \sigma \to \mathbb{R}$ be the linear map determined by the values of $f(e_i)$ via the formula

$$f\left(\sum_i t_i e_i\right) = \sum_i t_i f(e_i), \quad t_i \geq 0, \quad \sum_i t_i = 1, \qquad (7.8)$$

and let $t \in \mathbb{R}$ be such that $\sup_i f(e_i) > t > \inf_i f(e_i)$.

The map f and the number t determine two k-convex cells, σ_+ and σ_-, and a $(k-1)$-convex cell σ' as follows:

$$\begin{aligned}
\sigma_+ &= f^{-1}([t, \infty)) \cap \sigma, \\
\sigma_- &= f^{-1}((-\infty, t]) \cap \sigma, \\
\sigma' &= f^{-1}(t) \cap \sigma.
\end{aligned} \qquad (7.9)$$

An orientation $o(\sigma)$ on σ provides orientations $o(\sigma_+)$ and $o(\sigma_-)$ on σ_+ and σ_-, respectively, and induces an orientation $o'(\sigma')$ on σ', namely the unique orientation which followed by the direction provided by the vector field $\text{grad} f$ defines the orientation of σ. We have $I(\sigma_\pm, \sigma') = \pm 1$.

Recall that the map $f : X \to \mathbb{S}^1 \subset \mathbb{C}$ is simplicial if its infinite cyclic cover $\tilde{f}$ is simplicial.

STEP 1

For the reader's convenience we recall from Subsection 2.3.4 that:
The simplicial set X is encoded by

(1) The finite set $\mathcal{X}_0$ of vertices with a chosen total order.
(2) A specification of the subsets which define the collection $\mathcal{X}$ of simplices. Note that each simplex is just a collection of vertices. Implicit in the total order of vertices is

    (i) the total order of the simplices of $\mathcal{X}$ provided by the *lexicographic order* induced from the order of the vertices, and

    (ii) an orientation $o(\sigma)$ of each simplex, orientation provided by the relative ordering of the vertices of each simplex, and therefore

(iii) the incidence number $I(\sigma', \sigma)$ of any two simplexes σ' and σ in $\mathcal{X}$.

(3) The simplicial map f indicated by the collection of angles, the values of f on vertices.

The map f and the angle $\theta = e^{it}$ provide a decomposition of the set $\mathcal{X}$ as $\mathcal{X}' \sqcup \mathcal{X}''$, with $\mathcal{X}' := \{\sigma \in \mathcal{X} \mid \sigma \cap f^{-1}(\theta) \neq \emptyset\}$ and $\mathcal{X}'' := \mathcal{X} \setminus \mathcal{X}''$, and both $\mathcal{X}'$ and $\mathcal{X}''$ carry an induced total order.

We assume that f takes different values at different vertices: if not, one can achieve this by a small perturbation of the simplicial map, which does not change the homotopy class of f and consequently the monodromy. From these data one derives:

– first, the collection $\mathcal{Y}$ with the subcollections $\mathcal{Y}(1)$ and $\mathcal{Y}(2)$ of the cells of the complex $Y = \overline{X}_\theta^f$ and the subcomplexes $Y_1 = f^{-1}(\theta)$ and $Y_2 = f^{-1}(\theta)$,

– second, the incidence function on $\mathcal{Y} \times \mathcal{Y}$,

– third, a good order for the elements of $\mathcal{Y}$.

These data lead to the incidence matrix $\mathbb{I}(Y)$ written in a convenient form.

Description of the cells of Y: Each oriented simplex σ in $\mathcal{X}''$ provides a unique oriented cell σ in $\mathcal{Y}$.

Each oriented k-simplex σ in $\mathcal{X}'$ provides two oriented k-cells σ_+ and σ_- and two oriented $(k-1)$-cells $\sigma'(1)$ and $\sigma'(2)$, copies of the oriented cell σ'. So the cells of Y are of five types:

$$\mathcal{Y}_k'(1) = \mathcal{X}_{k+1}',$$
$$\mathcal{Y}_k'(2) = \mathcal{X}_{k+1}',$$
$$\mathcal{Y}_{k-}' = \mathcal{X}_k',$$
$$\mathcal{Y}_{k+}' = \mathcal{X}_k',$$
$$\mathcal{Y}_k'' = \mathcal{X}_k''.$$

Note that $\mathcal{Y}_{k+}'$ and $\mathcal{Y}_{k-}'$ are two copies of the same set $\mathcal{X}_k'$, and $\mathcal{Y}_k'(1)$ and $\mathcal{Y}_k'(2)$ are in bijective correspondence with the set $\mathcal{X}_{k+1}'$.

Inside the cell complex Y we have two sub complexes, Y_1 and Y_2, whose k-cells are $\mathcal{Y}_k'(1)$ and $\mathcal{Y}_k'(2)$, two copies of the same set $\mathcal{X}_{k+1}'$.

Incidence of the cells of $\mathcal{Y}$: The incidence of two cells in the same group (one of the five types indicated above) is the same as the incidence of the corresponding simplices. The incidence of a cell in $\mathcal{Y}(1)$ or $\mathcal{Y}(2)$ with a cell in $\mathcal{Y}$ is zero in case they are not derived from the same simplex σ, and the incidence $I(\sigma'(1), \sigma_+) = 1$ and $I(\sigma'(2), \sigma_-) = -1$. The remaining cells in $\mathcal{Y}$ are actually simplices in $\mathcal{X}''$ with the same incidence as in $\mathcal{X}$.

The good order: Start with the cells in $\mathcal{Y}(1)$, followed by the cells in $\mathcal{Y}(2)$, followed by the cells in $\mathcal{Y} \setminus (\mathcal{Y}(1) \sqcup \mathcal{Y}(2))$.

As a result we have the incidence matrix $\mathbb{I}(Y)$ which is of the form

$$
\begin{pmatrix}
\mathbb{I} & 0 & A \\
0 & \mathbb{I} & B \\
0 & 0 & C
\end{pmatrix},
\tag{7.10}
$$

with $\mathbb{I}$ the incidence matrix of Y_1 and Y_2, which is the same.

Running the persistence algorithm cf. [Cohen-Steiner, D., Edelsbrunner, H., Morozov, D. (2006)], [Zomorodian, A., Carlsson, G. (2005)] leads to the matrices representing $A_r : H_r(Y_1) \to H_r(Y)$ and $B_r : H_r(Y_2) \to H_r(Y)$ as follows.

One runs the persistence algorithm on the incidence matrix of $\mathcal{Y}$ and one calculates first a base for the homology of $H_r(Y_1) = H_r(Y_2)$; one continues the procedure by adding columns and rows to the matrix to obtain a base of $H_r(Y)$. It is straightforward to derive a matrix representation for the inclusion-induced linear maps $H_r(Y_i) \to H_r(Y)$, $i = 1, 2$.

Chapter 8

Applications

8.1 Relations with the classical Morse and Morse-Novikov theories

The key results of Morse theory on finite-dimensional manifolds are the construction of the Morse complex, also referred to as the Thom-Smale complex, derived from the critical points of a proper Morse real-valued function f and the instantons of a smooth vector field X which has f as a Lyapunov function, and the proof that its homology calculates the homology of the manifold (at least when f is bounded from below or from above). The vector field is supposed to satisfy the Morse-Smale condition. We will show that this complex tensored by a field is isomorphic to the AM complex defined in Chapter 4 in terms of barcodes.

The same holds true for the Novikov complex. The Novikov complex in this book is a chain complex which determines the Novikov homology of a pair (M, ξ_f), M closed (i.e., compact boundaryless) smooth manifold, and is defined using the critical points of a Morse angle-valued map f and the *instantons* of a Morse-Smale vector field which has f as a Lyapunov map. It is considerably more subtle because such vector fields might have infinitely many instantons and closed trajectories. A priori the Novikov complex is a chain complex of free $\mathbb{Z}[t^{-1}, t]]$-modules over the ring of Laurent power series with integer coefficients, which when tensored by a field κ becomes a chain complex of $\kappa[t^{-1}, t]]$-vector spaces. Recall that $\kappa[t^{-1}, t]]$ denotes Laurent power series with coefficients in the field κ, which is actually a field extending the ring of Laurent polynomials with coefficients in κ.

The Novikov theory in "post-Novikov" developments involves also the closed trajectories of the vector field (not present for vector fields with Lyapunov real-valued map) and relates them with the homological monodromy which is computable in terms of Jordan cells. This aspect is not elaborated in this book and is work in progress not quite in final form. For partial results the interested reader can consult the papers [Hutchings, M., Lee, Y.J. (1999)], [Hutchings, M. (2002)], [Pajitnov, A.V. (2006)], [Burghelea, D., Haller, S. (2008a)], [Burghelea, D. (2011)]. All these papers answer questions about the relation between the topology of the underling manifold and closed trajectories indirectly, via an invariant the topologists and geometers called "torsion" preceeded by names like Reidemeister, Milnor-Turaev, Ray-Singer, etc.

8.1.1 *The Morse complex*

In this section the space previously denoted by X is a smooth finite-dimensional manifold M^n and $f : M^n \to \mathbb{R}$ is a proper smooth map with all critical points nondegenerate. Such maps are referred to as *Morse functions.*

Recall that a point $x \in M$ is a *nondegenerate critical point* of f if one can find coordinates $t_1, t_2, \ldots, t_n$ (Morse coordinates) in a neighborhood of x such that x corresponds to $t_i = 0$, $i = 1, \ldots, n$ and

$$f(t_1, t_2, \cdots t_n) = f(x) - t_1^2 - t_2^2 - \cdots - t_k^2 + t_{k+1}^2 \cdots + t_n^2.$$

The integer k is called the *Morse index* of x and does not dependent on the choice of Morse coordinates.

Morse theory considers complete smooth vector fields[1] X on M^n which have a Morse function as Lyapunov function; that is, for any trajectory $\gamma(t)$ it holds that $df(\gamma'(t)) < 0$ iff $\gamma'(t) \neq 0$. The gradient of a Morse function with respect to a complete Riemannian metric is an example of such vector field. Given a complete smooth vector field, equivalently a smooth flow on M, and x a rest point of X, i.e., $X(x) = 0$, one denotes by $W_x^{\mp} := \{y \mid \lim_{t \to \mp \infty} \gamma_y(t) = x\}$ the corresponding unstable and stable set. Here $\gamma_y(t)$ denotes the trajectory with $\gamma_y(0) = y$. If X has f as a Lyapunov function, then x is a rest point for X iff is a critical point for f.

A complete smooth vector field with the properties

(1) for any rest point x the subsets $W_x^{\mp}$ are smooth submanifolds diffeomorphic to $\mathbb{R}^k$ and resepctively $\mathbb{R}^{n-k}$ for some integer k (called the *Morse index* of the rest point x and denoted by ind(x)), referred to as the stable and respectively unstable manifold, and

---

[1] Recall that a vector field is said to be complete if all its trajectories are defined on the entire real line $\mathbb{R}$. When the manifold M is closed any vector field on it is complete.

(2) for any two rest points x, y, the manifold W_x^- intersects transversally W_y^+, hence $W_x^- \cap W_y^-$ is a smooth submanifold of dimension $\text{ind}(x) - \text{ind}(y)$,

is called a Morse-Smale vector field. For such a vector field the rest points are isolated and hyperbolic. The set of Morse-Smale vector fields is generic[2] in the set of all smooth vector fields with respect to the C^r-topology for any $r \geq 1$.

For a Morse-Smale vector field a trajectory $\gamma : \mathbb{R} \to M$ from a rest point x to a rest point y is contained in the intersection $W_x^- \cap W_y^-$ and is isolated iff $\text{ind}(x) - \text{ind}(y) = 1$. Following [Bott, R. (1982)] we call such a trajectory an *instanton*.

The choice of an orientation O_x for each of the manifolds W_x^- permits to assign to the instanton γ a sign, $\epsilon(\gamma) = \pm 1$ [3].

In case the vector field has a proper real-valued map as Lyapunov function, the set of rest points and instantons confined to any compact region $f^{-1}([a, b])$ is finite, so for x, y rest points and a choice of orientations O_x and O_y, one can define $\mathbb{I}(x, y) := \sum_\gamma \epsilon(\gamma)$.

Denote by $\mathcal{X}_r$ the set of rest points of index r. Clearly $\mathbb{I}$ can be viewed as a collection of maps $\mathbb{I}_r^{O_X} : \mathcal{X}_r \times \mathcal{X}_{r-1} \to \mathbb{Z}$, referred to below as the *instanton counting matrix*.

Case 1: M closed manifold, $f : M \to \mathbb{R}$ Morse function.

A *Morse complex* is a complex of free abelian groups associated to a Morse-Smale vector field X which admits f as a Lyapunov function and to a collection $\mathcal{O}_X := \{O_x, x \in \bigsqcup_r \mathcal{X}_r\}$ of orientations. The component $C_r(X, \mathcal{O}_X)$ of the complex is the free abelian group generated by the rest points of index r,
$$C_r(X, \mathcal{O}_X) = Z[\mathcal{X}_r],$$
and the boundary map $\partial_r : C_r(X, \mathcal{O}_X) \to C_{r-1}(X, \mathcal{O}_X)$ is the $\mathbb{Z}$-linear maps defined by the $n_r \times n_{r-1}$ matrix with entries $\mathbb{I}_r^{\mathcal{O}_X}(x, y)$, $x \in \mathcal{X}_r$, $y \in \mathcal{X}_{r-1}$, the *instanton counting matrix*, where $n_r = \sharp \mathcal{X}_r$. It can be shown that $\partial_{r-1} \cdot \partial_r = 0$.

The main result about the Morse complex reads as follows.

Theorem 8.1. *Suppose M^n is a closed smooth manifold and $f : M \to \mathbb{R}$ is a smooth [4] Morse function.*

---

[2]i.e., is a residual set.

[3]For γ an instanton from x to y, we put $\epsilon(\gamma) = \pm 1$ if the orientation O_x transported along γ by the flow defined by the vector field agrees or not to O_y followed by the orientation of γ.

[4]Differentiable function of class C^2 suffices.

1. *Suppose X is a Morse-Smale vector field which has f as a Lyapunov function and $\mathcal{O}_X$ is a collection of orientations. The homology of the chain complex $C_*(X, \mathcal{O}_X)$ is isomorphic to the integral homology of the manifold M^n.*

2. *Given f as above, there exists Morse-Smale vector fields X which have f as Lyapunov functions. If X_1 and X_2 are two such vector fields and $\mathcal{O}_{X_1}$ and $\mathcal{O}_{X_2}$ are collections of orientations of the unstable manifolds for X_1 and X_2, then the complexes $C_*(X_1, \mathcal{O}_{X_1})$ and $C_*(X_2, \mathcal{O}_{X_2})$ are isomorphic.*

The statement remains true for compact manifolds with boundary provided f is constant on each component of the boundary as well as for noncompact manifolds with f a function bounded from below or above, but this cases are not considered in this book.

Consequently, up to an isomorphism, the Morse complex depends only on f and its isomorphism class can be denoted by $C_*(f)$ and referred to as the *Morse complex* of f. While the components are defined in terms of the critical points of f, the boundary maps ∂_r depend on the choice of X and orientations $\mathcal{X}$; however, different choices lead to boundary maps which are conjugated by a grading-preserving isomorphism of $\bigoplus_r C_r(f)$.

Item 2 (in the generality stated) above has first appeared in the work of Cornea and Ranicki, cf. [Cornea, O., Ranicki, A. (2003)] together with the additional result given by Theorem 8.2 below. While the linear maps ∂_r depend on the vector field X, their rank is actually an invariant of the pair (M, f) which will be associated with any pair consisting of a compact ANR and a tame map.

Theorem 8.2. *For f a Morse function defined on a closed manifold M there exists $\epsilon(f) > 0$ such that for any other Morse function g defined on M and satisfying $\sup_{x \in M} |f - g| < \epsilon(f)$, the Morse complex of f is a direct summand of the Morse complex of g.*

Note that if $f : M \to \mathbb{R}$ is a Morse function with the set of critical values

$$\cdots < c_{i-1} < c_i < c_{i+1} < \cdots$$

and f is a Lyapunov function for X, then the chain complex $C_*(X, \mathcal{O}_X)$ has a filtration

$$C_*(X, \mathcal{O})(i) \subseteq C_*(X, \mathcal{O})(i+1) \subseteq \cdots \subseteq C_*(X, \mathcal{O}),$$

with $C_r(X, \mathcal{O})(i)$ the free abelian group generated by $\{x \in \mathcal{X}_r \mid f(x) \leq c_i\}$. Indeed, since f is Lyapunov for X, for any instanton γ from $x \in \mathcal{X}_r$ to $y \in \mathcal{X}_{r-1}$ one has $f(x) > f(y)$. This implies that $\partial_r(C_r(X; \mathcal{O})(i)) \subseteq C_r(X, \mathcal{O})(i)$.

If one denotes $M_{\leq c_j} := f^{-1}((-\infty, c_j])$ and $\mathcal{X}_r(c_j) := \{x \in \mathcal{X}_r \mid f(x) = c_j\}$, then since all critical points are nondegenerate, $M_{\leq c_i}$ retracts by deformation to $M'_{\leq c_i}$ the union of $M_{\leq c_{i-1}}$ with the unstable manifoldes of the rest points $x \in \mathcal{X}_r(c_i)$. Precisely $M'_{\leq c_i} \setminus M_{\leq c_{i-1}}$ is homeomorphic to $\bigsqcup_{x \in \sqcup \mathcal{X}_r(c_i)} D_x^{\text{ind}(x)}$. Here D_x^k denotes the open ball of radius 1 in $\mathbb{R}^k$ indexed by the critical point x. As a consequence one has:

Observation 8.1.

(1) $H_r(M_{\leq c_i}, M_{\leq c_{i-1}}; \mathbb{Z})$ is isomorphic to $\mathbb{Z}[\mathcal{X}_r(c_i)]$, the free abelian group generated by the critical points of index r on the level $f^{-1}(c_i)$.

(2) $C_r(X, \mathcal{O}_X) = \bigoplus_k \mathbb{Z}[\mathcal{X}_r(c_k)]$ and $C_r(X, \mathcal{O}_X)(i) = \bigoplus_{k \leq i} \mathbb{Z}[\mathcal{X}_r(c_k)]$.

We call *Morse complex with coefficients in the field* κ the complex of κ-vector spaces $C^M(X, \mathcal{O}) \otimes \kappa$.

Case 2. M is not compact and f is a proper Morse function.

When M is not compact the sets $\mathcal{X}_r$ can be infinite and the matrix with entries $\mathbb{I}(x, y)$ might not define a linear $\mathbb{Z}$-map unless in each column all but finitely many entries are zero[5], in which case a Morse complex is defined; alternatively, sometimes one can pass to some type of completion of $C_r(X, \mathcal{O}_X)$ as in the case of the infinite cyclic cover $\widetilde{f} : \widetilde{M} \to \mathbb{R}$ of $f : M \to \mathbb{S}^1$, which ends up with the Novikov complex described below.

The above considerations can be recovered from any text book on Morse theory. The discussion above is close to what the reader can learn from [Milnor, J. (1965)], [Bott, R. (1982)].

8.1.2 *The Novikov complex*

For $f : M \to \mathbb{S}^1$ a Morse angle-valued map, X a Morse-Smale vector field which has f as Lyapunov map, i.e., $df(X)x)) < 0$ if $X(x) \neq 0$, $\mathcal{O}_X$ a collection of orientations of the unstable manifolds, and $\mathbb{Z}[t^{-1}, t]]$ the ring of Laurent power series with coefficients in $\mathbb{Z}$, the Novikov complex

---

[5]This is the case when f is bounded from below or from above

$(C_*^N(X, \mathcal{O}), \partial_*^N)$ is a chain complex of free $\mathbb{Z}[t^{-1}, t]]$-modules defined as follows.

Consider the vector field $\widetilde{X}$ on $\widetilde{M}$, the lift of X to $\widetilde{M}$ for the infinite cyclic cover $\widetilde{M} \to M$, and let $\widetilde{\mathcal{X}}_r$ be the set of rest points of $\widetilde{X}$, equivalently of critical points of $\widetilde{f}$, on which the group $\mathbb{Z}$ acts freely by deck transformations, with the quotient set $\widetilde{\mathcal{X}}_r / \mathbb{Z} = \mathcal{X}_r$.

Consider the collection $\mathcal{O}_{\widetilde{X}}$ of orientations for the unstable manifolds of the rest points in $\widetilde{\mathcal{X}}$ derived from the collection $\mathcal{O}_X$. Let $\mathbb{Z}[\widetilde{\mathcal{X}}_r]$ be the free abelian group generated by $\widetilde{\mathcal{X}}_r$, viewed as a $\mathbb{Z}[t^{-1}, t]$-module with the multiplication by t, the isomorphism induced by the action of the generator of the group of deck transformation on the rest points $\widetilde{\mathcal{X}}$ of $\widetilde{X}$.

The choice of one lift $\widetilde{x} \in \widetilde{\mathcal{X}}$ for each rest point $x \in \mathcal{X}$ specifies a basis of the free $\mathbb{Z}[t^{-1}, t]$-module for $\mathbb{Z}[\widetilde{\mathcal{X}}_r]$. With respect to this basis one observes (following Novikov) that the collections of integers $\mathbb{I}_r^{\mathcal{O}_X}(\widetilde{x}', \widetilde{y}')$, $\widetilde{x} \in \widetilde{\mathcal{X}}_r, \widetilde{y}' \in \widetilde{\mathcal{X}}_{r-1}$ can be interpreted as an $n_r \times n_{r-1}$ matrix ∂_r with entries not always in $\mathbb{Z}[t^{-1}, t]$, but in the larger ring $\mathbb{Z}[t^{-1}, t]]$. Therefore, taking $C_r^N(X, \mathcal{O}_X) := \mathbb{Z}[\widetilde{\mathcal{X}}_r] \otimes_{\mathbb{Z}[t^{-1}, t]} \mathbb{Z}[t^{-1}, t]] = \mathbb{Z}[t^{-1}, t]][\mathcal{X}_r]$ and interpreting the matrix ∂_r as a $\mathbb{Z}[t^{-1}, t]]$-linear map $\partial_r : C_r^N(X, \mathcal{O}_X) \to C_{r-1}^N(X, \mathcal{O}_X)$, we obtain, thanks to the property $\partial_{r-1} \cdot \partial_r = 0$, the so-called Novikov complex $C_*^N(X, \mathcal{O}_X)$, which calculates $H(\widetilde{X}) \otimes_{\mathbb{Z}[t^{-1}, t]} \mathbb{Z}[t^{-1}, t]]$. If κ is a field, then so is $\kappa[t^{-1}, t]]$, and tensoring the Novikov complex by κ one obtains a chain complex of finite dimensional $\kappa[t^{-1}, t]]$-vector spaces.

Observation 8.2.

(1) The free abelian group $\mathbb{Z}[\widetilde{\mathcal{X}}_r] = \bigoplus_{c \in \widetilde{f}(\widetilde{x})} H_r(\widetilde{f}^{-1}([c + \epsilon, c - \epsilon]), \widetilde{f}^{-1}(c - \epsilon); \mathbb{Z})$, with ϵ a small enough positive number,

(2) The free $\mathbb{Z}[t^{-1}, t]$-module structure on $\bigoplus_{c \in \widetilde{f}(\widetilde{x})} H_r(\widetilde{f}^{-1}([c + \epsilon, c - \epsilon]), \widetilde{f}^{-1}(c - \epsilon); \mathbb{Z})$ is provided by the deck transformation $\tau : \widetilde{X} \to \widetilde{X}$ which induces an isomorphism $H_r(\widetilde{f}^{-1}([c + \epsilon, c - \epsilon]), \widetilde{f}^{-1}(c - \epsilon); \mathbb{Z}) \to H_r(\widetilde{f}^{-1}([c + 2\pi + \epsilon, c + 2\pi - \epsilon]), \widetilde{f}^{-1}(c + 2\pi - \epsilon); \mathbb{Z})$, and then an isomorphism of $\bigoplus_{c \in \widetilde{f}(\widetilde{x})} H_r(\widetilde{f}^{-1}([c + \epsilon, c - \epsilon]), \widetilde{f}^{-1}(c - \epsilon); \mathbb{Z})$ to itself.

The discussion above can be recovered from [Novikov, S.P. (1991)], [Pajitnov, A.V. (2006)], [Farber, M. (2004)].

In what follows we will show that the Morse complex and the Novikov complex of X with f as a Lyapunov map, tensored by a field κ, can be recovered entirely from the closed, open, and closed-open barcodes of f. More precisely, the Morse complex tensored by a field κ is isomorphic to the AM complex, and the Novikov complex tensored by κ is isomorphic to the AN complex described in Chapter 4 Section 4.2.2. For this purpose one needs a few observations about chain complexes with coefficients in a field.

8.1.3 Chain complexes of vector spaces

A chain complex of finite-dimensional vector spaces is a finite sequence of vector spaces and linear maps

$$\{C_i, \partial_i : C_i \to C_{i-1}\}, \quad 0 \le i \le n,$$

with $\partial_0 = 0$ and $\partial_{i-1} \cdot \partial_i = 0, i \ge 1$.

$$0 \longrightarrow C_n \xrightarrow{\partial_n} C_{n-1} \xrightarrow{\partial_{n-1}} \cdots C_k \xrightarrow{\partial_k} C_{k-1} \cdots \xrightarrow{\partial_1} C_0 \xrightarrow{\partial_0} 0.$$

Clearly, then $\operatorname{img} \partial_{k+1} \subseteq \ker \partial_k$.

Introduce

(1) $C_r^- := C_r / \ker \partial_r$,
(2) $H_r := \ker \partial_r / \operatorname{img} \partial_{r+1}$,
(3) $C_r^+ := \operatorname{img} \partial_{r+1} \subseteq C_r$,

and denote

(1) $n_r = \dim C_r$,
(2) $n_r^- = \dim C_r^-$,
(3) $n_r^+ = \dim C_r^+$,
(4) $\beta_r = \dim H_r$.

Clearly, $n_{r+1}^- = n_r^+$ and ∂_{r+1} factors through $\overline{\partial}_{r+1} : C_{r+1}^- \to C_r^+$ as in the diagram

$$
\begin{array}{ccc}
C_{r+1} & \xrightarrow{\partial_{r+1}} & C_r \\
\downarrow{\scriptstyle \pi} & & \uparrow{\scriptstyle \subseteq} \\
C_{r+1}^- & \xrightarrow{\overline{\partial}_{r+1}} & C_r^+
\end{array}
$$

with the left vertical arrow a surjection, the right vertical arrow an injection, and $\underline{\partial}_{r+1}$ an isomorphism. An immediate consequence of this is

Proposition 8.1. (Hodge decomposition)

(1) *There exists isomorphisms* $\theta_r : C_r^- \oplus H_r \oplus C_r^+ \to C_r$ *such that the*

$$
\text{composition } \theta_{r-1}^{-1} \cdot \partial_r \cdot \theta_r = \begin{pmatrix} 0 & 0 & 0 \\ 0 & 0 & 0 \\ \overline{\partial}_r & 0 & 0 \end{pmatrix}.
$$

(2) *If $\kappa = \mathbb{R}$ or $\kappa = \mathbb{C}$ and each C_r is equipped with a Hermitian scalar product, then one can choose θ_r in a canonical manner, namely, realizing C_r^- as the orthogonal complement of $\ker \partial_r$ in C_r and H_r as the orthogonal complement of $\operatorname{img} \partial_{k+1}$ inside $\ker \partial_k$.*

In particular, the chain complex is isomorphic to the direct sum of the chain complex whose components are H_r and $\partial_r^H = 0 : H_r \to H_{r-1}$ and the acyclic complex $\underline{C}_r = C_r^- \oplus C_r^+$ with $\partial_r^C = \begin{pmatrix} 0 & 0 \\ \bar\partial_r & 0 \end{pmatrix}$. In particular, one has

Proposition 8.2. *Two chain complexes,*

$$\{C_i', \partial_i' : C_i' \to C_{i-1}'\}, \ 0 \le i \le n',$$

and

$$\{C_i'', \partial_i'' : C_i'' \to C_{i-1}''\}, \ 0 \le i'' \le n'',$$

are isomorphic iff $n' = n''$, $\operatorname{rank} \partial_r' = \operatorname{rank} \partial_r''$ and $\beta_r' = \beta_r''$, equivalently, $n_r' = n_r''$ and $\beta_r' = \beta_r''$, equivalently $\operatorname{rank} \partial_r' = \operatorname{rank} \partial_r''$ and $\beta_r' = \beta_r''$.

Proof of Propositions 8.1 and 8.2

Let $\operatorname{img} \partial_{r+1} = C_r^+ \subseteq \ker \partial_r = Z_r \subseteq C_r$ and let $\pi_r' : Z_r \to Z_r/C_r^+ = H_r$ be the canonical projection. Choose

$$s_r' : H_r \to Z_r,$$

a splitting for π_r'. Then

$$\theta_r' = (s_r' \oplus i_r^{C^+}) : H_r \oplus C_r^+ \to Z_r$$

is an isomorphism, with $i_r^{C^+} : C_r^+ \to Z_r$ the inclusion.

Let $\ker \partial_r = Z_r \subseteq C_r$ and let $\pi_r'' : C_r \to C_r/Z_r = C_r^-$ be the canonical projection. Choose

$$s_r'' : C_r^- \to C_r,$$

a splitting for π_r''. Then

$$\theta_r'' = (i_r^Z \oplus s_r'') : Z_r \oplus C_r^- \to C_r$$

is an isomorphism with $i_r^Z : Z_r \to C_r$ the inclusion.

This implies that the composition $\theta_r = \theta_r'' \cdot (\theta_r' \oplus Id) : C_r^- \oplus H_r \oplus C_r^+ \to C_r$ is an isomorphism.

Note that ∂_r factors through the injective map $\overline{\partial}_r : C_r/Z_r \to C_{r-1}$ with image C_{r-1}^+. The above choices imply that

$$\theta_{r-1}^{-1} \cdot \partial \cdot \theta_r = \begin{pmatrix} 0 & 0 & 0 \\ 0 & 0 & 0 \\ \overline{\partial}_r & 0 & 0 \end{pmatrix}.$$

If the components C_r carry a Hermitian scalar product, then the choices s_r', s_r'' above are canonical. Proposition 8.2 follows from Proposition 8.1.

8.1.4 *The* AM *and* AN *complexes for a Morse map*

Recall from Chapter 4 Section 4.2.2 that the AM complex $\mathcal{C}_*^M(f) = (C_*^M(f), \partial_*^M(f))$ and the AN complex $\mathcal{C}_*^N(f) = (C_*^N(f), \partial_*^N(f))$ [6] have as components $C_r^{\cdots}(f)$, the vector spaces generated by the sets

$$\mathcal{B}_r^c(f) \sqcup \mathcal{B}_{r-1}^o(f) \sqcup \mathcal{B}_r^{c,o}(f) \sqcup \mathcal{B}_{r-1}^{c,o}(f),$$

and the boundary maps $\partial_r(f)$ given by the block matrix

$$\begin{pmatrix} 0 & 0 & 0 & 0 \\ 0 & 0 & 0 & 0 \\ 0 & 0 & 0 & Id \\ 0 & 0 & 0 & 0 \end{pmatrix}.$$

The relevant field for the AM complex is κ (an arbitrary field) and for the AN complex, consistently with the classical Novikov complex, is $\kappa[t^{-1}, t]]$, the field of Laurent series with coefficients in an arbitrary field κ, although one could work with any other field.

In case of a real-valued Morse map with critical values

$$\cdots < c_{i-1} < c_i < c_{i+1} < \cdots$$

one has a natural filtration of $\mathcal{C}_*^M(f)$

$$\cdots \mathcal{C}^M(f)(i) \subseteq \mathcal{C}_*^M(f)(i+1) \subseteq \cdots \subseteq \mathcal{C}_*^M(f),$$

where the component $C_*^M(f)(i)$ is the vector space generated by the set

$$\begin{cases} I \in \mathcal{B}_r^c, \ \mathrm{l}(I) \leq c_i, \\ I \in \mathcal{B}_{r-1}^o, \ \mathrm{r}(I) \leq c_i, \\ I \in \mathcal{B}_r^{c,o}, \ \mathrm{l}(I) \leq c_i, \\ I \in \mathcal{B}_{r-1}^{c,o}, \ \mathrm{r}(I) \leq c_i, \end{cases}$$

---

[6] both denoted in Section 4.2.2 as $(C_*(f), \partial_*(f))$

where for an interval I one denotes by $l(I)$ and $r(I)$ the left and the right endpoint, respectively. Combining Proposition 8.2, Observation 8.1, Corollary 4.1 and Theorem 4.1 one obtains

Theorem 8.3.

1. *For any f real-valued Morse function on a closed manifold, κ a field, X a Morse-Smale vector field which has f as a Lyapunov function, and $\mathcal{O}$ a collection of orientations, it holds that*

$$(C_*^{\mathrm{M}}(X,\mathcal{O}),\partial_*^{\mathrm{M}}) \otimes \kappa \simeq (C^{\mathrm{M}}(f)_*,\partial_*^{\mathrm{M}}(f))$$

and

$$(C_*^{\mathrm{M}}(X,\mathcal{O})(i),\partial_*^{\mathrm{M}}) \otimes \kappa \simeq (C_*^{\mathrm{M}}(f)(i),\partial_*^{\mathrm{M}}(f)).$$

2. *For any f angle-valued Morse map on a closed manifold, κ a field, X a Morse-Smale vector field which has f as a Lyapunov map, and $\mathcal{O}$ collection of orientations, it holds that*

$$(C_*^{\mathrm{N}}(X,\mathcal{O}),\partial_*^{\mathrm{N}}) \otimes \kappa = (C_*^{\mathrm{N}}(X,\mathcal{O}),\partial_*^{\mathrm{N}}) \otimes_{\mathbb{Z}[t^{-1},t]} \kappa[t^{-1},t]] \simeq (C_*^{\mathrm{N}}(f),\partial_*^{\mathrm{N}}(f)).$$

Proof. Indeed, Observation 8.1 expresses the components of the Morse or Novikov chain complex tensored by the relevant field in terms of the homology of the pairs $(f^{-1}([c_i + \epsilon, c_i - \epsilon]), f^{-1}(c_{i-\epsilon}))$, or of the pairs $(f^{-1}([\theta_i + \epsilon, \theta_i - \epsilon]), f^{-1}(\theta_{i-\epsilon}))$ for $\epsilon > 0$ small enough. Precisely, for f real-valued

$$C_r(X,\mathcal{O}) \otimes \kappa = \bigoplus_i H_r(f^{-1}([c_i + \epsilon, c_i - \epsilon]), f^{-1}(c_i - \epsilon)),$$

with c_i all critical values of f, and for f angle-valued

$$C_r(\widetilde{X},\widetilde{\mathcal{O}}) \otimes \kappa = (\bigoplus_i H_r(f^{-1}([\theta_i + \epsilon, \theta_i - \epsilon]), f^{-1}(\theta_{i-\epsilon}))) \otimes_\kappa \kappa[t^{-1},t]],$$

with θ_i all critical angles of f.

By Corollary 4.1 Chapter 4, based on Observations 8.1 and 8.2 one calculates the dimensions of the components of the Morse complex $(C_*^{\mathrm{M}}(X,\mathcal{O}),\partial_*^{\mathrm{M}})$ and of the Novikov complex $(C_*^{\mathrm{N}}(X,\mathcal{O}),\partial_*^{\mathrm{N}})$ tensored by κ and $\kappa[t^{-1},t]]$, respectively, in terms of barcodes. One concludes that they are the same with the dimensions in the AM and AN complexes. By Theorem 4.1, one obtains the dimensions of the standard homology and of the Novikov homology in terms of barcodes and observes they are the same as the dimensions of the homology of the AM and AN complexes. Then the result follows from Proposition 8.2, which guarantees the claimed isomorphisms. $\square$

In case of real-valued map it is possible to produce an isomorphism $(C_*^M(X, \mathcal{O}), \partial_*^M) \otimes \kappa \to (C^M(f)_*, \partial_*^M)$ which preserves the filtration. The arguments are however longer.

The stability Theorems 5.2 and 5.4 in Chapter 5 and Theorem 6.2 in Chapter 6 imply

Proposition 8.3. *Let f be a Morse real- or angle-valued function. For any Morse function g satisfying $\sup_{x \in M} |f((x) - g(x)| < \epsilon(f)/2$, with $\epsilon(f) = \inf(c_{i+1} - c_i)$, the Morse complex of f tensored by a field κ, respectively the Novikov complex of f tensored by the field $\kappa[t^{-1}, t]]$, is isomorphic to a direct summand of the Morse, respectively Novikov complex of g.*

Cornea and Ranicki have established such result for the Morse complex and the Novikov complex defined in Section 8.1, cf. [Cornea, O., Ranicki, A. (2003)].

As a final note, if the real-valued map f is Lyapunov for the vector field X, then the real-valued map $-f$ is Lyapunov for the vector field $-X$, and $\mathcal{X}_r^f = \mathcal{X}_r^{-f}$. Also, any instanton for X remains an instanton for $-X$ with the specification that if it goes from the critical value c to the critical value c' for f then, it goes from the critical value $-c'$ to the critical value $-c$ for $-X$. These observations permit to derive the Poincaré Duality Theorems 6.2 and 6.4, as well as item (1) of Theorem 5.3 and 5.6 from Theorem 8.3 for smooth manifolds, cf. E.4 in Section 8.3.

8.2 A few computational applications

8.2.1 *Novikov-Betti numbers, standard and twisted Betti numbers in relation with Jordan cells*

For a pair $(X, \xi \in H^1(X; \mathbb{Z}))$, with X a compact ANR, and κ a fixed field, the relevant numerical invariants derived from various homologies are

   (i) The Betti numbers $\beta_r(X)$,
   (ii) For $u \in \bar{\kappa} \setminus 0$, the twisted Betti numbers $\beta_r(X; (\xi, u)) := \dim H_r(X; (\xi, u))$,
   (iii) The Novikov-Betti numbers $\beta_r^N(X; \xi)$.

If $\kappa = \mathbb{R}$ or $\mathbb{C}$ and $\widetilde{X} \to X$ is an infinite cyclic cover defined by ξ, then the Novikov-Betti number $\beta_r^N(X; \xi)$ is the same as the L_2-Betti number $\beta_r^{L_2}(\widetilde{X})$, that is, the von Neumann dimension of the $L^\infty(\mathbb{S}^1)$-Hilbert module

$H_r^{L_2}(\widetilde{X})$, the L_2-homology of $\widetilde{X}$.

The calculations in terms of barcodes and Jordan cells of $f : X \to \mathbb{S}^1$ representing ξ given by Theorem 4.1 in Chapter 4 yield the following equalities:

$$\boxed{\beta_r(X; (\xi, u)) = \beta_r^{\mathrm{N}}(X, f) + \sharp \mathcal{J}_r(f)(1/u) + \sharp \mathcal{J}_{r-1}(f)(u)} \qquad (8.1)$$

and

$$\boxed{\beta_r^{\mathrm{N}}(X; \xi_f) = \beta_r(X; \kappa) - \sharp \mathcal{J}_r(f)(1) - \sharp \mathcal{J}_{r-1}(f)(1)} \qquad (8.2)$$

where $\mathcal{J}_r(f)(u) := \{J = (u, k) \in \mathcal{J}_r(f) \mid u \in \kappa \setminus 0, k \in \mathbb{Z}_{\leq 1}\}$ and $\sharp$ denotes cardinality. We remind the reader that if $u \in \kappa \setminus 0$, then any Jordan blocks $J = (T, V)$ such that T has u as an eigenvalue is a Jordan cell.

While the twisted Betti numbers $\beta_r(X; (\xi, u))$ and the Novikov-Betti numbers $\beta_r^{\mathrm{N}}(X; \xi)$ are defined using an infinite cyclic cover, a computer unfriendly mathematical object, the formulae (8.2) and (8.1) show that they are actually computer friendly invariants, since both the standard Betti numbers and the Jordan cells are computer friendly invariants.

We also have the following mild extension of the Poincaré Duality Theorem 5.5 in Chapter 5.

Proposition 8.4. *If $(M^n, \partial M^n)$ is an orientable compact manifold with boundary and $\xi \in H^1(M; \mathbb{Z})$ is such that the Novikov-Betti numbers of $(\partial M, \xi|_{\partial M})$ vanish, then*

$$\beta_r^{\mathrm{N}}(M; \xi) = \beta_{n-r}^{\mathrm{N}}(M; \xi). \qquad (8.3)$$

Proof. Suppose $X = X_1 \cup X_2$ and $Y = X_1 \cap X_2$, with X_1, X_2, Y closed subsets of X and X, X_1, X_2, Y all compact ANRs. Let $\xi \in H^1(X; \mathbb{Z})$ and let ξ_1, ξ_2, ξ_0 be the pull-backs of ξ to X_1, X_2, Y. Let $\widetilde{X}, \widetilde{X}_1, \widetilde{X}_2, \widetilde{Y}$ be the infinite cyclic covers corresponding to ξ, ξ_1, ξ_2, ξ_0. Note that $\widetilde{X} = \widetilde{X}_1 \cup \widetilde{X}_2$ and $\widetilde{X}_1 \cap \widetilde{X}_2 = \widetilde{Y}$, and then the long exact sequence in homology

$$\cdots \longrightarrow H_r(\widetilde{Y}) \longrightarrow H_r(\widetilde{X}_1) \oplus H_r(\widetilde{X}_1) \longrightarrow H_r(\widetilde{X}) \longrightarrow H_{r-1}(\widetilde{Y}) \longrightarrow \cdots$$

is a sequence of $\kappa[t^{-1}, t]$-modules with all arrows $\kappa[t^{-1}, t]$-linear. If $H^{\mathrm{N}}(Y; \xi_0) = 0$, then $\beta_r^{\mathrm{N}}(X_1; \xi_1) + \beta_r^{\mathrm{N}}(X_2; \xi_2) = \beta_r^{\mathrm{N}}(X; \xi)$.

We apply this to the *double of M*, $X = DM = M_1 \cup_{\partial M} M_2$, with M_1 equal to M and M_2 the manifold M with the opposite orientation and consider $\xi_D \in H^1(DM; \mathbb{Z})$, the cohomology class obtained from ξ by the foregoing doubling construction. Clearly, $\beta_r^N(DM; \xi_D) = 2\beta_r^N(M; \xi)$. Since DM is closed and orientable, and consequently satisfies $\beta_r^N(DM; \xi_D) = \beta_{n-r}^N(DM; \xi_D)$, the statement follows. $\qquad\square$

As a consequence of the above one has the following computation with pleasant algebraic geometric applications.

Theorem 8.4. *Suppose* $(M^n, \partial M)$ *is a connected compact* $\kappa-$*orientable manifold with boundary and* $\xi \in H^1(M; \mathbb{Z})$ *is such that* $\beta_r^N(\partial M, \xi|_{\partial M}) = 0$. *Suppose further that* M *retracts by deformation to a simplicial complex of dimension* $\leq [n/2]$, *where* $[n/2]$ *denotes the integer part of* $n/2$ *and* $\chi(M)$ *is the Euler-Poincaré characteristic of* M *over the field* κ. *Then the following statements hold true.*

(i) *If* $n = 2k$, *then one has:*

$$\beta_r^N(X; \xi) = \begin{cases} 0, & \text{if } r \neq k, \\ (-1)^k \chi(M), & \text{if } r \neq k, \end{cases}$$

$$\beta_r(X) = \begin{cases} \mathcal{J}_{r-1}(\xi)(1) + \mathcal{J}_r(\xi)(1), & \text{if } r \neq k, \\ \mathcal{J}_{k-1}(\xi)(1) + \mathcal{J}_k(\xi)(1) + (-1)^k \chi(M), & \text{if } r = k, \end{cases}$$

$$\beta_r(X; (\xi, u)) = \begin{cases} \mathcal{J}_{r-1}(\xi)(u) + \mathcal{J}_r(\xi)(1/u), & \text{if } r \neq k, \\ \mathcal{J}_{k-1}(\xi)(u) + \mathcal{J}_k(\xi)(1/u) + (-1)^k \chi(M), & \text{if } r = k. \end{cases}$$
$$\tag{8.4}$$

(ii) *If* $n = 2k + 1$, *then one has:*

$$\beta_r^N(X; \xi) = 0,$$
$$\beta_r(X) = \mathcal{J}_{r-1}(\xi)(1) + \mathcal{J}_r(\xi)(1), \tag{8.5}$$
$$\beta_r(X; (\xi, u)) = \mathcal{J}_{r-1}(\xi)(1/u) + \mathcal{J}_r(\xi)(u).$$

(iii) *If* $V^{2n-1} \subset M^{2n}$ *is compact proper submanifold (i.e.,* $V \pitchfork \partial M$ [7] *and* $V \cap \partial M = \partial V$) *with Poincaré Dual class* $\xi \in H^1(M; \mathbb{Z})$ [8] *and* $H_r(V) = 0$, *then the set of Jordan cells* $J_r(M, \xi)$ *is empty.*

Items (i) and (ii) follow from (8.2), (8.1) and (8.3). To check item (iii), note that one can construct a smooth function $f : M \to \mathbb{S}^1$ with θ regular value, such that $V = f^{-1}(\theta)$ and $\xi = \xi_f$. Since the geometric monodromy can be calculated using the cut at θ with respect to the map f from the relation $R_\theta^f(r) : H_r(V) \rightsquigarrow H_r(V)$, and $H_r(V) = 0$ the statement follows.

---

[7] $\pitchfork$ = transversal

[8] The Poincaré Dual of the fundamental class $[V] \in H_{n-1}(M, \partial M; \mathbb{Z})$ representing the submanifold V

The complement of a complex algebraic hypersurface V **in** $\mathbb{C}^p$ which is "regular at infinity" in the sense considered by L. Maxim, cf. [Maxim, L. (2014)], or tame at infinity (cf [Némethi, A., Zaharia, A. (1992)]) has compactifications to compact manifolds with boundary which satisfy the hypotheses of Theorem 8.4, item (i). This holds in view of the following folklore theorem implicit in the above references.

Theorem 8.5. *Let* $V \subset \mathbb{C}^p$ *be a complex hypersurface given by the polynomial equation* $f(z_1, z_2, \ldots, z_p) = 0$ *and* ξ *be the integral cohomology class represented by* $h = f/|f| : \mathbb{C}^p \setminus V \to \mathbb{S}^1$. *Suppose* V *is in general position at* ∞[9] *cf.* [Maxim, L. (2014)] *or is tame at* ∞ *cf.* [Némethi, A., Zaharia, A. (1992)].

Denote

$$D_R := \{(z_1, z_2, \ldots, z_p) \in \mathbb{C}^p \mid \sum |z_i|^2 \leq R\}$$

and

$$N_\epsilon := \{(z_1, z_2, \ldots, z_p) \in \mathbb{C}^p \mid |f(z_1, z_2, \ldots, z_p)| < \epsilon\}.$$

Then for ϵ small enough $M^{2p} := D_{1/\epsilon} \setminus N_\epsilon$ *is a compact topological manifold with boundary whose interior is diffeomorphic to* $\mathbb{C}^p \setminus V$, *and the restriction of* $h = f/|f|$ *to* ∂M^{2p} *is a fibration over* $\mathbb{S}^1$.

Well known results in complex algebraic geometry imply that $\mathbb{C}^p \setminus V$ retracts by deformation to a subspace homeomorphic to a finite simplicial complex of dimension $\leq p$; moreover, the fibration property of the restriction of h to ∂M^{2p} stated in Theorem 8.5 implies that the Novikov-Betti numbers $\beta_r^N(\partial M, \xi|_{\partial M})$ vanish.

Let ξ be the cohomology class represented by $f/|f| : \mathbb{C}^p \setminus V \to \mathbb{S}^1$. Then $\mathbb{C}^p \setminus V$ is diffeomorphic to the interior of a compact manifold $(M^{2p}, \partial M^{2p})$ which satisfies the hypotheses of Theorem 8.4 and therefore the conclusions of Theorem 8.4 (i) hold.

The assertion in item (i) of Theorem 8.4 in the case of the complement of a hypersurface are not new. The first statement about Novikov-Betti number in terms of L_2-Betti numbers is due to Maxim [Maxim, L. (2014)]. Maxim's results concern a considerably larger class of L_2-Betti numbers or Novikov-Betti numbers, cf. [Maxim, L. (2014)], [Friedl, S., Maxim, L. (2016)][10]. The remaining part of item (i) for the complement of a hypersurface is also apparently known, at least for $\kappa = \mathbb{C}$.[11]

---

[9]i.e., its projective completion intersects the hyperplane at infinity transversally.

[10]But are still consequences of analogues of the Poincaré Duality, i.e., of the equality (2.29) for more sophisticated Novikov homologies.

[11]As I understand, these formulae can be recovered from [Dimca, A. (1992)].

8.2.2 *Alexander polynomial of a knot, generalizations*

A *knot* $K \subset \mathbb{S}^3$ is a locally flat embedding of $\mathbb{S}^1$ in the three-dimensional sphere $\mathbb{S}^3$, cf. Subsection 2.3.6. Consider $X = \mathbb{S}^3 \setminus N$, where N is an open tubular neighborhood of K in $\mathbb{S}^3$. The compact space X is a deformation retract of $\mathbb{S}^3 \setminus K$. The Alexander dual of the generator $u \in H_1(K; \mathbb{Z})$ is an integral cohomology class $\xi \in H^1(X; \mathbb{Z})$.

The Alexander polynomial of the knot K, is a polynomial with integral coefficients

$$P(t) = a_r t^r + \cdots + a_1 t + a_0,$$

with $a_0 \neq 0$ and $a_n \geq 0$, and is defined as the only generator of the principal ideal $(P(t))$ defined by the isomorphism $H_r(\widetilde{X}; \mathbb{Z}) \equiv \mathbb{Z}[t^{-1}, t]/(P(t))$, where as before, $\mathbb{Z}[t^{-1}, t]$ denotes the ring of Laurent polynomial with coefficients in $\mathbb{Z}$.

For detailed definitions and examples we recommend [Rolfsen, D. (1976)][12]. As established first by Milnor, cf. [Milnor, J. (1968)], the monic polynomial $(1/a_n)P(t)$ can be calculated as the characteristic polynomial of the homological 1-monodromy of (X, ξ), and is exactly

$$\prod_{J=(\lambda_J, n_J),\, J \in \mathcal{J}_1(\xi)} (z - \lambda_J)^{n_J}. \qquad (8.6)$$

The characteristic polynomial of the geometric monodromy $[T^{(X,\xi)}(1)]$ for the pair $(X; \xi)$, $X = \mathbb{S}^3 \setminus N$, N an open tube around an embedded oriented circle (knot) and $\xi \in H^1(S^3 \setminus K; \mathbb{Z})) = H^1(S^3 \setminus N; \mathbb{Z})$ the class defined above, as defined in 7.2, is exactly the monic Alexander polynomial of the knot.

Section 7.3 permits to propose a new algorithm to calculate the monic Alexander polynomial[13].

The alternating product of the characteristic polynomials $P_r(z)$ of the monodromies $[T^{X;\xi}(r)]$,

$$A(X; \xi)(z) = \prod P_r(z)^{(-1)^r},$$

---

[12]For example, for the familiar figure-eight knot on has $P(t) = t^2 - 3t + 1$, cf. [Rolfsen, D. (1976)] page 166, and for the torus knot $(4, 7)$ one has $P(t) = t^{18} - t^{17} + t^{14} - t^{13} + t^{11} - t^9 + t^7 - t^5 + t^4 - t + 1$, cf. [Rolfsen, D. (1976)] page 178.

[13]A more detailed discussion on the calculation of Alexander polynomials of knots and links, including knots in higher dimensions, and of the role of the algorithm provided in Section 7.3 is in preparation. In a similar vein, important cases of the Milnor-Turaev torsion of (M, ξ), a rational function defined on $GL(n, \mathbb{C})$, when regarded as the variety of rank-n complex representations of $\mathbb{Z}$ can be calculated by algorithms which determine the Jordan cell for monodromies.

known to topologists as the Alexander rational function, calculates an important part of the Reidemeister torsion of X equipped with the degree-one representation of the fundamental group $\pi_1(X)$ defined by ξ, and the complex number $z \in \mathbb{C}$, $z \neq 0$. This was pointed out first by J. Milnor and refined by V. Turaev, cf. [Turaev, V. (1986)], so the Jordan cells can be used for the calculation of this invariant as well. Given the need of additional background and definitions, the relations between *Alexander function* and *torsion* will not be addressed in this book. Section 7.3 provides algorithms to calculate the Alexander function and then Reidemeister torsion for rank one representations. The precise relation between these two is "work in progress".

8.3   Exercises

E.1 Let M^n be a closed smooth manifold.

Is it possible for a real-valued or angle-valued Morse map to have a closed barcode of the form $[a, a]$? But for a continuous map? Explain your answer or construct examples.

E.2 Suppose that $f : M^n \to \mathbb{R}$ is a perfect Morse function with respect to some fixed field (i.e., the number of critical points of index r is equal to the Betti numbers $\beta_r(M)$ for all r).

a) Show that f has no mixed barcodes.

b) Show that every Morse function whose gradient with respect with some Riemannian metric has no instantons is a perfect Morse function.

E.3 Show that for any Morse function $f : M^n \to \mathbb{R}, n \geq 2$, M^n closed smooth manifold, any $\epsilon > 0$, any open neighborhood U of a point x, and any $0 \leq k < n$, one can produce a new Morse function arbitrarily closed to f which has exactly one additional mixed barcode in dimension k and one in dimension $n - k - 1$.

E.4 Use Morse-Novikov theory to establish the Poincaré Duality for the configurations δ_r^f and γ_r^f.

E.5 Consider the real projective space $\mathbb{R}P^n = \mathbb{S}^n / \sim$, where $\mathbb{S}^n = \{\bar{x} = (x_1, x_2, \ldots, x_{n+1}) \in \mathbb{R}^{n+1} \mid \sum_i x_i^2 = 1\}$ and one puts $\bar{x} \sim \bar{y}$ iff $\bar{x} = -\bar{y}$, and denote by $[x_1, x_2, \ldots, x_{n+1}] \in \mathbb{R}P^n$ the point represented by

$\overline{x} = (x_1, x_2, \ldots, x_{n+1})$. For any $\overline{\lambda} = (\lambda_1, \lambda_2, \ldots, \lambda_{n+1})$ define

$$F_{\overline{\lambda}}([x_1, x_2, \ldots, x_{n+1}]) = \sum_i \lambda_i x_i^2.$$

Calculate the critical values and determine the barcodes of the real-valued map $F_{\overline{\lambda}}$ for the fields $\kappa = \mathbb{Z}_2, \ \mathbb{Z}_3, \ \mathbb{R}$ and for any $\overline{\lambda} \in \mathbb{R}^{n+1}$.

Chapter 9

Comments

9.1 Relation to Persistence Theory

9.1.1 *Persistence Theory, a summary*

The barcodes, equivalently the supports of the configurations δ_r^f and γ_r^f, and the Jordan cells $\mathcal{J}_r(f)$ discussed in this book, were first introduced as invariants for topological persistence of real- and circle-valued maps (level or zig-zag persistence and persistence for circle-valued maps cf. [Carlsson, G., de Silva, V., Morozov, D. (2009)] and [Burghelea, D., Dey, T (2013)]). They were proposed as invariants refining the *Persistent Homology*[1] introduced in [Edelsbrunner, H., Letscher, D., Zomorodian, A. (2002)]. In this section we present a brief overview of *Persistence Theory*, following closely [Burghelea, D., Dey, T (2013)]. A similar but more detailed presentation is contained in [Dong Du (2012 and 2014)].

We denote by $H_r(X)$ the homology vector space of X in dimension r with coefficients in a fixed field κ. Departing a little from the notations in the previous chapters, for a real-valued tame map $f : X \to \mathbb{R}$ we denote

$$X_t := f^{-1}(t), \; X_{[a,b]} := f^{-1}([a,b]), \; t, a, b \in \mathbb{R}, \; a < b,$$

and for a tame angle-valued map we denote

$$X_\theta := f^{-1}(\theta), \; X_{[\theta,\theta']} := f^{-1}([\theta, \theta']), \; 0 < \theta < \theta' \leq 2\pi.$$

Sublevel persistence

Persistent homology, introduced in [Edelsbrunner, H., Letscher, D., Zomorodian, A. (2002)] and further developed in [Zomorodian, A., Carls-

---

[1]For a survey on the basic definitions on persistent homology the reader can consult [Edelsbrunner, H., Harer, J. (2010)].

son, G. (2005)], considers real-valued tame maps $f : X \to \mathbb{R}$ and is concerned with the following questions:

Q1. Does the homology class $x \in H_r(X_{(-\infty,t]})$ originate from $H_r(X_{(-\infty,t'']})$ for $t'' < t$? Does the homology class $x \in H_r(X_{(-\infty,t]})$ vanish in $H_r(X_{(-\infty,t']})$ for $t < t'$? *Originate from* t' means that x is in the image of the induced linear map $H_r(X_{(-\infty,t']}) \to H_r(X_{(-\infty,t]})$, and *vanishes at* t'' means that x is in the kernel of the induced linear map $H_r(X_{(-\infty,t']}) \to H_r(X_{(-\infty,t]})$.

Q2. What are the smallest t' and t'' for which this happens?

This information is contained in the linear maps $H_r(X_{(-\infty,t]}) \to H_r(X_{(-\infty,t']})$, where $t' \geq t$, and is known as *topological persistence*, better called *homological persistence*. Since the involved subspaces are sublevel sets, one refers to this persistence as *sublevel persistence*. When f is tame, the persistence for each $r = 0, 1, \ldots, \dim X$, is determined by a finite collection of invariants referred to as *barcodes* [Zomorodian, A., Carlsson, G. (2005)]. For sublevel persistence the barcodes provide a collection of *closed intervals* of the form $[s, s']$ or $[s, \infty)$, with s, s' being the critical values of f. From these barcodes one can derive the Betti numbers of $X_{(-\infty,a]}$, the dimension of $\mathrm{img}(H_r(X_{(-\infty,t]}) \to H_r(X_{(-\infty,t']}))$ and get answers to the questions Q1 and Q2. For example, the number of r-barcodes which contain the interval $[a, b]$ is the dimension of $\mathrm{img}(H_r(X_{(-\infty,a]}) \to H_r(X_{(-\infty,b]}))$. The number of r-barcodes which identify to the interval $[a, b]$ is the maximal number of linearly independent homology classes born exactly in $X_{(-\infty,a]}$, but not before, and dead exactly in $H_r(X_{(-\infty,b]})$, but not before. These intervals give an idea on the topological changes of the sublevels of f when t varies and therefore qualitative information about the shapes under considerations, and they are as computable as the Betti numbers by algorithms of comparable complexity.

Such information is used in data analysis cf [Carlsson, G. (2009)], [Ghrist (2008)].

Level persistence

If instead of sublevels of $f : X \to \mathbb{R}$ we use levels, we obtain what we call it level persistence. Level persistence was first considered in [Dey, T.K., Wenger, R. (2007)], but was better formulated and understood computationally when the *zigzag persistence* was introduced in [Carlsson, G., de Silva, V., Morozov, D. (2009)]. Level persistence is concerned with the homology $H_r(X_t)$ of the fibers X_t and addresses questions of the following type:

Q1. Does the image of $x \in H_r(X_t)$ vanish in $H_r(X_{[t,t']})$, where $t' > t$, or in $H_r(X_{[t'',t]})$, where $t'' < t$?

Q2. Can x be detected in $H_r(X_{t'})$, where $t' > t$, or in $H_r(X_{t''})$, where $t'' < t$? The precise meaning of "detection" is explained below.

Q3. What are the smallest t' resp. the largest t'' for which the answers to Q1 and Q2 are affirmative?

To answer such questions one has to record information about the maps

$$H_r(X_t) \rightarrow H_r(X_{[t,t']}) \leftarrow H_r(X_{t'}). \tag{9.1}$$

The *level persistence* is the information provided by this collection of vector spaces and linear maps (9.1) for all t, t'.

We say that $x \in H_r(X_t)$ is *dead* at t', $t' > t$, if its image under the map $H_r(X_t) \rightarrow H_r(X_{[t,t']})$ vanishes. Similarly, x is *dead* at t'', $t'' < t$, if its image under the map $H_r(X_t) \rightarrow H_r(X_{[t'',t]})$ vanishes.

We say that $x \in H_r(X_t)$ is *detectable* at t', $t' > t$, respectively t'', $t'' < t$, if its image in $H_r(X_{[t,t']})$, respectively in $H_r(X_{[t'',t]})$, is nonzero and is contained in the image of $H_r(X_{t'}) \rightarrow H_r(X_{[t,t']})$, respectively $H_r(X_{t''}) \rightarrow H_r(X_{[t'',t]})$.

In the case of a tame map the collection of the vector spaces and linear maps is determined up to coherent isomorphisms by a collection of invariants called *barcodes for level persistence* which are intervals of the form $[s, s'], (s, s'), (s, s'], [s, s')$ with s, s' critical values, as opposed to the *barcodes for sublevel persistence*, which are intervals of the form $[s, s'], [s, \infty)$ with s, s' critical values. The open end of an interval signifies the death of a homology class at that end (left or right), whereas a closed end signifies that a homology class cannot be detected beyond this level (left or right) but it can be at that level. Note that in the case of sublevel persistence the left end signifies *birth* while the right *death*.

Two tame maps $f : X \rightarrow \mathbb{R}$ and $g : Y \rightarrow \mathbb{R}$ which are fiberwise homotopy equivalent have the same associated barcodes.

Level persistence provides considerably more information than the sublevel persistence. The barcodes of the sublevel persistence (for a tame map) can be recovered from the ones of level persistence. Precisely, a level barcode $[s, s']$ gives a sublevel barcode $[s, \infty)$ and a level barcode $[s, s')$ gives a sublevel barcode $[s, s']$; the sublevel persistence does not see any of the level barcodes (s, s') or $(s, s']$.

In Figure 9.1, we indicate the barcodes for both sublevel and level persistence for some simple map $f : X \to \mathbb{R}$ in order to illustrate their differences. The space X is the one-end-open tube and f is the "horizontally directed" height function.

In Figure 9.1 the class consisting of the sum of two circles at level t is not detected on the right, but is detected at all levels on the left up to (but not including) the level t'.

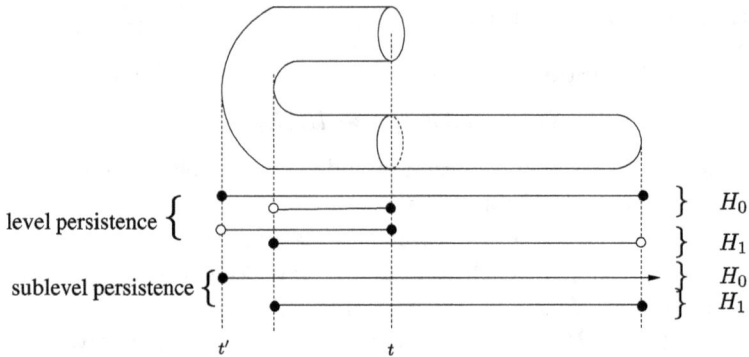

Fig. 9.1   Barcodes for level and sublevel persistence

Persistence for (circle-) angle-valued maps

Let $f : X \to \mathbb{S}^1$ be an angle-valued map. The sublevel persistence for such a map cannot be defined since circularity in values prevents defining sublevels. Even level persistence cannot be defined as per se, since the intervals may repeat over values. To overcome this difficulty one considers the infinite cyclic cover map $\tilde{f} : \tilde{X} \to \mathbb{R}$ for f, defined by the pull-back diagram

$$
\begin{array}{ccc}
\tilde{X} & \xrightarrow{\ \tilde{f}\ } & \mathbb{R} \\
{\scriptstyle p}\big\downarrow & & \big\downarrow{\scriptstyle \pi} \\
X & \xrightarrow{\ f\ } & \mathbb{S}^1
\end{array}
$$

The map $\pi : \mathbb{R} \to \mathbb{S}^1$ is the universal covering of the circle (the map which assigns to the number $t \in \mathbb{R}$ the angle $\theta = t \pmod{2\pi}$) and p is the pullback of π by the map f. One takes as barcodes of f the barcodes of $\tilde{f}$, intervals whose left end is in $[0, 2\pi)$. Notice that $X_\theta = \tilde{X}_t$ if $\pi(t) = \theta$. If

$x \in H_r(X_\theta) = H_r(\widetilde{X}_t)$, $\pi(t) = \theta$, the questions Q1, Q2, Q3 for f and X can be formulated in terms of the level persistence for $\widetilde{f}$ and $\widetilde{X}$.

Suppose that $x \in H_r(\widetilde{X}_t) = H_r(X_\theta)$ is detected in $H_r(\widetilde{X}_{t'})$ for some $t' \geq t + 2\pi$. Then x returns to $H_r(X_\theta)$ going along the circle $\mathbb{S}^1$ one or more time. When this happens, the class x may change in some respect. This gives rise to new questions that were not encountered in sublevel or level persistence.

Q4. When $x \in H_r(X_\theta)$ returns, how does the "returned class" compare with the original class x? It may disappear after going around the circle a number of times, or it might never disappear, and if so how does this class change after its return.

To answer Q1–Q4 one has to record information about $H_r(\widetilde{X}_t) \to H_r(\widetilde{X}_{[t,t']}) \leftarrow H_r(\widetilde{X}_{t'})$ with $t \in [0, 2\pi)$. If the barcode has $t = \theta, t' = \theta' + 2\pi k$ with $\theta, \theta' \in (0, 2\pi]$ this means that x has disappeared after at most $k+1$ runs along the circle. if it never disappear, hence always returns then x generates a Jordan cell.

The persistence for (angle-) circle-valued maps is the right persistence to get qualitative information about the complexity of shapes observed from a central location, or data collected from observations in angular directions, like the observation of a city from an observation tower.

When f is tame, in addition to sublevel and level persistence for real-valued maps, the invariants include structures other than barcodes called *Jordan cells* which, as shown in Chapter 7, describe the homological monodromy.

The barcodes for $f : X \to \mathbb{S}^1$ can be inferred from $\widetilde{f} : \widetilde{X}_{[a,b]} \to \mathbb{R}$, with $[a,b]$ being any large enough interval. The Jordan cells cannot be derived from $\widetilde{f} : \widetilde{X} \to \mathbb{R}$ or any of its restrictions $\widetilde{f} : \widetilde{X}_{[a,b]} \to \mathbb{R}$.

9.1.2 A few observations about δ_r^f and γ_r^f

Poincaré duality for barcodes

The Poincaré Duality for mixed barcodes in case M^n is a manifold and f a tame map is largely the consequence of the Poincaré Duality for the regular fibers of f, which are manifolds of dimension $n-1$ if M is a manifold of dimension n. This explains *complementarity* with respect to $n-1$ for the configurations γ_r^f rather than with respect to n, which holds for the configurations δ_r^f. A proof on these lines should be provided.

Algebrization via modules over principal ideal domains

In the paper [Zomorodian, A., Carlsson, G. (2005)], for a real-valued tame map $f : X \to \mathbb{R}$ with X compact and critical values $c_1 < c_2 < c_3 < \cdots$, one considers $\bigoplus_{1 \leq i} H_r(X_{(-\infty, c_i]}$ as a f.g $\kappa[t]$-module, $\kappa[t]$ being the ring of polynomials, with the multiplication by t being by $t(x_1, x_2, x_3, \ldots) = (0, i_1(x_1), i_2(x_2), \ldots)$, where i_k is the inclusion-induced linear map $i_k : H_r(X_{(-\infty c_k]}) \to H_r(X_{(-\infty, c_{k+1}]})$.

Since $\kappa[t]$ is a principal ideal domain, this module is a direct sum of indecomposable components which are free submodules of rank one and torsion components, modules of form $\kappa[t]/t^r \kappa[t]$, which provide the torsion. The torsion is responsible for the finite sublevel barcodes, each component providing a finite barcode, while the free components are responsible for the infinite sublevel barcodes. If the generator of a torsion component is the element $x \in H_r(X_{(-\infty, c_l]})$ which dies in $H_r(X_{(-\infty, c_{l+i]}})$, then it corresponds to a barcode is $[c_l, c_{l+i}]$, while if the element $x \in H_r(X_{(-\infty, c_l]})$ is the generator of a free component, then it corresponds to the barcode is $[c_l, \infty)$. For more details the reader should consult [Zomorodian, A., Carlsson, G. (2005)] or [Ghrist (2008)].

As shown in Chapter 3, for a tame map $f : X \to \mathbb{S}^1$ with $\widetilde{f} : \widetilde{X} \to \mathbb{R}$ its infinite cyclic cover, $H_r(\widetilde{X})$ is a f.g. $\kappa[t^{-1}, t]$-module. Again, $\kappa[t^{-1}, t]$ is a principal ideal domain and, as explained in Section 3.3 Chapter 3, the indecomposable torsion components are responsible for Jordan blocks or Jordan cells, while the free part is responsible for closed r-barcodes and open $r - 1$-barcodes. The role of closed-open and open-closed barcodes is however not visible in this picture.

There is an alternative algebrization, which relies on closed, open, and closed-open or open-closed barcodes. This involves a different type of algebraic structure, namely chain complexes of finite-dimensional vector spaces, referred to (in Chapter 4) as the AM complex and the AN complex. In this structure the closed-open or the open-closed barcodes are responsible for the boundary maps in the complex. These complexes, described and discussed in Chapter 4 Subsection 4.2.2 and Chapter 8 Section 8.1, provide different descriptions of the Morse complex and respectively the Novikov complex when f is a Morse real-, respectively angle-valued map, descriptions that make sense for any tame map.

Barcodes, Jordan blocks, and the Leray spectral sequence

The barcodes in both cases, real-valued map $f : X \to \mathbb{R}$ and angle-valued map $f : X \to \mathbb{S}^1$, can be recovered from the differential of the Leray spectral sequence of the map f. This is known to many experts cf. [Edelsbrunner, H., Harer, J. (2010)]. As spectral sequence tools are not considered in this book, and this aspect is not discussed.

From this point of view, the persistence theory is implicit in the work of R. Deheuvels [Deheuvels, R. (1955)].

Analogy with spectral theory of linear maps in finite-dimensional vector spaces

The most familiar example of a configuration of points with multiplicity in the complex plane $\mathbb{C}$ and a refinement of such a configuration to a configuration of vector spaces is provided by a linear map $T : V \to V$, where V is a finite-dimensional complex vector space, say of dimension $\dim V = n$.

In this case the eigenvalues provide a finite collection of complex numbers with multiplicities of total cardinality n, hence an element $\delta^T \in \mathrm{Conf}_n(\mathbb{C})$, and the assignment

$$\mathbb{C} \ni z \rightsquigarrow \text{the generalized eigenspace of } z,$$

defines an element $\widehat{\delta}^T \in \mathrm{CONF}_V(\mathbb{C})$ s.t. $\dim \widehat{\delta}^T(z) = \delta^T(z)$. They satisfy the following properties:

(1) Both assignments $T \rightsquigarrow \delta^T$ and $T \rightsquigarrow \widehat{\delta}^T$ are continuous maps from the space of linear maps with the norm topology to the space of configurations with any of the collision topologies.

(2) For an open dense set of linear maps T one has $\delta^T(z) = 0$ or 1.

(3) $\delta^T(z) = \delta^{T^*}(\overline{z})$.

These properties are analogues of the results stated in Theorems 5.1, 5.2, 5.4, 5.6 about the pair δ^f, $\widehat{\delta}^f$.

While there is no apparent common structure behind these configurations of vector points and their refinements to configurations of vector subspaces, it might be useful to pursue the analogies between the properties of the pair δ_r^f, $\widehat{\delta}_r^f$ and the pair δ^T, $\widehat{\delta}^T$.

9.2 A measure-theoretic aspect of the configurations δ_r^f, γ_r^f

In the plane $\mathbb{R}^2$ a **Box** is determined by the cartesian product of two intervals $(a, b) \times (c, d)$, $a < b, c < d$, to which one adds one vertex, called the *relevant vertex*, and the open sides adjacent to the relevant vertex. The boxes with relevant vertex can be of three types,

(1) **Box of type 1**: the relevant vertex is upper right.
(2) **Box of type 2**: the relevant vertex is lower left.
(3) **Box of type 3**: the relevant vertex is lower right.

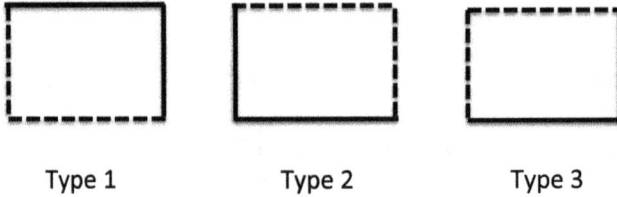

Type 1   Type 2   Type 3

Fig. 9.2   Box types 1, 2, 3

We will also refer to the opposite vertex, as the *opposite vertex*. Consider the collection boxes of a specified type contained in $\mathbb{R}^2$ or $(\mathbb{R}^2 \setminus \Delta)^{\pm}$, κ a fixed field and denote by VECT the class of finite dimensional κ-vector spaces.

A *vector-space valued measure*[2] on the σ-algebra generated by the collection of boxes of a type described above is an assignment $B \rightsquigarrow M(B) \in$ VECT together with

(i) a linear map $\pi_B^{B'} : M(B) \to M(B')$ for any two boxes $B' \subset B$ with the same relevant vertex,
(ii) a linear map $i_B^{B'} : M(B) \to M(B')$ for any two boxes $B' \subset B$ with the same opposite vertex,

which satisfies the following properties:

(1) if $B'' \subset B' \subset B$ all have the same relevant vertex, then
$$\pi_{B''}^{B} = \pi_{B'}^{B} \cdot \pi_{B''}^{B'};$$

(2) if $B'' \subset B' \subset B$ all have the same opposite vertex, then
$$i_{B''}^{B} = i_{B'}^{B} \cdot i_{B''}^{B'},$$

---

[2]Apparently, this concept is not found in the literature.

(3) if $B = B_1 \sqcup B_2$ with B_2 and B having the same relevant vertex (hence B_1 and B have the same opposite vertex), then the following sequence is exact

$$0 \longrightarrow M(B_1) \xrightarrow{\ i^B_{B_2}\ } M(B) \xrightarrow{\ \pi^B_{B_2}\ } M(B_2) \longrightarrow 0.$$

To such vector-space valued measure one associates an integer-valued measure $m(B) := \dim M(B)$ as well as the densities $\widehat{\delta} : \mathbb{R}^2 \rightsquigarrow \mathrm{VECT}$ and $\delta : \mathbb{R}^2 \rightsquigarrow \mathbb{Z}$ defined as follows.

For any a, b consider the box $B(a, b; \epsilon)$ with relevant vertex (a, b) and sides of length ϵ (for example, the box above diagonal of type (1), $B(a, b; \epsilon) = (a - \epsilon, a] \times (b - \epsilon, b]$, the box below diagonal of type (2), $B(a, b; \epsilon) = [a, a + \epsilon) \times [b, b + \epsilon)$, and the box of type (3), $B(a, b; \epsilon) = (a - \epsilon, a] \times [b, b + \epsilon))$, and then the direct system

$$\pi^{B(a,b;\epsilon)}_{B(a,b;\epsilon')} : M(B(a, b; \epsilon)) \to M(B(a, b; \epsilon'))$$

for $\epsilon > \epsilon'$. Define the densities

$$\widehat{\delta}(a, b) := \varinjlim_{\epsilon \to 0} M(B(a, b; \epsilon))$$

and

$$\delta(a, b) := \dim \varinjlim_{\epsilon \to 0} M(B(a, b; \epsilon)).$$

Clearly, both $\widehat{\gamma}^f_r$ and $\widehat{\delta}^f_r$ appear in this way and ultimately the stability property and Poincaré Duality can be formulated and proved for such configuration. The reader will recognize the configurations $\widehat{\gamma}^f_r$ and γ^f_R as densities of such a vector-space valued measure and an integer-valued measure provided by the collections of boxes of type (1) (when above diagonal) and type (2) (when below diagonal), and of the configurations $\widehat{\delta}^f_r$ and δ^f_r as densities for a vector-space valued measure and an integer-valued measure for boxes of type (3). This perspective can be applied to other vector-space valued functors in algebraic topology and extended to higher-dimensional boxes.

9.3 An invitation

An analytic description of barcodes and Jordan cells when X is a Riemannian manifold is a challenging research problem. It is justified by the observation that the Morse and Novikov complexes of a Morse real- and respectively angle-valued map can be recovered analytically via the

Witten deformation defined by a Morse function, respectively by a Morse angle-valued function and thanks to the relations between various types of analytic torsion and the Alexander rational function, as discussed in [Burghelea, D., Haller, S. (2008b)].

Bibliography

Allili, M., Corriveau, D. (2007). Topological analysis of shapes using morse theory, *Computer Vision and Image understanding* **105**, pp. 188–199.

Ball, B. (1975). Alternative approaches to proper shape theory, in: *Studies in topology (Proc. Conf., Univ. North Carolina, Charlotte, N.C., 1974; dedicated to Math. Sect. Polish Acad. Sci.), Academic Press, New York*, pp. 1–27.

Benson, J.D. (1998). *Representations and Cohomology. I.* (Cambridge University Press).

Borel, A., Moore, J.C. (1960). Homology theory for locally compact spaces, *Mich. Math. J.* **7**, pp. 137–159.

Bott, R. (1982). Lectures on Morse theory, old and new, *Bull. Amer. Math. Soc.* **7**, pp. 331–358.

Bucur, I., Deleanu, A. (1968). *Introduction to the Theory of Category and Functors,* With the collaboration of Peter J. Hilton and Nicolae Popescu. Pure and Applied Mathematics, Vol. XIX (John Wiley and Sons, A Wiley Interscience Publication).

Burghelea, D. (2011). Dynamics, spectral geometry and topology, in: *"Alexandru Myller" Mathematical Seminar, 3548, AIP Conf. Proc., 1329, Amer. Inst. Phys., Melville, NY* **1329**, pp. 35–48.

Burghelea, D. (2015a). Linear relations, monodromy and Jordan cells of a circle valued map, *arXiv:1501.02486.*

Burghelea, D. (2015b). A refinement of Betti numbers in the presence of a continuous function (i), *arXiv:1501.01012.*

Burghelea, D. (2016a). A refinement of Betti numbers and homology in the presence of a continuous function ii (the case of an angle valued map, *arXiv:1603.01861.*

Burghelea, D. (2016b). Refinement of Novikov-Betti numbers and of Novikov homology provided by an angle valued map, *Fundamentalnaya i prikladnaya matematika = Fundamental and Applied Mathematics; arXiv:1509.0773.*

Burghelea, D., Dey, T (2013). Persistence for circle-valued maps, *Discrete and Comput. Geom.* **50**, pp. 69–98.

Burghelea, D., Haller, S. (2008a). Dynamics, Laplace transform and spectral geometry, *Journal of Topology*, pp. 115–151.

Burghelea, D., Haller, S. (2008b). Torsion, as a function on the space of representations, in: *C*-algebras and elliptic theory II, Trends in Mathematics, Birkhäuser, Basel*, pp. 41–66.

Burghelea, D., Haller, S. (2015). Topology of angle valued maps, bar codes and Jordan blocks, *arXiv:1303.4328*.

Carlsson, G. (2009). Topology and data, *Bull. Amer. Math. Soc.* **46**, pp. 255–308.

Carlsson, G., de Silva, V., Morozov, D. (2009). Zigzag persistent homology and real-valued functions, in: *Proc. of the 25th annual symposium on computational geometry*, pp. 247–256.

Chapman, T.A. (1976). Lectures on Hilbert cube manifolds, *CBMS Regional Conference Series in Mathematics* American Mathematical Society, Providence, R. I. **28**.

Chapman, T.A. (1977). Simple homotopy theory for ANR's, *General Topology and Appl.* **7**, p. 165174.

Cohen-Steiner, D., Edelsbrunner, H., Harer, J. (2007). Stability of persistence diagrams, *Discrete Comput. Geom.* **37**, pp. 103–120.

Cohen-Steiner, D., Edelsbrunner, H., Morozov, D. (2006). Vines and vineyards by updating persistence in linear time, *Computational geometry (SCG'06), ACM, New York*, pp. 119–126.

Cornea, O., Ranicki, A. (2003). Rigidity and gluing for Morse and Novikov complexes, *J. Eur. Math. Soc. (JEMS)* **5**, pp. 343–394.

Daverman, R.J., Walsh, J.J. (1981). A ghastly generalized *n*-manifold, *Illinois Journal of Mathematics* **25**, pp. 555–576.

Deheuvels, R. (1955). Topologie d'une fonctionelle, *Annals of Mathematics* **61**, pp. 13–72.

Dey, T.K., Wenger, R. (2007). Stability of critical points with interval persistence, *Discrete Comput. Geom.* **38**, pp. 479–512.

Dimca, A. (1992). *Singularities and Topology of Hypersurfaces* (Universitex, Springer Verlag, New York).

Dong Du (2012 and 2014). Contributions to persistence theory, *arXiv: 12103092* and *An. Univ. Vest Timi. Ser. Mat.-Inform.* **52**, pp. 13–95.

Donovan, P., Freislich, M.R. (1973). *The representation theory of finite graphs and associated algebras*, Vol. 5 (Carleton Mathematical Lecture Notes, Carleton University, Ottawa, Ont.).

Edelsbrunner, H., Harer, J. (2010). *Computational topology: An introduction* (American Mathematical Society, Providence, RI).

Edelsbrunner, H., Letscher, D., Zomorodian, A. (2002). Topological persistence and simplification, *Discrete Comput. Geom.* **28**, pp. 511–533.

Farber, M. (2004). *Topology of closed one-forms* Mathematical Surveys and Monographs, Vol. 108 (American Mathematical Society, Providence, RI).

Forman, R. (1998). Morse theory for cell complexes, *Adv. Math.* **134**, pp. 90–145.

Forman, R. (2002). Combinatorial Novikov-Morse theory, *Internat. J. Math.* **13**, pp. 333–368.

Friedl, S., Maxim, L. (2016). Twisted Novikov homology of complex hypersurface complements, *arXiv:1602.04943*.

Gelfand, I. M. (1961). *Lectures on Linear Algebra* (Interscience Publishers Inc., New York).

Ghrist, R. (2008). Barcodes: the persistent topology of data, *Bull. Amer. Math. Soc.* **45**, pp. 61–75.

Goresky, M., MacPherson, R. (1988). *Stratified Morse Theory*, Vol. 14 (Ergebnisse der Mathematik und Ihrer Grenzgebiete (3), Springer-Verlag, Berlin).

Hatcher, A. (2002). *Algebraic Topology* (Cambridge University Press, Cambridge).

Hu, S.T. (1965). *Theory of Retracts* (Wayne State University Press, Detroit, MI).

Hutchings, M. (2002). Reidemeister torsion in generalized Morse theory, *Forum Math.* **14**, pp. 209–244.

Hutchings, M., Lee, Y.J. (1999). Circle-valued Morse theory, Reidemeister torsion, and Seiberg-Witten invariants of 3-manifolds, *Topology* **38**, pp. 861–888.

Lang, S. (2002). *Algebra,* Revised third edition (Graduate Texts in Math. 211, Springer-Verlag).

Maxim, L. (2014). l^2-Betti numbers of hypersurface complements, *Int. Math. Res. Not. IMRN*, pp. 4665–4678.

Milnor, J. (1959). On spaces having the homotopy type of a CW-complex, *Trans. Amer. Math. Soc* **90**, pp. 272–280.

Milnor, J. (1965). *Lectures on the h-Cobordism Theorem,* Notes by L. Siebenmann and J. Sondow (Princeton Univ. Press).

Milnor, J. (1966). Whitehead torsion, *Bull. Amer. Math. Soc.* **72**, pp. 358–426.

Milnor, J. (1968). Infinite cyclic coverings, *Conference on the Topology of Manifolds (Michigan State Univ., E. Lansing, Mich., 1967)*, Prindle, Weber and Schmidt, Boston, Mass., pp. 115–133.

Munkres, J. (1984). *Elements of Algebraic Topology* (Addison Wesley Publishing Company, Menlo Park, CA).

Nazarova, L.A. (1973). Representations of quivers of infinite type (russian), *Izv. Akad. Nauk SSSR Ser. Mat.* [English Translation in: Math. USSR-Izv. 7 (1973), 749792] **37**, pp. 752–791.

Némethi, A., Zaharia, A. (1992). Milnor fibration at infinity, *Indag. Math.*, pp. 323–335.

Novikov, S.P. (1991). Quasiperiodic structures in topology, in: *Topological methods in modern mathematics, (Stony Brook, NY, 1991) Publish or Perish, Houston, TX*, pp. 223–233.

Pajitnov, A.V. (2006). *Circle-valued Morse theory*, Vol. 32 (De Gruyter Studies in Mathematics, Walter de Gruyter and Co., Berlin).

Rolfsen, D. (1976). *Knots and Links* (Publish or Perish, Inc.,Berkeley, CA).

Turaev, V. (1986). Reidemeister torsion in knot theory, *Uspekhi Mat. Nauk* [English translation: *Russian Math. Surveys* 41 (1986), no. 1, 119182] **41**, pp. 97–147.

Warfield, R.B. (1969). A Krull-Schmidt theorem for infinite sums of modules, *Proc. Amer. Math. Soc.* **22**, pp. 460–465.

Zomorodian, A., Carlsson, G. (2005). Computing persistent homology, *Discrete Comput. Geom.* **33**, pp. 249–274.

Index

k-minor, 14

Alexander function, 8
Alexander polynomial, 10, 59
Alexander rational function, 59
ANR, 44

barcode, 71
barcodes, 5
Borel-Moore homology, 61
bottleneck topology, 54

canonical divisor, 7, 15
cell, 51
cell complex, 51
collision topology, 54, 56
compatible splittings, 108, 161
computer friendly, 7, 59
configuration of points, 54
critical value, 45
critical values, 44

elementary divisors, 15
equivalent matrices, 14

fiberwise weak homotopy
equivalence, 96
Fredholm cross-ratio, 23
Fredholm map, 22

generalized simplex, 51

good ANR, 45
good map, 195
good total order, 126
graph $\mathcal{Z}$, 65
graph G_{2m}, 65

homological critical value, 47
homological persistence, 228
homological monodromy, 59
homologically regular value, 46
homologically weakly tame, 46

index, 22
infinite cyclic cover, 49
infinite cyclic cover of map, 49
instanton, 1, 211
instanton counting matrix, 211

Jordan block, 17, 70
Jordan canonical form, 18
Jordan cell, 7, 17
Jordan decomposition, 18

knot, 223

leading vector, 18
level persistence, 228
linear relation, 30
locally conservative vector
field, 1
locally polynomial, 45